AF360723

Power Converter of Electric Machines, Renewable Energy Systems, and Transportation

Power Converter of Electric Machines, Renewable Energy Systems, and Transportation

Editors

Adolfo Dannier
Gianluca Brando
Marino Coppola

MDPI • Basel • Beijing • Wuhan • Barcelona • Belgrade • Manchester • Tokyo • Cluj • Tianjin

Editors
Adolfo Dannier
University of Napoli Federico II
Italy

Gianluca Brando
University of Napoli Federico II
Italy

Marino Coppola
University of Napoli Federico II
Italy

Editorial Office
MDPI
St. Alban-Anlage 66
4052 Basel, Switzerland

This is a reprint of articles from the Special Issue published online in the open access journal *Energies* (ISSN 1996-1073) (available at: https://www.mdpi.com/journal/energies/special_issues/ Power_Converter_Electric_Machines_Renewable_Energy_Systems_Transportation).

For citation purposes, cite each article independently as indicated on the article page online and as indicated below:

LastName, A.A.; LastName, B.B.; LastName, C.C. Article Title. *Journal Name* **Year**, *Volume Number*, Page Range.

ISBN 978-3-0365-1170-2 (Hbk)
ISBN 978-3-0365-1171-9 (PDF)

© 2021 by the authors. Articles in this book are Open Access and distributed under the Creative Commons Attribution (CC BY) license, which allows users to download, copy and build upon published articles, as long as the author and publisher are properly credited, which ensures maximum dissemination and a wider impact of our publications.

The book as a whole is distributed by MDPI under the terms and conditions of the Creative Commons license CC BY-NC-ND.

Contents

About the Editors

Adolfo Dannier (Associate Professor) Adolfo Dannier, born in Naples in 1976, graduated in 2003 in Electrical Engineering at University Federico II of Naples, where he obtained his Ph.D. in Electrical Engineering in 2008, with a dissertation on multilevel converters with fault-tolerant structures. Since 2020, he has been an Associate Professor at University Federico II in "Converters, Electrical Machines and Drives" and teaches "Modelling of Electric Machines and Converters", "Power Electric Converter" and "Dynamics of electromechanical systems" (UNICAMPANIA). Active in the research topics of his scientific sector, he collaborated with national and international research projects, as attested by numerous and significant scientific publications. The research interest is mainly targeted at high-performance dynamic drives with PM motors, conversion equipment and electrical energy storage for electric vehicles and for the integration of renewable energy sources.

Gianluca Brando (Associate professor) received his MS (cum laude) and Ph.D. in electrical engineering from the University of Naples Federico II, Naples, Italy, in 2000 and 2004. From 2000 to 2012, he was a Postdoctoral Research Fellow with the Department of Electrical Engineering, University of Naples Federico II. Since 2020, he has been an Assistant Professor of Electrical Machines and Drives in the Department of Electrical Engineering and Information Technology, University of Naples Federico II, where he teaches "Machines and Electrical Drives" and "Converters and Storage Systems for Electrical Drives". He has authored several scientific papers published in international journals and conference proceedings. His research interests include control strategies for power converters and electrical drives, modulation techniques for multilevel converters and the advanced modelling of multiphase electrical machines.

Marino Coppola (Postdoctoral Researcher). Marino Coppola earned his Laurea degree cum laude in electronic engineering and his Ph.D. in electronic and telecommunication engineering from the University of Napoli Federico II, Italy, in 2008 and 2012, respectively. From 2012 to 2015, he was a Postdoctoral Researcher with the Department of Electrical Engineering and Information Technologies at the University of Napoli Federico II. From 2015 to 2018, he was a Senior Researcher with PNP Lab (a spinoff company of the University of Napoli Federico II) dealing with EU project ASPIRE (Advanced Smart-grid Power dIstRibution systEm). Since 2019, he has been a Postdoctoral Researcher with the Department of Electrical Engineering and Information Technologies at the University of Napoli Federico II. He has been a Visiting Researcher at Department of Energy Technology (Aalborg University, Denmark) and at Laplace Laboratories (Institut National Polytechnique of Toulouse, France). He has co-authored 39 papers in refereed international journals and conference proceedings. His initial research activities covered the area of design of highperformance CMOS standard cells, and design of high-speed analog-to-digital converters. His current research interests include the design and control of power converters for photovoltaics and distributed power generation systems, integration of renewable energy resources in electrical systems, modulation of multilevel inverters, design and control of high efficiency DC-DC converters, SiC-based converters, and the design of digital circuits on FPGA.

Preface to "Power Converter of Electric Machines, Renewable Energy Systems, and Transportation"

Nowadays, energy is becoming more electrical in each field of engineering application, thus power converters have assumed an increasingly relevant role for electric machines, renewable energy and transportation systems. The converters' design and control are critical, as they are evolving into an essential component to interface and integrate different power systems. In this evolution, power converters' topology and technology play an enabling role in the advancement of electrical machines performances, renewable energy integration and emerging transport applications. The use of reliable and efficient electric machines is crucial in allowing the higher penetration of renewable energy in electrical systems, as well as in the ongoing change transport sector. Moreover, energy storage systems can be the way towards higher system flexibility and reliability. Consequently, it is necessary to develop innovative systems "devoted" to the specific application under study. Indeed, in recent years, research has been very active in working to improve each electric subsystem to achieve advantages in terms of system performances, robustness, easy installation, and maintenance. It is necessary to continue on this path in order to obtain significant technological advances.

Adolfo Dannier, Gianluca Brando, Marino Coppola
Editors

Review

Transformerless Multilevel Voltage-Source Inverter Topology Comparative Study for PV Systems

Adyr A. Estévez-Bén [1,†], Alfredo Alvarez-Diazcomas [2,†] and Juvenal Rodríguez-Reséndiz [2,*,†]

[1] Facultad de Química-Facultad de Ingeniería, Universidad Autónoma de Querétaro, Cerro de las Campanas, Las Campanas, Querétaro 76010, Mexico; aestevez05@alumnos.uaq.mx

[2] Facultad de Ingeniería, Universidad Autónoma de Querétaro, Cerro de las Campanas, Las Campanas, Querétaro 76010, Mexico; aalvarez78@alumnos.uaq.mx

* Correspondence: juvenal@uaq.edu.mx; Tel.: +52-442-192-1200

† These authors contributed equally to this work.

Received: 2 June 2020; Accepted: 18 June 2020; Published: 24 June 2020

Abstract: At present, renewable energies represent 25% of the global power generation capacity. The increase in clean energy facilities is mainly due to the high levels of pollution generated by the burning of fossil fuels to satisfy the growing electricity demand. The global capacity of generating electricity from solar energy has experienced a significant increase, reaching 505 GW in 2018. Today, multilevel inverters are used in PV systems to convert direct current into alternating current. However, the use of multilevel inverters in renewable energies applications presents different challenges; for example, grid-connected systems use a transformer to avoid the presence of leakage currents. The grid-connected systems must meet at least two international standards analyzed in this work: VDE 0126-1-1 and VDE-AR-N 4105, which establish a maximum leakage current of 300 mA and harmonic distortion maximum of 5%. Previously, DC/AC converters have been studied in different industrial applications. The state-of-the-art presented in the work is due to the growing need for a greater use of clean energy and the use of inverters as an interface between these technologies and the grid. Also, the paper presents a comparative analysis of the main multilevel inverter voltage-source topologies used in transformerless PV systems. In each scheme, the advantages and disadvantages are presented, as well as the main challenges. In addition, current trends in grid-connected systems using these schemes are discussed. Finally, a comparative table based on input voltage, switching frequency, output levels, control strategy used, efficiency, and leakage current is shown.

Keywords: DC/AC converter; voltage-source; multilevel inverter; PV systems; neutral point clamped inverter; flying capacitor inverter; cascaded inverter; renewable energy systems

---

1. Introduction

In the 21st century, Renewable Energy Sources (RES) have acquired an unprecedented role [1]. Governments are increasingly betting on clean energy to comply with international agreements. For example, India seeks the installation of 40,000 MW electricity generation capacity from renewable energy sources by 2022. In Argentina, one of the largest solar plants in Latin America is being built, the project will provide the grid 300 MW of power [2]. The main RES are: hydraulic, wind and solar. In 2018, solar energy experienced an increase of 103.2 GW in global electricity production capacity over the previous year, for a total of 505 GW. However, electrical capacity from hydraulic and wind power only increased by 20 GW and 51.3 GW, respectively. A detailed description of the current status of RES is presented in [3]. The greater use of Photovoltaic (PV) systems is, essentially, due to the decrease in cell production costs, which means that the return on investment occurs in less time. At present, solar energy is consolidated as a competitive option for both the industrial and residential sectors [4]. Figure 1 shows the increase in electrical production per year of the main clean sources.

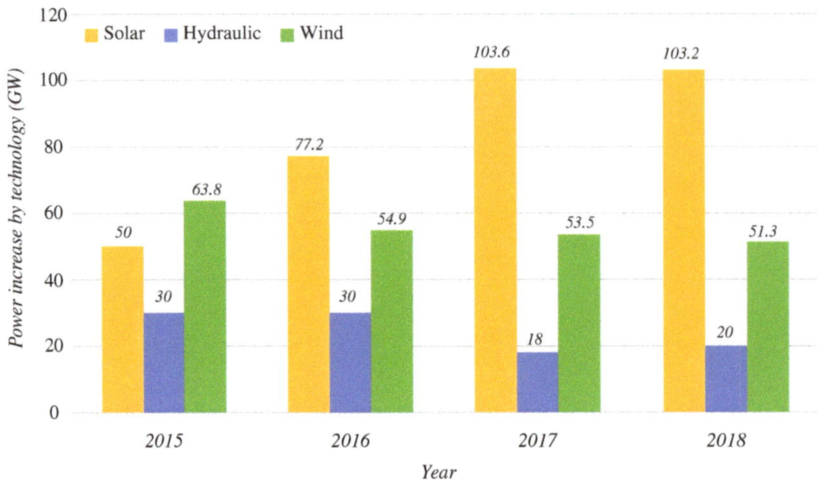

Figure 1. Annual increase in electricity generation capacity from RES by technology.

The advance in PV systems worldwide has caused a reduction in the cost of investments to generate electricity by this means. Gatta et al. in [5] analyze the replacement of diesel generators by hybrid RES plants in Italy, where 500 kWp (Peak Power) PV power plant and a 1000 kW/500 kWh lithium-ion Battery Energy Storage System (BESS) were installed. However, to achieve widespread use of PV systems worldwide, this technology must be competitive in terms of cost compared to conventional methods. Certain estimates assure that the price of PV systems will decrease from USD 0.18/kWh in 2016 to USD 0.05/kWh in 2030 [6]. This price actually contrasts with an average of USD 0.12/kWh for conventional energy sources.

Currently, the integration of BESS and PV systems has been efficiently achieved in certain applications. The authors in [7] present a traction system based on obtaining solar energy stored in a BESS. The research proposes a solution based on PV for one of the most polluting sectors in the world. However, it is important to mention that the energy obtained from other RES is currently expanding. Wind farms are an alternative to the absence of sunlight at night. In this regard, the research [8] offers a broad panorama of this technology use.

The efficiency and lifetime of the system are two aspects that directly affect the reduction of costs. Currently, there is a trend to oversize the PV array; thus, the nominal power of the array is greater than the nominal inverter power [9,10] and the converter gets more energy in the same period. The system as a whole will capture more energy during production periods. It is possible to oversize the panels considering that the price decreases approximately 13% [11] each year; therefore, the cost of a greater PV string will be compensated with the energy captured.

The inverters are: Current-Source Inverter (CSI), Voltage-Source Inverter (VSI), or Impedance-Source Inverter (ZSI) [12]. The VSI and CSI differ in the type of input element (capacitor or inductor) [13]. The use of the inverters cover a wide range of applications, from power supplies to high-power industrial applications, certain examples are found in the literature [14–18]. The DC/AC converters are also employed as intermediate stage between RES and the grid since they transform the Direct Current (DC) into Alternating Current (AC).

Sahan et al. in [19] present a comparative study between VSI and CSI. Other classifications are line-switched or auto-switched; inverters for autonomous systems or inverters for grid-connected systems and single-phase (<5 kW) or three-phase (>5 kW). Traditional DC/AC converters, also called conventional inverters, are limited to only two output levels and require specific characteristics to achieve an adequate signal. For example, the two-level H-bridge topology uses a high-switching frequency to obtain low harmonic distortion at the output [20]. The main disadvantages of traditional topologies are shown in Table 1.

Table 1. Main conventional inverters limitations.

Parameters	Description
High-switching frequency	Require fast switching and stray inductance should be minimized with the proper circuit.
High dv/dt	The energy injected into the load must be a sinusoidal signal. When intermediate energy levels are not used, the load must support high dv/dt stress.
Power loss	The fast switching causes a temperature increase in the semiconductor devices, which requires an adequate heat dissipation system.
Electromagnetic Interference (EMI)	Electromagnetic interference problems increase with the switching frequency of semiconductors.

Multilevel topology has emerged to remove the limitations of traditional DC/AC converters. The main multilevel voltage-source schemes are: Neutral Point Clamped (NPC) [21], Flying Capacitor (FC) [22] and Cascaded (CMLI) [23], although other topologies are also used to a lesser extent, such as Hexagram [24] and Hybrid [25]. The main difference between conventional topologies and multilevel inverters is the number of output levels, while traditional converters have only two levels of power at their output, multilevel inverters deliver more than two levels.

The development of multilevel inverters evolves together with the different control strategies. Nowadays, Model Predictive Control (MPC) is commonly used in the control area of these converters [26]. Its effectiveness has been proved in various power converter topologies [27]. Conventional control techniques such as Proportional-Integral (PI) [28] or Proportional-Resonant (PR) controller [29,30] are also used. However, the control strategy depends on the topology and the application.

Although inverters have been previously studied in industrial applications [31], the use of these converters in RES is a subject that presents its own challenges, trends and problems. Precisely, due to the current need for a greater use of RES and inverters as an interface for the injection of energy into the grid, the current state-of-the-art research on multilevel inverters in these applications is presented. The paper addresses NPC, FC, and CMLI topologies and focuses on establishing benchmarks, including the latest research on the topologies mentioned. Figure 2 shows an inverter classification scheme and highlights the topologies that are addressed in the work. The comparative analysis is performed taking into account several aspects of importance in PV inverters such as: number of elements and power supplies, leakage current, fault tolerance capacity, compliance with international standards and the complexity of the control and modulation strategy developed. Therefore, the work presents the following contributions:

- Presents an overview of the current integration of RES with energy injection systems to the grid.
- Provides an evaluation and comparison between three voltage-source multilevel inverter topologies.
- Discusses about the modulation strategy in NPC inverters.
- Presents future trends and research opportunities to contribute to the field.
- Presents the challenges and issues concerning the interconnection between the inverters and the grid.
- Summarizes more than 20 inverter application works in PV systems.

The work is structured as follows: In Section 2, the basic concepts of multilevel inverters, the advantages of their use, as well as main standards of grid-connected systems are presented.

Section 3 shows the NPC topology, highlighting the different modulation strategies used for the correct balance of the input capacitors. Similarly, FC-based topology is addressed in Section 4. Section 5 present the scheme based on cascade inverters, addressing the issues of fault tolerance. Finally, Section 6 presents a summary table of the most recent works reported in the literature, allowing comparison points based on the type of converter, input voltage, switching frequency, control strategy, efficiency and leakage current. A list of the acronyms used in this paper is presented at the end of the document.

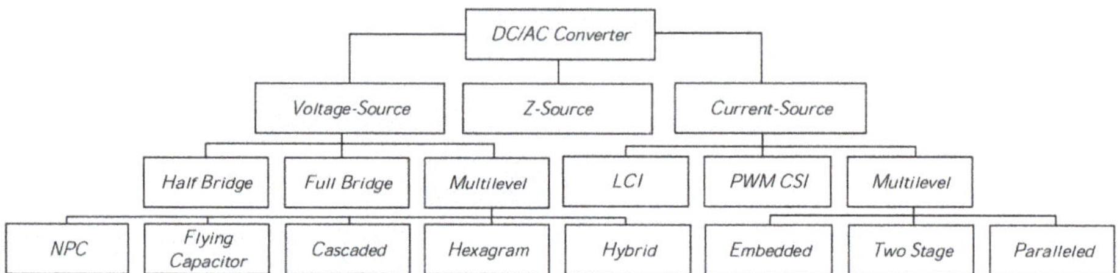

Figure 2. Multilevel inverters classification presented in [32].

2. Multilevel Voltage-Source Inverters

The multilevel inverter generate various levels of voltage or current at the output and obtain their energy from different DC sources, to deliver it with the use of lower-rated switches. In general, the power is obtained from capacitors, batteries, or other conventional storage, including RE sources [18]. Different MLI topologies have been studied [33–35]. Figure 3 present a comparison between the output signal of traditional inverters and MLI. It is observed that the MLI shows a more sinusoidal waveform that traditional two level inverters, which allows obtaining the characteristics presented in Table 2.

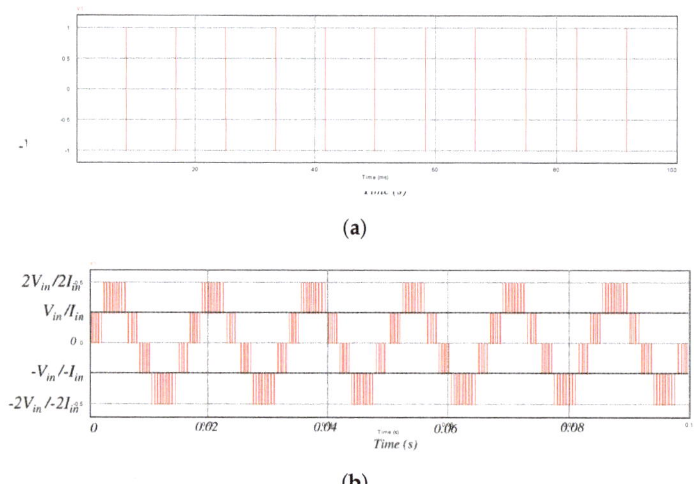

(a)

(b)

Figure 3. Output comparison: (a) Traditional inverters, (b) Multilevel inverters.

Table 2. Main multilevel inverters advantages.

Parameters	Description
Low-switching frequency	The switching frequency is lower, since generally more switches are used to generate the scaled output levels.
Low dv/dt of output voltages	The voltage stress is lower in each switch since the output levels are distributed among a greater number of semiconductors, thus obtaining a lower dv/dt of output voltage.
Structure	Modular structure that allows increasing the number of input sources and output power.
Power	High-output power without increasing the rating of the topology switches.
Total Harmonic Distortion (THD)	Low THD due to a more sinusoidal signal.
Reduced losses	Switching and conduction losses are low.
Fault tolerant operation	Using an adequate control strategy and state redundancy.

The development of MLI has been marked by the progress of semiconductor materials technology (IGBT, MOSFET, etc.) and the evident evolution of digital processors (microprocessors, DSP and FPGA). In this sense, in [36], an interesting investigation is presented, and the most important conclusion could be that the use of 4H-SiC constitutes one of the most important aspects that enabled the current development of power converters. Hence, the multilevel inverters are rapidly emerging as a promising alternative in photovoltaic systems for high-power/medium-voltage DC/AC conversion. Within multilevel inverters, the Multilevel Voltage-Source Inverter (MVSI) inverter has got attention for a better quality power supply. In MVSI, the amplitude of the output voltage generally is less than the input, in these cases, the inverter behaves as a buck converter. Therefore, an intermediate boost stage is generally required since, in RES applications, the panel voltage is low. Nevertheless, by including this stage, the complexity of the system control increases, and the efficiency decreases.

There is a close compromise between THD, inverter output levels and filter size. When the levels increases, THD and filter size decrease, but a higher number of components is required. Nowadays, an important challenge is to design schemes with a reduced number of components. Authors in [37] introduce a novel topology, which seeks to reduce the number of switches as shown in Figure 4. This scheme has three power supplies and ten semiconductors to deliver fifteen output levels.

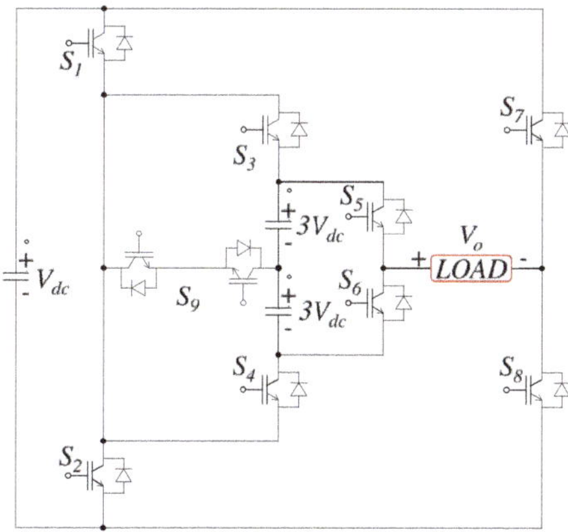

Figure 4. Multilevel topology proposed in [37].

The authors propose that the maximum output voltage by using this configuration is:

$$V_{o,max} = V_{dc} + 2V_{dc} = 7V_{dc} \tag{1}$$

where: V_{dc} is the input power supply and $V_{o,max}$ is the output voltage in the load.

The document presented in [38] summarizes the main schemes that make it possible to reduce the number of elements. The work concludes that the reduction of components causes the use of more expensive devices, by raising the voltage rating of semiconductor circuit breakers. Other aspects, such as the increase in the number of energy sources and more complex control schemes are also pointed out. The above approach shows the compromise that exists between the number of energy levels that the converter delivers and the complexity of its control, presenting a proportional relationship between both variables [39].

The grid-connected PV systems must comply with certain standards such as VDE 0126-1-1, which regulates the maximum allowed of leakage current in the system. Leakage current flows when the terminals have high-frequency voltage transitions.This systems generally use transformers to ensure the isolation of the PV system. Hence, it is avoided the appearance of leakage current between the stray capacitances of the panel and the grid [40], causing EMI problems, increased harmonic distortion and possible damage to health. The use of an isolation stage transformer increases as the weight, cost and volume of the system. In addition, this element causes losses in efficiency of around 3%. The newer topologies seek to eliminate this element. The main advantages are higher efficiency, adequate power density, and lower cost [41]. Also, the performance of the control would be affected according to the winding settings [42]. For this reason, transformerless PV inverters capture the interest of the scientific community [43].

The European Network Code "Requirements for Generators" or VDE-AR-N 4105 is aimed at low voltage systems. The code establishes the grid connection standards for generation systems in Germany. The main parameters to monitor are the capabilities for frequency stabilization and the provision of reactive power. A comparative summary of the aforementioned standards is presented in Table 3. In [44], the principal regulations that the grid-connected systems must comply with are summarized.

Table 3. German code VDE comparison [45].

Parameters		VDE 0126-1-1	VDE-AR-N 4105
Leakage current		RMS Value	The use of the leakage current protection devices is inevitable. The standard IEC 60755 defines the detail requirements for the leakage current protection devices.
		i > 300 mA	
		Δi > 30 mA	
		Δi > 60 mA	
		Δi > 150 mA	
Grid frequency monitor	50.2 < f < 51.5	Disconnected from the grid within 0.2 s	Adjustable generation systems must reduce (for f increase) or increase (for f decrease) the Active Power (P_M) generated instantaneously with a gradient of 40% of P_M by Hertz
	f > 51.5 or f < 47.5		Disconnected from the grid within 0.2 s
Active power		None	The generation systems (>100 kW) could reduce their active power to set point provided by the network operator.
Reactive power		None	The generation systems should output required reactive power in accordance with the characteristic curve provided by the network operator.

3. Neutral Point Clamped Based Topologies

In NPC inverters, multiple DC sources are generated by dividing the input bus voltage using a capacitors bank as shown in Figure 5a. The topology is recognized as one of the most popular schemes among MLI [46]. Table 4 shows the allowed, potentially destructive, and destructive switching patterns that could be implemented in the basic structure of the Three-Level NPC (3L-NPC) inverters. The NPC converters are mainly employed in high and medium-power range [47,48]. This DC/AC converter have low dv/dt, low THD [49] and can remove common-mode current making it attractive for PV applications [50]. There are certain high-power applications in which NPC inverters allow a higher DC-link voltage and also avoid the series connection of semiconductors in the same branch [51]. This devices have a large number of clamping diodes, unbalance problems in the DC-bus capacitors and a non-uniform distribution of losses in the switches. In standard operating conditions the values of the capacitors must have a value similar voltage. In [52], a new PWM modulation strategy is proposed to control the output voltage and balance the DC bus capacitors for converters.

An alternative to the traditional NPC topology is the Active-NPC (ANPC) scheme. This variant arises to eliminate the previously mentioned disadvantages. In ANPC design, instead of using clamping diodes, bidirectional semiconductors are used, as shown in Figure 5b. In both schemes (NPC and ANPC), it is possible to reduce the leakage current by connecting the middle point of the DC-bus to the grid ground. In this way, the value of $DC+$ or $DC-$ will depend on the sign of the output current. Therefore, the stray capacitance voltage remains constant and no leakage current arises. The conduction losses can be reduced by using different paths in the zero state. Currently, certain modifications of the traditional PWM have been implemented in ANPC topologies to achieve correct operation of the inverter, reducing conduction losses and achieving a correct balance of the capacitors.

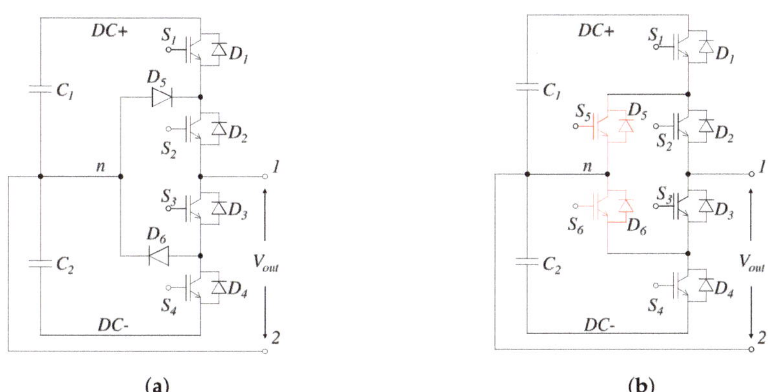

(a)　　　　　　　　　　　　　　　**(b)**

Figure 5. Comparison of basic configurations: (**a**) 3L-NPC and (**b**) 3L-ANPC.

Table 4. Switching states 3L-NPC.

Switches	Allowed						Potentially Destructive					Destructive				
S_1	0	0	0	1	0	0	1	0	1	1	0	1	1	1	0	1
S_2	0	1	0	1	1	0	0	0	0	0	1	1	1	0	1	1
S_3	0	0	1	0	1	1	0	0	0	1	0	1	0	1	1	1
S_4	0	0	0	0	0	1	0	1	1	0	1	0	1	1	1	1

The techniques for eliminating the leakage current in PV inverters are grouped into two categories. The first introduces one switch to isolate the grid from the panels in freewheeling times. The second category maintains a neutral connection from the grid to the midpoint of the input capacitors and ensures low-voltage variations. Considering an NPC inverter, the common-mode voltage is defined from Equation (2). From the mathematical point of view the reduction of the leakage current is achieved when there is no variation in the terms V_{cm-dm} and V_{cm}. Figure 6 shows a simplified electrical diagram, where the influence of the above terms in the emergence of leakage current is observed.

$$V_{cm} = \frac{V_{1n} - V_{2n}}{2} \tag{2}$$

$$V_{dm} = V_{1n} - V_{2n} \tag{3}$$

where: the common voltage and the differential voltage are defined as V_{cm} and V_{dm}, the voltage at the inverter output at (1) and (2) with respect to the neutral point (n) is defined as V_{1n} and V_{2n} respectively.

By considering Equations (2) and (3), V_{1n} and V_{2n} can be presented as:

$$V_{1n} = \frac{V_{cm} + V_{dm}}{2} \tag{4}$$

$$V_{2n} = \frac{V_{cm} - V_{dm}}{2} \tag{5}$$

$$V_{cm-dm} = -\frac{V_{dm}}{2} \tag{6}$$

where: the relationship between the common mode and the differential mode is determined by V_{cm-dm}.

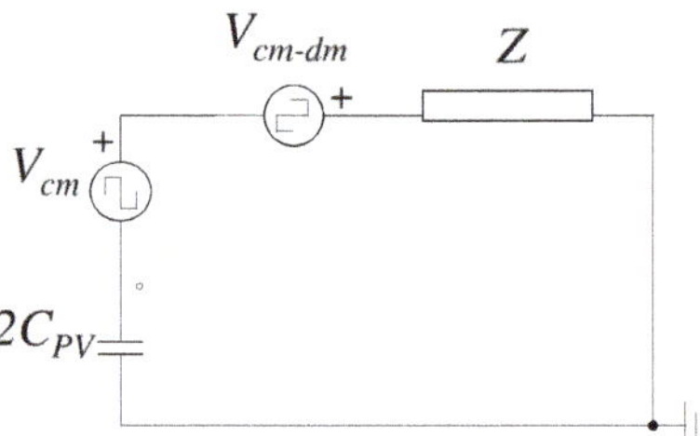

Figure 6. Common-mode represented based on simplified electrical circuit diagram.

The control strategy in this type of converter is divided into current loop based controllers and Direct Power Control (DPC). A good dynamic inverter response as well as a simple control scheme are two characteristics present in the DPC technique. The main disadvantage of this control strategy is the presence of a variable-switching frequency. Today this technique is combined with others such as Space Vector Modulation (SVM) [53,54] and MPC to achieve a fixed-switching frequency [55–58]. Control schemes should also include an appropriate modulation strategy, with particular emphasis on capacitor balancing. In this way, safe operation of the switches is achieved, avoiding over-voltage conditions. Table 5 summarizes the works on the modulation techniques for NPC inverters. There are three factors that must be considered for the selection of the modulation strategy [59]:

- The redundant switching states.
- The direction of the output current.
- The influence on the instantaneous value of the capacitors.

Table 5. Summary of recent works on modulation strategies in NPC inverters.

Ref./Year	Modulation Strategy	Contribution to the Field
[54]/2019	SVPWM	The authors present a modulation strategy (SVPWM) for 3L-NPC buck-boost inverters, which can be used with impedance sources.
[60]/2019	SPWM/DPWM	This article proposes an adaptive modulation technique for multilevel inverters. This strategy adjusts its switching states to provide a seamless transition from SPWM to DPWM and vice versa.
[61]/2019	DPWM	The proposed modulation scheme mitigates the imbalance in the capacitor voltages even during transients.
[62]/2019	PWM	The work proposes a optimized PWM highlighting its convergence ability for each operating condition.
[63]/2019	DPWM	The paper exposes a novel pulse sequence DPWM to reduce the switching losses of semiconductor devices, in addition, it controls the neutral point voltage.
[64]/2020	DPWM	A modulation strategy is proposed to optimize four types of DPWM. Optimization is performed according to the modified spatial vector.
[65]/2020	SPWM	This article analyzes the implementation of a MSCMM-SPWM to make easier the use of grid-connected inverters.
[66]/2020	PWM	A technique that ensures the stability and efficiency of the system is proposed. The strategy can select different modulation methods: unipolar, dipolar or partial-dipolar. The selection will correspond to the fundamental frequency and the inverter output current.
[67]/2020	PWM	The novelty of the work lies in the proposal of a method to reduce CMV and THD from 3L-NPC schemes.

Most of the methods [68] that reduce the common-mode base their principle on selecting the vectors corresponding to Common-Mode Voltage (CMV) lower or zero without considering the oscillation of the neutral point voltage. In [69], a novel virtual SVM where a zero NP current average and a low CMV in one control cycle is achieved. Martinez et al. in [70], present a comparative analysis on different modulation techniques used in PV inverters. The results of the work throw certain conclusions that are interesting:

- Phase Shifted-Pulse Width Modulation (PS-PWM) is a suitable solution for power filters, controlled rectifiers, etc., but this technique is not recommended for transformerless inverter applications.
- Two-Sectors Hybrid-PWM (2SH-PWM) is easy to implement, reduces leakage ground current and is more efficient than 3L-PWM.
- Six-Sectors Hybrid PWM (6SH-PWM) is capable of halving leakage ground current spikes compared to 2SH-PWM.
- Three-Level PWM (3L-PWM), the 2SH-PWM, and the 6SH-PWM are three modulation strategies that achieve the correct operation of the transformerless grid-connected systems.

An analysis of the lifetime of inverters for photovoltaic applications is carried out in [71], where an NPC based topology and a T-type inverter are compared. The authors conclude that inverters based on the NPC topology have a longer lifetime than T-type inverters. This conclusion exposes the durability and the use of this type of inverter in RES applications [72,73]. In this sense, authors in [74] summarize a group of inverter topologies used in RE applications, highlighting the presence of a low leakage current in each of them. Ma et al. in [72] propose a new PWM strategy for ANPC topologies, the scheme is illustrated in Figure 7. The cited work present a modulation strategy based on an adjustable losses distribution that offers excellent performance and an increase the efficiency of the topology of the 97%. The switching pattern is presented in Table 6.

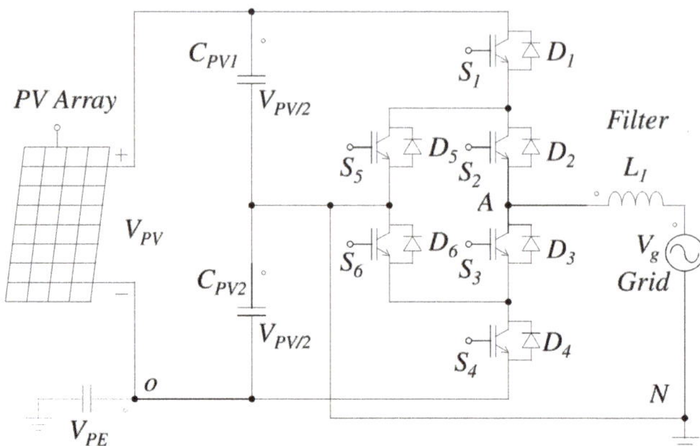

Figure 7. ANPC Half Bridge proposed in [72].

Table 6. Switches states of adjustable losses distribution of ANPC Half-Bridge inverter proposed in [72].

Output Voltage	S_1	S_2	S_3	S_4	S_5	S_6
Positive	1	1	0	0	0	1
0^{+In}	1	0	1	0	0	1
0^{+Out}	0	1	1	0	0	1
0^+	0	0	1	0	0	1
0^-	0	1	0	0	1	0
0^{-Out}	0	1	1	0	1	0
0^{-In}	0	1	1	1	1	0
Negative	0	0	1	1	1	0

Wang et al. in [75] propose a grid-connected 6S-5L-ANPC inverter. The topology reduces the number of switches since eight switches are generally used. This advantage reduces conduction and switching losses. An important comparison with traditional ANPC topologies considering the stress of semiconductor devices, the switching frequency, the switching losses, the conduction losses and the system volume is presented. In PV applications, special attention should be paid to THD. Therefore, the authors select the phase disposition PWM scheme as modulation strategy. This method directly affects the balance of flying capacitors. The proposal achieves a correct balance of the capacitors, also using a selection method to limit its voltage ripple. Figure 8 present the aforementioned topology and Table 7 shows the switching states of the inverter. In total, there are eight possible states. One of the most important results is a THD of 1.6%. The authors also present the equations for sizing capacitors in active and reactive power conditions. Equation (7) establishes the capacitor value under the condition of unit power factor and Equation (8) under reactive power condition.

$$C_{fc} = \frac{I_{pk}}{2\Delta V_{fc}f_s M} \tag{7}$$

$$C_{fc} = \frac{\sum_{n=1}^{N}\Delta Q_{fc}}{\Delta V_{fc}} = \frac{2MI_{pk}}{\Delta V_{fc}f_s}\sum_{n=1}^{\frac{\varphi f_s}{2\pi f_{Line}}}sin^2(n\frac{f_{Line}}{f_s})2\pi \tag{8}$$

where: f_{Line} represents the line, f_s is the switching frequency frequency, I_{pk} is the peak value of the output, ΔV_{fc} is the voltage drop, ΔQ_{fc} is the electric charge, the modulation index is defined as M and N is the number of switching cycles.

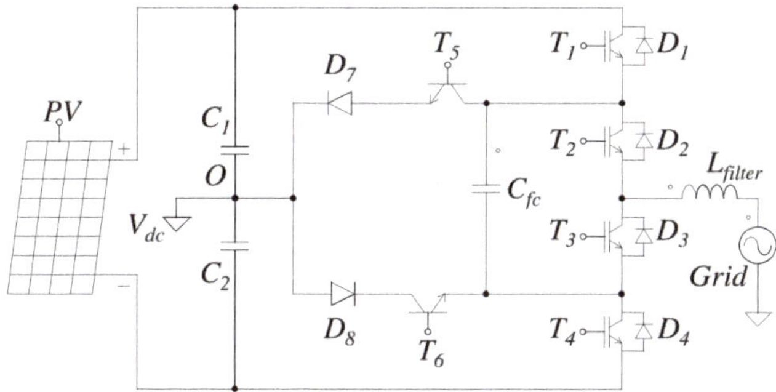

Figure 8. Proposed ANPC inverter topology in [75].

Table 7. Operation modes of the proposed topology in [75].

No	T_1	T_2	T_3	T_4	T_5	T_6	V_{out}	$i_{out} > 0$	$i_{out} < 0$
	Active Switch State							**Flying Capacitor** C_{fc}	
A	1	1	0	0	0	1	+2	-	-
B	1	0	1	0	0	1	+1	Charge	Discharge
C	0	1	0	0	0	1	+1	Discharge	Charge
D	0	0	1	0	0	1	+0	-	-
E	0	1	0	0	1	0	−0	-	-
F	0	0	1	0	1	0	−1	Charge	Discharge
G	0	1	0	1	1	0	−1	Discharge	Charge
H	0	0	1	1	1	0	−2	-	-

4. Flying Capacitor Based Topologies

The FC concept was first introduced in 1992; this type of inverter uses different capacitors to deliver various levels of power at the converter output [76]. The topology benefits include attractive properties in different power ranges, however they are more suitable for medium-voltage applications. Another advantage of topology is the possibility of using natural self-balancing. Furthermore, it has an equitable distribution of voltage stress between switches [77]. Also, as in the case of NPC, a single source can be used to generate multiple voltage levels.In commercial applications, the use of FC with more than three output levels are more common than the NPC alternative [78]. The presented topology is generally not used in PV applications. The scheme is more suitable for use in electric vehicles. However, it was decided to include it in work, since it is part of the most widely used voltage-source topologies. Figure 9 illustrates a Five-Level FC (5L-FC) inverter and Table 8 the operation modes are shown.

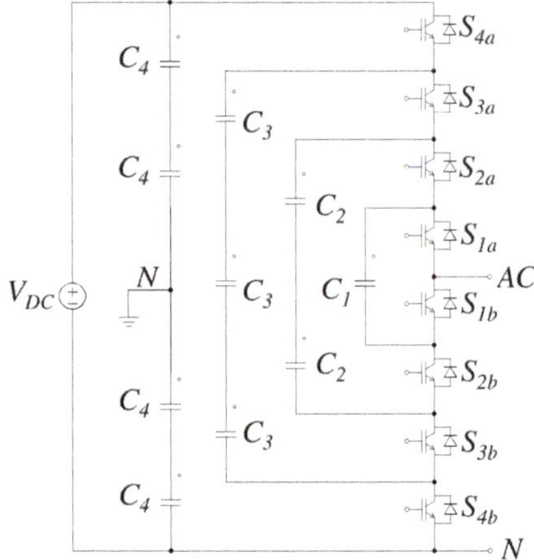

Figure 9. Basic configuration of 5-level inverter based on FC.

Table 8. 5L-FC operation modes.

Output Voltage	S_{4a}	S_{3a}	S_{2a}	S_{1a}	S_{1b}	S_{2b}	S_{3b}	S_{4b}
$V_{DC}/2$	1	1	1	1	0	0	0	0
$V_{DC}/4$	1	1	1	0	1	0	0	0
0	1	1	0	1	1	1	0	0
$-V_{DC}/4$	1	0	0	1	1	1	1	0
$-V_{DC}/2$	0	0	0	0	1	1	1	1

Despite the advantages mentioned, FC inverters have certain limitations that are addressed in current works. For example, capacitor banks reduce the life of the system, and sometimes the balance of floating capacitors can be complex [79]. The problem of capacitor voltage balance is the main limitation of the use of the FC topology. Consequently, the scheme has not been generally used in PV applications. In the last decade, investigations related have been reported, for example, in [80], using a D_A-D_B duty cycle mismatch measurement between two groups of Three-Level Flying Capacitor (3L-FC) topology switches to control the system without any additional detection. Table 9 summarizes several works between the years 2015–2019.

Table 9. Recent work on voltage balancing in FC.

Ref./Year	V_{in}/P_{in}	Output Level	Balancing Method
[81]/2015	100 V	5	Phase-Disposition Pulse Width Modulation (PD-PWM)
[82]/2017	600 V	5	Logic-Form Equations
[83]/2019	120 V	3	Proportional-Integral
[84]/2019	200 V	4	Valley Current Detection
[85]/2019	100 V	3	Time-Domain Power Averaging-Based Approach

In [86] a novel converter is proposed, which has certain advantages, for example, reduced voltage stress on semiconductors, a wide voltage gain and a common grounded scheme. These characteristics can be commonly found in this type of inverter, however, FC-based converters are not generally used in RES, since they require a large number of input capacitors that increase the complexity of the techniques for balancing them [87].

THD is reduced with more energy levels at the inverter output. In the case of FC, requires a large number of capacitors. A derivation of this configuration has been presented in [88], where cross-connecting capacitors have achieved a higher voltage level at the output through additional switches.Another of the biggest challenges in this topology is to provide the necessary energy to activate the large number of switches that the scheme has. Ye et al. in [89] present the comparison of five methods that reduce space and increase the efficiency of gate drive power supply circuits. Also, the operation of a multilevel FC inverter where an additional circuit is provided to avoid the defective cell, if it exists, is presented in [90]. However, in this topology, the elements have to be oversized to operate at full-power level when the failure of one of the cells is detected.

The FC design must have several considerations. Various design methodologies are found in the bibliography. For example, authors in [91] propose a methodology based on harmonic representation of the switching functions. The advantage of the proposed methodology lies in the possibility of being extrapolated to any FC-based scheme. Currently the uses of the treated scheme are very varied. There are certain applications in which a DC-bus is used due to voltage variations. Large capacitors are connected in parallel to the bus to avoid such voltage variations. Some of this applications are back-to-back converters, Power Factor Compensators (PFC), and uninterruptible power supply. The FC topology is chosen as the infinite virtual capacitor converter, which is a nonlinear capacitor where the voltage dependence of the load has a flat region and the voltage remains constant [92].

5. Cascaded Based Topologies

The CMLI integrates multiple H-bridge schemes to generate a multilevel voltage [93]. The scheme has certain advantages compared to NPC and FC topologies, for example, they do not employ clamping diodes, in addition, a greater number of energy sources making it more suitable for specific applications such as electric vehicle [94] and PV applications [95]. Another advantage of the CMLI scheme is that, if any device fails in the bridge, the converter will continue to operate although it will deliver less energy. Therefore, this configuration is, to some extent, fault-tolerant. Also, its modularity and smaller filter size make it more attractive for high and medium-power PV applications. Figure 10 present a basic configuration of the CMLI topology, and Table 10 shows the switching pattern for the five output levels.

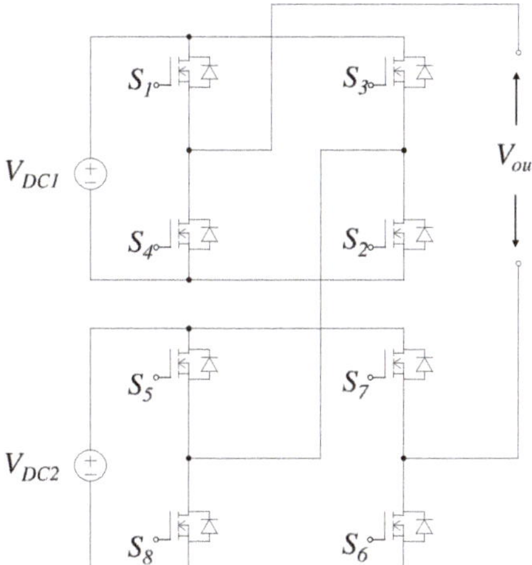

Figure 10. 5L-CMLI basic configuration.

Table 10. 5L-CMLI switching table.

Output Voltage	S1	S2	S3	S4	S5	S6	S7	S8
$2V_{DC}$	1	1	0	0	1	1	0	0
V_{DC}	1	1	0	0	0	1	0	1
0	0	1	0	1	0	1	0	1
$-V_{DC}$	0	0	1	1	0	1	0	1
$-2V_{DC}$	0	0	1	1	0	0	1	1

Despite the aforementioned advantages, the topology has certain limitations. The main disadvantage is the use of isolated DC sources for each H-bridge. This problem was solved in the FC and NPC topology, but the voltage adjustment of the capacitors is complex [96]. Also, during partial shading the energy captured by the system is reduced. In certain investigations have presented various studies about the partial shading of PV modules, but most of the schemes are complex designs that generally cause a decrease in the efficiency of the system and an increase in the cost of the inverter [97,98].

The selection of the controller in CMLI depends on the topology and the application. Each controller has favorable characteristics in certain systems, ranging from less complexity to a desired dynamic response. Various types of controllers are used with the scheme discussed. The most widely used are PR controller and PI. When used LC or LCL filters using traditional controllers such as PI is not appropriate because it does not completely eliminate the steady state error [99]. Control systems that employ proportional-resonant controllers eliminate steady-state error. These controllers provide infinite gain in resonance frequency. Equation (9) defines the ideal PR control. In [100] a new technique to synchronize MLI with the grid using PR controller is shown. In the presented design, the control scheme has a lower error between the real power and the reference compared to the PI controller.

$$G(S) = k_P + \frac{k_i S}{S^2 + w_0^2} \tag{9}$$

where: w_0 is the fundamental frequency (grid frequency), k_p and k_i represent proportional and resonant gains respectively.

Grid-connected systems can be classified according to the Maximum Power Point Tracking (MPPT) method used. The two classifications are centralized or distributed. The distributed technique reports better efficiency in the literature, but is more complex and has a larger volume than the centralized MPPT methods [101]. In the case of cascade inverters, the implementation of a distributed method requires a large number of sensors and considerably increases the cost of the system. Figure 11 illustrates the most common architectures in the distributed MPPT method. In micro-inverters of the Figure 11a, the energy generated by the different modules is injected directly into the grid. In front-end DC optimizers presented in Figure 11b, the converters perform the MPPT separately. Its output is connected in series; thus, the power that is injected into the grid is the sum of each module. If a module has low efficiency, it does not affect the rest of the converters, since each module provides its power separately.

Authors in [102] present a low cost and straightforward distributed MPPT method for energy optimizers in CMLI-based photovoltaic systems using front-end DC optimizers. In [103] a simplified feedforward distributed MPPT method for grid-connected CMLI is presented. The authors use the method as a *"... superior solution for PV system grid integration due to its simple implementation, signal stage power conversion, no added complexity with increasing the number of connected modules, and it eliminates the need for individual control loop for each module..."* It is important to mention that most conventional techniques do not achieve a distributed MPPT, which decreases the efficiency of the system. Goetz et al. in [104] propose a modular Double Cascading H-bridge (CHB^2) topology.

The scheme reduces the output of the inverter filter and achieves a fast dynamic response. The MPPT is carried out in each module, which incorporates a battery for energy storage. The use of batteries in PV applications is avoided since these elements are highly polluting, then they break with the concept of clean energy. One of the essential advantages of the topology is the possibility of being extended to CHB^2 circuits in general.

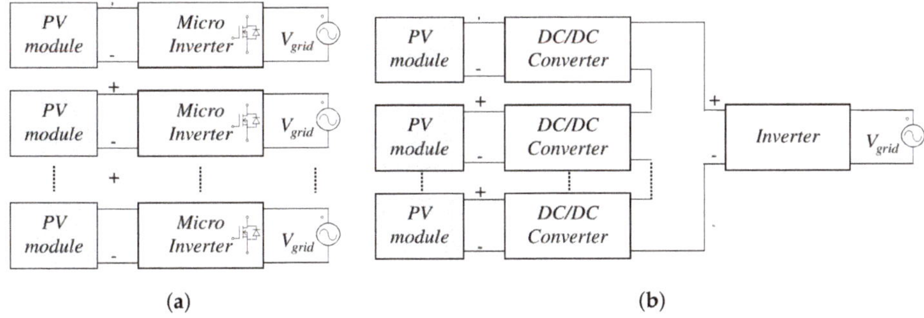

(a)   (b)

Figure 11. Distributed MPPT architectures: (**a**) micro-inverters and (**b**) front-end DC optimizers.

A brief bibliographic review shows the use of topology as a fault-tolerance scheme. In photovoltaic applications, faults distort the output voltage, degrading the power supplied. A fault diagnosis scheme must detect the problem in the shortest possible time to avoid serious failures in the system, and each design requires its strategy. A common problem is considering that the system is in open circuit to monitor its status [105]. Shao et al. in [106], present a technique for detecting faults using Sliding Mode Observer (SMO) able to locate the fault element in the system. The authors in [107] present a detailed review of various faults in photovoltaic systems. The work identifies the main faults as line-line, earth, arc, shadow and others, and proposes its protection strategy. Various strategies are employed for fault detection in PV systems, using standard protection devices or offline/real-time testing of PV systems. Table 11 shows a summary of typical failure cases and respective protection/detection devices.

Table 11. Typical fault occurrences and respective protection/detection devices present in [107].

Fault	Severity	Occurrence	Protection Devices
Single ground faults	High	Common	GFDI, RCD, IMD
Double ground faults	Very high	Rare	GFDI, OCPD, RCD, IMD
Line-line faults	High	Common	OCPD
Series arc faults	Very high	Rare	AFCI, AFD
Parallel arc faults	Very high	Rare	Not Available
Temporary shading	Low	Frequent	Not Available
Permanent shading	High	Frequent	Not Available
Open circuit faults	Low	Rare	ECM, Line Checker

Note: Over Current Protection Devices (OCPDs), Ground Fault Detection and Interruption (GFDI) fuses/Ground Fault Protection Devices (GFPDs), Arc Fault Circuit Interrupters (AFCIs), Residual Current Monitoring Device (RCD), Insulation Monitoring Device (IMD), Earth Capacitance Measurement (ECM).

Currently, there is a trend towards the combined use of PV energy and batteries as a storage medium and certain studies analyze its feasibility. For example, authors in [108] conducts extensive research concluded in 2019, in Finland. The work considers the profitability of BESS investments between the years 2018 and 2035. The authors conclude that, these systems would not be profitable for RES applications at present, although a decrease in costs is estimated from 1270 to 1370 euros/kWh in 2018 to 830–930 euros/kWh in 2035. However, sometimes an uninterrupted flow of energy is required and the use of batteries is essential due to PV power intermittent nature. The BESS output is controllable and the system can be treated as a controllable load [109].

The grid-connected schemes in this typology, as in the previous ones, still have certain challenges. Decreasing the leakage current in transformerless systems is one of the main aspects to consider. In this regard, some research has been carried out, such as [110]. In this work an analysis of the behavior of the leakage current of the different modes of operation of the basic structure CMLI is presented. The analysis considers the common-mode inductor in each switching state and can be simpler if pole voltages are used. Also, the authors propose two schemes of suppression of the leakage current. The first solution uses low-capacitance common-mode capacitors and stray capacitors as part of the output filter. It is included that this solution is suitable for inverters operated at high-frequencies. Solution two is appropriate when the converter uses a switching frequency of less than 1.5 kHz.

Sonti et al. in [111] present a PWM technique to eliminate or reduce leakage current in CMLI-based schemes. The work integrates the applied MPPT and PWM algorithm. In this way, it is possible to reduce the high-frequency transitions of voltage and the CMV. Figure 12 shows the proposed architecture and Table 12 present the switching states. The switches S_{w4}, S_{w5} and S_{w6}, S_{w7} operate in a complementary way. Hence, there are three pairs of switches $[S_{w1}, S_{w2}, S_{w3}]$, $[S_{w4}, S_{w5}]$ and $[S_{w6}, S_{w7}]$. Modulation proposal isolates PV array and grid during freewheeling states, operating similarly to an H5 topology. The authors achieve leakage current reduction with the presence of low-frequency transitions in the PV terminal voltage.

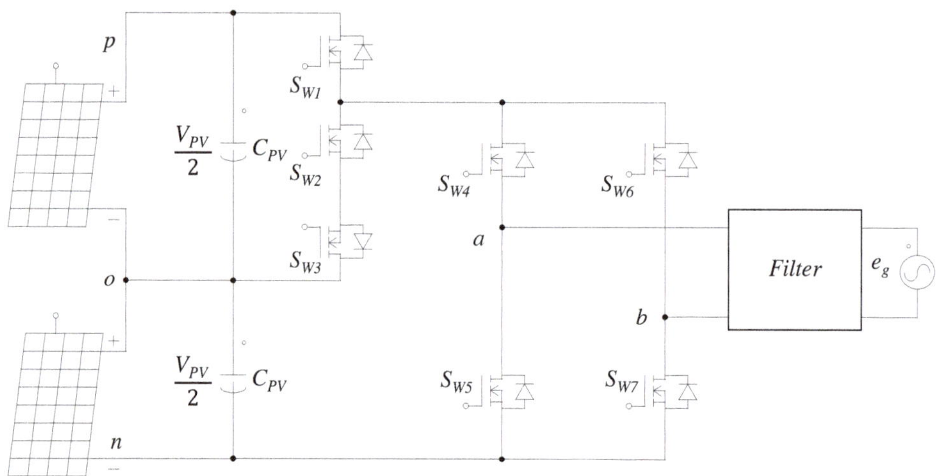

Figure 12. Five-level CMLI proposed in [111].

Table 12. Operation mode of the topology proposed in [111].

S_{w1}	S_{w2}	S_{w4}	S_{w6}	Output Voltage
1	0	1	0	$+V_{PV}$
0	1	1	0	$+V_{PV}/2$
0	0	1	1	0
0	1	0	1	$-V_{PV}/2$
1	0	0	1	$-V_{PV}$

6. Comparative Study

The section present a comparative table between NPC and CMLI schemes. As mentioned above, at present, FC-based topology is not generally used in PV applications. The comparison takes into account different aspects such as input voltage value, switching frequency, control strategy, efficiency and leakage current and only includes recent works. Other recent works establish different comparisons according to the topic they address. The authors in [112] show five control methods based on SMC, the comparison is made considering the topology, the modes of operation and the number of sensors required. Lee in [113], establishes a comparative analysis of recent topologies based on cascade inverters, which reveals that the S^3CM proposed in the document achieves the reduction of the switch count.

It can be appreciated that the switching frequencies rarely exceed 15 kHz. Increasing the switching speed causes an increase in the system losses and the THD. There is a compromise between the number of output levels and the speed of the switches. Increasing the number of power levels imply slower switching frequency.

All the works presented comply with the two most important grid connection standards. These standards establish a maximum leakage current and THD of 300 mA and 5%, respectively. Compliance with standards largely depends on the strategy of modulation employed. From the data presented, it is observed that the SVPWM technique and its variants are the most recurrent when it is required to decrease the voltage in common-mode topologies.

Another element that is noted is the use of inverters with 3 output levels. The authors note a compromise between the number of levels and the number of elements in their topology. Generally, with 3 levels, satisfactory results are obtained, as showed in Table 13.

Table 13. Multilevel inverters used on PV applications.

Ref./Year	V_{in}/P_{in}	f_s	Output Levels	Strategy	Leakage Current/THD	Eff./Power Loss
[114]/2015	200-450 V	5 kHz	5	Proportional Resonant Control	2%	97.3%
[115]/2015	400 V	16 kHz	3	SPWM	80 mA	97%
[116]/2017	200 V/1 kW	10 kHz	3	Active NPC Method	≈0 mA	≈95.5%
[117]/2017	190 V	2.2 kHz	3	Model Predictive Control	≈0 mA	≈95.8%
[118]/2017	400 V/500 W	24 kHz	3	Proportional Resonant Control	≈0 mA	97.4%
[119]/2017	400 V	8 kHz	3	SVPWM	1%	-
[120]/2018	150 V	10 kHz	5	A novel modulation strategy	<300 mA	-
[121]/2018	1 kW	2–3 kHz	3	Model Predictive Control	≈25 mA	-
[122]/2019	102 V	10.02 kHz	5	-	2.85%	-
[123]/2019	450 V	60 Hz	21	Proportional Integral Control	27 mA/4.6%	98.5%
[112]/2019	200 V	2.5 kHz	3	Sliding Mode Control	2.1%	90%
[70]/2019	220 V	10 kHz	5	Three-Level PWM	7.18 mA	≈96%
[124]/2019	200 V	5 kHz	5	PWM modified	170 mA	-
[125]/2019	220 V	5 kHz	3	SV-PWM	100 mA	95%
[65]/2020	350–600 V	10 kHz	5	Multimodulation SPWM	<5%	-
[126]/2020	168 V	5 kHz	7	CB-PWM	2.03%	-

Table 13. *Cont.*

Ref./Year	V_{in}/P_{in}	f_s	Output Levels	Strategy	Leakage Current/THD	Eff./Power Loss
[127]/2020	200 V	5 kHz	5	COPWM	4.32%	-
[128]/2020	320 V	10 kHz	3	Strategy based on duty-cycle function	0.7%	90.9%
[129]/2020	260 V	15 kHz	3	SVPWM	2.35%	-
[130]/2020	500 V	9.4 kHz	3	MPC-based virtual vector modulation	3.5%	-

7. Conclusions

Nowadays, inverter technology is achieving efficiencies above 98%, leaving little room for improvement for future work. It is important to note that the use of each of the schemes is conditioned by the requirements to be met in each application. There are several interesting points to highlight:

- In general, the multilevel inverters schemes focus on reducing the total cost of ownership and the number of the switches.
- There is a close relationship among the efficiency, cost, and complexity, a relationship that is evident when analyzing the main parameters obtained in each design, such as THD, leakage current and efficiency.
- The NPC topology has high efficiency and low leakage current in transformerless schemes, making it attractive in RE applications.
- Certain disadvantages of the NPC topology were detected, for example, it uses additional numbers of clamping diodes to achieve a higher number of output levels. Neutral-point voltage balancing problem is the main challenge of the topology. Different factors such as the modulation index, load current, and fundamental frequency must be considered to achieve a correct balance of the capacitors. In addition, it presents an unequal distribution of its losses.
- The ANPC topology is used to overcome these drawbacks, which directly affect the useful life of the system and, therefore, the investment cost.
- An important aspect of the ANPC scheme is the possibility of using various modulation schemes to obtain the lowest loss distribution. Thus, ANPC topology is more suitable for applications of high-power transformerless PV systems.
- The FC-based topology is not commonly used in RES applications. It should be noted that the voltage balancing of flying capacitors in each PWM cycle, which guarantees the safe operation of the converter, is a crucial topic in these topologies. Furthermore, research trying to reduce the cost and volume of floating capacitors, especially when the number of cells increases.
- The CMLI topology is emerging as an excellent interface between different RES sources and the grid, offering high efficiency and fault tolerance capabilities. This topology is suitable when energy needs to be obtained from several RES.
- In this sense, within the current challenges of the CMLI is the ability to respond to fluctuations and the drop of some of its DC sources (generally, 20% and 80%, respectively). Furthermore, the researches seek to integrate various RES.

Author Contributions: Conceptualization, A.A.E.-B. and A.A.-D.; methodology, A.A.E.-B.; formal analysis, A.A.E.-B.; investigation, A.A.E.-B. and A.A.-D.; resources, J.R.-R.; data curation, A.A.E.-B., and A.A.-D. and J.R.-R.; writing–original draft preparation, A.A.E.-B. and A.A.-D.; writing–review and editing, A.A.E.-B. and A.A.-D.; supervision, J.R.-R. All authors have read and agreed to the published version of the manuscript.

Funding: This research received funding from PRODEP and CONACYT.

Conflicts of Interest: The authors declare no conflict of interest.

Abbreviations

The following abbreviations are used in this manuscript:

RES	Renewable Energy Sources
PV	Photovoltaic
DC	Direct Current
AC	Alternating Current
BESS	Battery Energy Storage System
CSI	Current-Source Inverter
VSI	Voltage-Source Inverter
ZSI	Impedance-Source Inverter
PI	Proportional-Integral
PR	Proportional-Resonant
MPC	Model Predictive Control
NPC	Neutral-Point-Clamped
NP	Neutral Point
ANPC	Active Neutral-Point-Clamped
3L-NPC	Three-Level Neutral-Point-Clamped
3L-ANPC	Three-Level Active Neutral-Point-Clamped
MVSI	Multilevel Voltage-Source Inverter
FC	Flying Capacitor
CMLI	Cascaded Multilevel Inverter
MLI	Multilevel Inverter
EMI	Electromagnetic Interference
THD	Total Harmonic Distortion
PWM	Pulse With Modulation
DSP	Digital Signal Processor
FPGA	Field-Programmable Gate Array
DPC	Direct Power Control
SVM	Space Vector Modulation
SVPWM	Space Vector PWM
SPWM	Senoidal PWM
DPWM	Digital PWM
MSCMM-SPWM	Modified Single-Carrier and Multimodulation Sine PWM
PS-PWM	Phase Shifted-PWM
6S-5L-ANPC	Six Switches Five Levels ANPC
2SH-PWM	Two-Sectors Hybrid-PWM
3L-PWM	Three-Level PWM
CMV	Common-Mode Voltage
PS-PWM	Phase Shifted-PWM
2SH-PWM	Two-Sectors Hybrid-PWM
3L-PWM	Three-Level PWM
CMV	Common-Mode Voltage
MPPT	Maximum Power Point Tracking
CHB	Cascading H-Bridge
OCPDs	Over Current Protection Devices
GFDI	Ground Fault Detection and Interruption
GFPDs	Ground Fault Protection Devices
AFCIs	Arc Fault Circuit Interrupters
RCD	Residual Current Monitoring Device
IMD	Insulation Monitoring Device
ECM	Earth Capacitance Measurement

References

1. Aleem, S.A.; Hussain, S.M.; Ustun, T.S. A review of strategies to increase PV penetration level in smart grids. *Energies* **2020**, *13*, 636. [CrossRef]
2. Kumar, N.; Saxena, V.; Singh, B.; Panigrahi, B.K. Intuitive control technique for grid connected partially shaded solar PV-based distributed generating system. *IET Renew. Power Gener.* **2020**, *14*, 600–607. [CrossRef]
3. Raturi, A.K. *Renewables 2019 Global Status Report*; REN21 Secretariat: Paris, France, 2019.
4. Kumar, A.; Sharma, S.; Verma, A. Optimal sizing and multi-energy management strategy for PV-biofuel-based off-grid systems. *IET Smart Grid* **2020**, *3*, 83–97. [CrossRef]
5. Gatta, F.M.; Geri, A.; Lauria, S.; Maccioni, M.; Palone, F.; Portoghese, P.; Buono, L.; Necci, A. Replacing diesel generators with hybrid renewable power plants: Giglio smart island project. *IEEE Trans. Ind. Appl.* **2019**, *55*, 1083–1092. [CrossRef]
6. Sangwongwanich, A.; Yang, Y.; Sera, D.; Blaabjerg, F.; Zhou, D. On the impacts of PV array sizing on the inverter reliability and lifetime. *IEEE Trans. Ind. Appl.* **2018**, *54*, 3656–3667. [CrossRef]
7. Zhong, Z.; Zhang, Y.; Shen, H.; Li, X. Optimal planning of distributed photovoltaic generation for the traction power supply system of high-speed railway. *J. Clean. Prod.* **2020**, *263*, 121394. [CrossRef]
8. Naderi, E.; Pourakbari-Kasmaei, M.; Lehtonen, M. Transmission expansion planning integrated with wind farms: A review, comparative study, and a novel profound search approach. *Int. J. Electr. Power Energy Syst.* **2020**, *115*, 105460. [CrossRef]
9. Khatib, T.; Mohamed, A.; Sopian, K. A review of photovoltaic systems size optimization techniques. *Renew. Sustain. Energy Rev.* **2013**, *22*, 454–465. [CrossRef]
10. Hussin, M.Z.; Omar, A.M.; Shaari, S.; Sin, N.D.M. Review of state-of-the-art: Inverter-to-array power ratio for thin—Film sizing technique. *Renew. Sustain. Energy Rev.* **2017**, *74*, 265–277. [CrossRef]
11. Wirth, H.; Schneider, K. *Recent Facts about Photovoltaics in Germany*; Fraunhofer ISE: Freiburg, Germany, 2015; Volume 92.
12. Vázquez, N.; Baeza, E.; Perea, A.; Hernández, C.; Vázquez, E.; López, H. "Z" and "qZ" source inverters as electronic ballast. *IEEE Trans. Power Electron.* **2016**, *31*, 7651–7660. [CrossRef]
13. Tapia Hector Jua, L.; Rodriguez Jose Juan, A.; Gonzalez Aurelio, D.; Resendiz Juvenal, R. Eight levels multilevel voltage source inverter modulation technique. *IEEE Lat. Am. Trans.* **2018**, *16*, 1121–1127. [CrossRef]
14. Yong, J.; Li, X.; Xu, W. Interharmonic source model for current-source inverter-fed variable frequency drive. *IEEE Trans. Power Deliv.* **2017**, *32*, 812–821. [CrossRef]
15. Komurcugil, H.; Altin, N.; Ozdemir, S.; Sefa, I. An extended lyapunov-function-based control strategy for single-phase UPS inverters. *IEEE Trans. Power Electron.* **2015**, *30*, 3976–3983. [CrossRef]
16. Esteve, V.; Jordán, J.; Sanchis-Kilders, E.; Dede, E.J.; Maset, E.; Ejea, J.B.; Ferreres, A. Enhanced pulse-density-modulated power control for high-frequency induction heating inverters. *IEEE Trans. Ind. Electron.* **2015**, *62*, 6905–6914. [CrossRef]
17. Wen, X.; Fan, T.; Ning, P.; Guo, Q. Technical approaches towards ultra-high power density SiC inverter in electric vehicle applications. *CES Trans. Electr. Mach. Syst.* **2017**, *1*, 231–237. [CrossRef]
18. Mondol, M.H.; Tur, M.R.; Biswas, S.P.; Hosain, M.K.; Shuvo, S.; Hossain, E. Compact three phase multilevel inverter for low and medium power photovoltaic systems. *IEEE Access* **2020**, *8*, 60824–60837. [CrossRef]
19. Sahan, B.; Araujo, S.V.; Noding, C.; Zacharias, P. Comparative evaluation of three-phase current source inverters for grid interfacing of distributed and renewable energy systems. *IEEE Trans. Power Electron.* **2011**, *26*, 2304–2318. [CrossRef]
20. Beig, A.R.; Dekka, A. Experimental verification of multilevel inverter-based standalone power supply for low-voltage and low-power applications. *IET Power Electron.* **2012**, *5*, 635–643. [CrossRef]
21. Aqeel Anwar, M.; Abbas, G.; Khan, I.; Awan, A.B.; Farooq, U.; Saleem Khan, S.; Majeed, R. An impedance network-based three level quasi neutral point clamped inverter with high voltage gain. *Energies* **2020**, *13*, 1261. [CrossRef]
22. Rana, R.A.; Patel, S.A.; Muthusamy, A.; Lee, C.W.; Kim, H.J. Review of multilevel voltage source inverter topologies and analysis of harmonics distortions in FC-MLI. *Electronics* **2019**, *8*, 1329. [CrossRef]
23. Yuan, W.; Wang, T.; Diallo, D.; Delpha, C. A fault diagnosis strategy based on multilevel classification for a cascaded photovoltaic grid-connected inverter. *Electronics* **2020**, *9*, 429. [CrossRef]

24. Devi, G.R.; Rajesh, P.; Sathish, S.; Sivaraman, S.; Fayaz, S.M. Performance investigation of hexagram inverter for high power applications. In Proceedings of the 2019 IEEE International Conference on System, Computation, Automation and Networking (ICSCAN), Pondicherry, India, 29–30 March 2019; pp. 1–8. [CrossRef]
25. Jiao, L.; Qiu, D.; Zhang, B.; Chen, Y. A hybrid nine-arm high-voltage inverter with DC-fault blocking capability. *Energies* **2019**, *12*, 3850. [CrossRef]
26. Rodriguez, J.; Kazmierkowski, M.P.; Espinoza, J.R.; Zanchetta, P.; Abu-Rub, H.; Young, H.A.; Rojas, C.A. State of the art of finite control set model predictive control in power electronics. *IEEE Trans. Ind. Inform.* **2013**, *9*, 1003–1016. [CrossRef]
27. Panten, N.; Hoffmann, N.; Fuchs, F.W. Finite control set model predictive current control for grid-connected voltage-source converters with LCL filters: A study based on different state feedbacks. *IEEE Trans. Power Electron.* **2016**, *31*, 5189–5200. [CrossRef]
28. Yang, S.; Lei, Q.; Peng, F.Z.; Qian, Z. A robust control scheme for grid-connected voltage-source inverters. *IEEE Trans. Ind. Electron.* **2011**, *58*, 202–212. [CrossRef]
29. Kumar, N.; Saha, T.K.; Dey, J. Control, implementation, and analysis of a dual two-level photovoltaic inverter based on modified proportional-resonant controller. *IET Renew. Power Gener.* **2018**, *12*, 598–604. [CrossRef]
30. Revana, G.; Kota, V.R. Simulation and implementation of resonant controller based PV fed cascaded boost-converter three phase five-level inverter system. *J. King Saud Univ. Eng. Sci.* **2019**. [CrossRef]
31. Gupta, K.K.; Ranjan, A.; Bhatnagar, P.; Sahu, L.K.; Jain, S. Multilevel inverter topologies with reduced device count: A review. *IEEE Trans. Power Electron.* **2016**, *31*, 135–151. [CrossRef]
32. Estévez-Bén, A.A.; López Tapia, H.J.C.; Carrillo-Serrano, R.V.; Rodríguez-Reséndiz, J.; Vázquez Nava, N. A new predictive control strategy for multilevel current-source inverter grid-connected. *Electronics* **2019**, *8*, 902. [CrossRef]
33. Rodriguez, J.; Lai, S.-J.; Peng, F.Z. Multilevel inverters: a survey of topologies, controls, and applications. *IEEE Trans. Ind. Electron.* **2002**, *49*, 724–738. [CrossRef]
34. Fazel, S.S.; Bernet, S.; Krug, D.; Jalili, K. Design and comparison of 4-kV neutral-point-clamped, flying-capacitor, and series-connected H-bridge multilevel converters. *IEEE Trans. Ind. Appl.* **2007**, *43*, 1032–1040. [CrossRef]
35. Cheng, Y.; Qian, C.; Crow, M.L.; Pekarek, S.; Atcitty, S. A comparison of diode-clamped and cascaded multilevel converters for a STATCOM with energy storage. *IEEE Trans. Ind. Electron.* **2006**, *53*, 1512–1521. [CrossRef]
36. Araujo, S.V.; Zacharias, P. Analysis on the potential of Silicon Carbide MOSFETs and other innovative semiconductor technologies in the photovoltaic branch. In Proceedings of the 2009 13th European Conference on Power Electronics and Applications, Barcelona, Spain, 8–10 September 2009; pp. 1–10.
37. Siddique, M.D.; Mekhilef, S.; Shah, N.M.; Sarwar, A.; Iqbal, A.; Memon, M.A. A new multilevel inverter topology with reduce switch count. *IEEE Access* **2019**, *7*, 58584–58594. [CrossRef]
38. Omer, P.; Kumar, J.; Surjan, B.S. A review on reduced switch count multilevel inverter topologies. *IEEE Access* **2020**, *8*, 22281–22302. [CrossRef]
39. Rawa, M.; Siddique, M.D.; Mekhilef, S.; Mohamed Shah, N.; Bassi, H.; Seyedmahmoudian, M.; Horan, B.; Stojcevski, A. Dual input switched-capacitor-based single-phase hybrid boost multilevel inverter topology with reduced number of components. *IET Power Electron.* **2020**, *13*, 881–891. [CrossRef]
40. Xiao, H.F.; Liu, X.P.; Lan, K. Zero-voltage-transition full-bridge topologies for transformerless photovoltaic grid-connected inverter. *IEEE Trans. Ind. Electron.* **2014**, *61*, 5393–5401. [CrossRef]
41. Vosoughi, N.; Hosseini, S.H.; Sabahi, M. Single-phase common-grounded transformer-less grid-tied inverter for PV application. *IET Power Electron.* **2020**, *13*, 157–167. [CrossRef]
42. Mahmood, H.; Jiang, J. Modeling and control system design of a grid connected VSC considering the effect of the interface transformer type. *IEEE Trans. Smart Grid* **2012**, *3*, 122–134. [CrossRef]
43. Guo, X.; Wang, N.; Wang, B.; Lu, Z.; Blaabjerg, F. Evaluation of three-phase transformerless DC-bypass PV inverters for leakage current reduction. *IEEE Trans. Power Electron.* **2020**, *35*, 5918–5927. [CrossRef]
44. Estévez-Bén, A.A.; Alvarez-Diazcomas, A.; Macias-Bobadilla, G.; Rodriguez-Resendiz, J. Leakage current reduction in single-phase grid-connected inverters—A review. *Appl. Sci.* **2020**, *10*, 2384. [CrossRef]

45. Li, W.; Gu, Y.; Luo, H.; Cui, W.; He, X.; Xia, C. Topology review and derivation methodology of single-phase transformerless photovoltaic inverters for leakage current suppression. *IEEE Trans. Ind. Electron.* **2015**, *62*, 4537–4551. [CrossRef]

46. Lin, H.; Leon, J.I.; Luo, W.; Marquez, A.; Liu, J.; Vazquez, S.; Franquelo, L.G. Integral sliding-mode control-based direct power control for three-level NPC converters. *Energies* **2020**, *13*, 227. [CrossRef]

47. Choudhury, A.; Pillay, P.; Williamson, S.S. Comparative analysis between two-level and three-level DC/AC electric vehicle traction inverters using a novel DC-link voltage balancing algorithm. *IEEE J. Emerg. Sel. Top. Power Electron.* **2014**, *2*, 529–540. [CrossRef]

48. Madhusoodhanan, S.; Mainali, K.; Tripathi, A.; Patel, D.; Kadavelugu, A.; Bhattacharya, S.; Hatua, K. Harmonic analysis and controller design of 15 kV SiC IGBT-based medium-voltage grid-connected three-phase three-level NPC converter. *IEEE Trans. Power Electron.* **2017**, *32*, 3355–3369. [CrossRef]

49. Khojakhan, Y.; Choo, K.M.; Won, C.Y. Stator inductance identification based on low-speed tests for three-level NPC inverter-fed induction motor drives. *Electronics* **2020**, *9*, 183. [CrossRef]

50. Zhang, L.; Sun, K.; Feng, L.; Wu, H.; Xing, Y. A family of neutral point clamped full-bridge topologies for transformerless photovoltaic grid-tied inverters. *IEEE Trans. Power Electron.* **2013**, *28*, 730–739. [CrossRef]

51. Wang, K.; Zheng, Z.; Li, Y. Topology and control of a four-level ANPC inverter. *IEEE Trans. Power Electron.* **2020**, *35*, 2342–2352. [CrossRef]

52. Cervone, A.; Brando, G.; Dordevic, O.; Pizzo, A.D.; Meo, S. An adaptive multistep balancing modulation technique for multipoint-clamped converters. *IEEE Trans. Ind. Appl.* **2020**, *56*, 465–476. [CrossRef]

53. Guzman-Guemez, J.; Laila, D.S.; Sharkh, S.M. State-space approach for modelling and control of a single-phase three-level NPC inverter with SVPWM. In Proceedings of the 2016 IEEE Power and Energy Society General Meeting (PESGM), Boston, MA, USA, 17–21 July 2016; pp. 1–5. [CrossRef]

54. Shults, T.E.; Husev, O.; Blaabjerg, F.; Roncero-Clemente, C.; Romero-Cadaval, E.; Vinnikov, D. Novel space vector pulsewidth modulation strategies for single-phase three-level NPC impedance-source inverters. *IEEE Trans. Power Electron.* **2019**, *34*, 4820–4830. [CrossRef]

55. Malinowski, M.; Jasinski, M.; Kazmierkowski, M.P. Simple direct power control of three-phase PWM rectifier using space-vector modulation (DPC-SVM). *IEEE Trans. Ind. Electron.* **2004**, *51*, 447–454. [CrossRef]

56. Hu, J.; Zhu, Z.Q. Improved voltage-vector sequences on dead-beat predictive direct power control of reversible three-phase grid-connected voltage-source converters. *IEEE Trans. Power Electron.* **2013**, *28*, 254–267. [CrossRef]

57. Scoltock, J.; Geyer, T.; Madawala, U.K. Model predictive direct power control for grid-connected NPC converters. *IEEE Trans. Ind. Electron.* **2015**, *62*, 5319–5328. [CrossRef]

58. Kahia, B.; Bouafia, A.; Abdelrahem, M.; Zhang, Z.; Chaoui, A.; Krama, A.; Kennel, R. A predictive direct power control strategy for three-level npc rectifier. In Proceedings of the 2017 5th International Conference on Electrical Engineering, Boumerdes, Algeria, 29–31 October 2017; pp. 1–5. [CrossRef]

59. Dargahi, V.; Sadigh, A.K.; Corzine, K.A.; Enslin, J.H.; Rodriguez, J.; Blaabjerg, F. A new control technique for improved active-neutral-point-clamped (I-ANPC) multilevel converters using logic-equations approach. *IEEE Trans. Ind. Appl.* **2020**, *56*, 488–497. [CrossRef]

60. Mukherjee, S.; Giri, S.K.; Banerjee, S. A flexible discontinuous modulation scheme with hybrid capacitor voltage balancing strategy for three-level NPC traction inverter. *IEEE Trans. Ind. Electron.* **2019**, *66*, 3333–3343. [CrossRef]

61. Mukherjee, S.; Kumar Giri, S.; Kundu, S.; Banerjee, S. A generalized discontinuous PWM scheme for three-level NPC traction inverter with minimum switching loss for electric vehicles. *IEEE Trans. Ind. Appl.* **2019**, *55*, 516–528. [CrossRef]

62. Giri, S.K.; Banerjee, S.; Chakraborty, C. An improved modulation strategy for fast capacitor voltage balancing of three-level NPC inverters. *IEEE Trans. Ind. Electron.* **2019**, *66*, 7498–7509. [CrossRef]

63. Jiang, W.; Li, L.; Wang, J.; Ma, M.; Zhai, F.; Li, J. A novel discontinuous PWM strategy to control neutral point voltage for neutral point clamped three-level inverter with improved PWM sequence. *IEEE Trans. Power Electron.* **2019**, *34*, 9329–9341. [CrossRef]

64. Zhang, G.; Wan, Y.; Wang, Z.; Gao, L.; Zhou, Z.; Geng, Q. Discontinuous space vector PWM strategy for three-phase three-level electric vehicle traction inverter fed two-phase load. *World Electr. Veh. J.* **2020**, *11*, 27. [CrossRef]

65. Luo, S.; Wu, F.; Zhao, K. Modified single-carrier multilevel SPWM and online efficiency enhancement for single-phase asymmetrical NPC grid-connected inverter. *IEEE Trans. Ind. Inform.* **2020**, *16*, 3157–3167. [CrossRef]

66. Jung, J.; Ku, H.; Im, W.; Kim, J. A carrier-based PWM control strategy for three-level NPC inverter based on bootstrap gate drive circuit. *IEEE Trans. Power Electron.* **2020**, *35*, 2843–2860. [CrossRef]

67. Pham, K.D.; Nguyen, N.V. A reduced common-mode-voltage pulsewidth modulation method with output harmonic distortion minimization for three-level neutral-point-clamped inverters. *IEEE Trans. Power Electron.* **2020**, *35*, 6944–6962. [CrossRef]

68. Ramasamy, P.; Krishnasamy, V. SVPWM control strategy for a three phase five level dual inverter fed open-end winding induction motor. *ISA Trans.* **2020**. [CrossRef] [PubMed]

69. Jiang, W.; Wang, P.; Ma, M.; Wang, J.; Li, J.; Li, L.; Chen, K. A novel virtual space vector modulation with reduced common-mode voltage and eliminated neutral point voltage oscillation for neutral point clamped three-level inverter. *IEEE Trans. Ind. Electron.* **2020**, *67*, 884–894. [CrossRef]

70. Martinez-Garcia, J.F.; Martinez-Rodriguez, P.R.; Escobar, G.; Vazquez-Guzman, G.; Sosa-Zuñiga, J.M.; Valdez-Fernandez, A.A. Effects of modulation techniques on leakage ground currents in a grid-tied transformerless HB-NPC inverter. *IET Renew. Power Gener.* **2019**, *13*, 1250–1260. [CrossRef]

71. Choi, U.M.; Lee, J.S. Comparative evaluation of lifetime of three-level inverters in grid-connected photovoltaic systems. *Energies* **2020**, *13*, 1227. [CrossRef]

72. Ma, L.; Kerekes, T.; Rodriguez, P.; Jin, X.; Teodorescu, R.; Liserre, M. A new PWM strategy for grid-connected half-bridge active NPC converters with losses distribution balancing mechanism. *IEEE Trans. Power Electron.* **2015**, *30*, 5331–5340. [CrossRef]

73. Saridakis, S.; Koutroulis, E.; Blaabjerg, F. Optimization of SiC-based H5 and conergy-NPC transformerless PV inverters. *IEEE J. Emerg. Sel. Top. Power Electron.* **2015**, *3*, 555–567. [CrossRef]

74. Zhou, L.; Gao, F.; Xu, T. A family of neutral-point-clamped circuits of single-phase PV inverters: Generalized principle and implementation. *IEEE Trans. Power Electron.* **2017**, *32*, 4307–4319. [CrossRef]

75. Wang, H.; Kou, L.; Liu, Y.; Sen, P.C. A new six-switch five-level active neutral point clamped inverter for PV applications. *IEEE Trans. Power Electron.* **2017**, *32*, 6700–6715. [CrossRef]

76. Dargahi, V.; Khoshkbar Sadigh, A.; Abarzadeh, M.; Pahlavani, M.R.A.; Shoulaie, A. Flying capacitors reduction in an improved double flying capacitor multicell converter controlled by a modified modulation method. *IEEE Trans. Power Electron.* **2012**, *27*, 3875–3887. [CrossRef]

77. Sadigh, A.K.; Hosseini, S.H.; Sabahi, M.; Gharehpetian, G.B. Double flying capacitor multicell converter based on modified phase-shifted pulsewidth modulation. *IEEE Trans. Power Electron.* **2010**, *25*, 1517–1526. [CrossRef]

78. Feng, C.; Liang, J.; Agelidis, V.G. Modified Phase-Shifted PWM control for flying capacitor multilevel converters. *IEEE Trans. Power Electron.* **2007**, *22*, 178–185. [CrossRef]

79. Lee, S.-G.; Kang, D.-W.; Lee, Y.-H.; Hyun, D.-S. The carrier-based PWM method for voltage balance of flying capacitor multilevel inverter. In Proceedings of the 2001 IEEE 32nd Annual Power Electronics Specialists Conference (IEEE Cat. No.01CH37230), Vancouver, BC, Canada, 17–21 June 2001; Volume 1, pp. 126–131. [CrossRef]

80. Abdelhamid, E.; Corradini, L.; Mattavelli, P.; Bonanno, G.; Agostinelli, M. Sensorless stabization technique for peak current mode controlled three-level flying-capacitor converters. *IEEE Trans. Power Electron.* **2020**, *35*, 3208–3220. [CrossRef]

81. Ghias, A.M.Y.M.; Pou, J.; Agelidis, V.G.; Ciobotaru, M. Optimal switching transition-based voltage balancing method for flying capacitor multilevel converters. *IEEE Trans. Power Electron.* **2015**, *30*, 1804–1817. [CrossRef]

82. Sadigh, A.K.; Dargahi, V.; Corzine, K.A. New active capacitor voltage balancing method for flying capacitor multicell converter based on logic-form-equations. *IEEE Trans. Ind. Electron.* **2017**, *64*, 3467–3478. [CrossRef]

83. Chen, H.; Lu, C.; Lien, W.; Chen, T. Active capacitor voltage balancing control for three-level flying capacitor boost converter based on average-behavior circuit model. *IEEE Trans. Ind. Appl.* **2019**, *55*, 1628–1638. [CrossRef]

84. Stillwell, A.; Candan, E.; Pilawa-Podgurski, R.C.N. Active voltage balancing in flying capacitor multi-level converters with valley current detection and constant effective duty cycle control. *IEEE Trans. Power Electron.* **2019**, *34*, 11429–11441. [CrossRef]

85. Penczek, A.; Mondzik, A.; Stala, R.; Ruderman, A. Simple time-domain analysis of a multilevel DC/DC flying capacitor converter average aperiodic natural balancing dynamics. *IET Power Electron.* **2019**, *12*, 1179–1186. [CrossRef]
86. Elsayad, N.; Moradisizkoohi, H.; Mohammed, O. A new three-level flying-capacitor boost converter with an integrated LC2D output network for fuel-cell vehicles: Analysis and design. *Inventions* **2018**, *3*, 61. [CrossRef]
87. Vijeh, M.; Rezanejad, M.; Samadaei, E.; Bertilsson, K. A general review of multilevel inverters based on main submodules: Structural point of view. *IEEE Trans. Power Electron.* **2019**, *34*, 9479–9502. [CrossRef]
88. Chaudhuri, T.; Rufer, A. Modeling and control of the cross-connected intermediate-level voltage source inverter. *IEEE Trans. Ind. Electron.* **2010**, *57*, 2597–2604. [CrossRef]
89. Ye, Z.; Lei, Y.; Liu, W.; Shenoy, P.S.; Pilawa-Podgurski, R. Improved bootstrap methods for powering floating gate drivers of flying capacitor multilevel converters and hybrid switched-capacitor converters. *IEEE Trans. Power Electron.* **2020**, *35*, 5965–5977. [CrossRef]
90. Lezana, P.; Pou, J.; Meynard, T.A.; Rodriguez, J.; Ceballos, S.; Richardeau, F. Survey on fault operation on multilevel inverters. *IEEE Trans. Ind. Electron.* **2010**, *57*, 2207–2218. [CrossRef]
91. Bressan, M.V.; Rech, C.; Batschauer, A.L. Design of flying capacitors for n-level FC and n-level SMC. *Int. J. Electr. Power Energy Syst.* **2019**, *113*, 220–228. [CrossRef]
92. Lin, J.; Weiss, G. Multilevel converter with variable flying capacitor voltage used for virtual infinite capacitor. In Proceedings of the 2017 International Symposium on Power Electronics (Ee), Novi Sad, Serbia, 19–21 October 2017; pp. 1–4. [CrossRef]
93. Guerriero, P.; Coppola, M.; Napoli, F.D.; Brando, G.; Dannier, A.; Iannuzzi, D.; Daliento, S. Three-phase PV CHB inverter for a distributed power generation system. *Appl. Sci.* **2016**, *6*, 287. [CrossRef]
94. Zhang, C.; Gao, Z. A cascaded multilevel inverter using only one battery with high-frequency link and low-rating-voltage MOSFETs for motor drives in electric vehicles. *Energies* **2018**, *11*, 1778. [CrossRef]
95. Gopal, Y.; Birla, D.; Lalwani, M. Selected harmonic elimination for cascaded multilevel inverter based on photovoltaic with fuzzy logic control maximum power point tracking technique. *Technologies* **2018**, *6*, 62. [CrossRef]
96. Banaei, M.R.; Khounjahan, H.; Salary, E. Single-source cascaded transformers multilevel inverter with reduced number of switches. *IET Power Electron.* **2012**, *5*, 1748–1753. [CrossRef]
97. Uno, M.; Shinohara, T. Module-integrated converter based on cascaded quasi-Z-source inverter with differential power processing capability for photovoltaic panels under partial shading. *IEEE Trans. Power Electron.* **2019**, *34*, 11553–11565. [CrossRef]
98. Kumar, A.; Verma, V. Performance enhancement of single-phase grid-connected PV system under partial shading using cascaded multilevel converter. *IEEE Trans. Ind. Appl.* **2018**, *54*, 2665–2676. [CrossRef]
99. Ye, T.; Dai, N.; Lam, C.; Wong, M.; Guerrero, J.M. Analysis, design, and implementation of a quasi-proportional-resonant controller for a multifunctional capacitive-coupling grid-connected inverter. *IEEE Trans. Ind. Appl.* **2016**, *52*, 4269–4280. [CrossRef]
100. Majumder, M.G.; Patra, M.; Kasari, P.R.; Das, B.; Chakraborti, A. Photovoltaic array based grid connected cascaded multilevel inverter using PR controller. In Proceedings of the 2017 Innovations in Power and Advanced Computing Technologies (i-PACT), Vellore, India, 21–22 April 2017; pp. 1–5. [CrossRef]
101. Li, Y.; Wang, Y.; Li, B.Q. Generalized theory of phase-shifted carrier PWM for cascaded H-bridge converters and modular multilevel converters. *IEEE J. Emerg. Sel. Top. Power Electron.* **2016**, *4*, 589–605. [CrossRef]
102. Elmelegi, A.; Aly, M.; Ahmed, E.M.; Alhaider, M.M. An efficient low-cost distributed MPPT method for energy harvesting in grid-tied three-phase PV power optimizers. In Proceedings of the 2019 21st International Middle East Power Systems Conference (MEPCON), Cairo, Egypt, 17–19 December 2019; pp. 1042–1047. [CrossRef]
103. Elmelegi, A.; Aly, M.; Ahmed, E.M.; Alharbi, A.G. A simplified phase-shift PWM-based feedforward distributed MPPT method for grid-connected cascaded PV inverters. *Sol. Energy* **2019**, *187*, 1–12. [CrossRef]
104. Goetz, S.M.; Wang, C.; Li, Z.; Murphy, D.L.K.; Peterchev, A.V. Concept of a distributed photovoltaic multilevel inverter with cascaded double H-bridge topology. *Int. J. Electr. Power Energy Syst.* **2019**, *110*, 667–678. [CrossRef]
105. Boutasseta, N.; Ramdani, M.; Mekhilef, S. Fault-tolerant power extraction strategy for photovoltaic energy systems. *Sol. Energy* **2018**, *169*, 594–606. [CrossRef]

106. Shao, S.; Wheeler, P.W.; Clare, J.C.; Watson, A.J. Fault detection for modular multilevel converters based on sliding mode observer. *IEEE Trans. Power Electron.* **2013**, *28*, 4867–4872. [CrossRef]

107. Pillai, D.S.; Rajasekar, N. A comprehensive review on protection challenges and fault diagnosis in PV systems. *Renew. Sustain. Energy Rev.* **2018**, *91*, 18–40. [CrossRef]

108. Kuleshov, D.; Peltoniemi, P.; Kosonen, A.; Nuutinen, P.; Huoman, K.; Lana, A.; Paakkonen, M.; Malinen, E. Assessment of economic benefits of battery energy storage application for the PV-equipped households in Finland. *J. Eng.* **2019**, *2019*, 4927–4931. [CrossRef]

109. Alhaider, M.; Fan, L. Planning energy storage and photovoltaic panels for demand response with heating ventilation and air conditioning systems. *IEEE Trans. Ind. Inform.* **2018**, *14*, 5029–5037. [CrossRef]

110. Zhou, Y.; Li, H. Analysis and suppression of leakage current in cascaded-multilevel-inverter-based PV systems. *IEEE Trans. Power Electron.* **2014**, *29*, 5265–5277. [CrossRef]

111. Sonti, V.; Jain, S.; Bhattacharya, S. Analysis of the modulation strategy for the minimization of the leakage current in the PV grid-connected cascaded multilevel inverter. *IEEE Trans. Power Electron.* **2017**, *32*, 1156–1169. [CrossRef]

112. Bayhan, S.; Komurcugil, H. A sliding-mode controlled single-phase grid-connected quasi-Z-source NPC inverter with double-line frequency ripple suppression. *IEEE Access* **2019**, *7*, 160004–160016. [CrossRef]

113. Lee, S.S. Single-stage switched-capacitor module (S3CM) topology for cascaded multilevel inverter. *IEEE Trans. Power Electron.* **2018**, *33*, 8204–8207. [CrossRef]

114. Wu, F.; Li, X.; Feng, F.; Gooi, H.B. Modified cascaded multilevel grid-connected inverter to enhance european efficiency and several extended topologies. *IEEE Trans. Ind. Inform.* **2015**, *11*, 1358–1365. [CrossRef]

115. Cui, W.; Luo, H.; Gu, Y.; Li, W.; Yang, B.; He, X. Hybrid-bridge transformerless photovoltaic grid-connected inverter. *IET Power Electron.* **2015**, *8*, 439–446. [CrossRef]

116. Zhou, L.; Gao, F.; Xu, T. Implementation of active NPC circuits in transformer-less single-phase inverter with low leakage current. *IEEE Trans. Ind. Appl.* **2017**, *53*, 5658–5667. [CrossRef]

117. Rojas, C.A.; Aguirre, M.; Kouro, S.; Geyer, T.; Gutierrez, E. Leakage current mitigation in photovoltaic string inverter using predictive control with fixed average switching frequency. *IEEE Trans. Ind. Electron.* **2017**, *64*, 9344–9354. [CrossRef]

118. Ardashir, J.F.; Sabahi, M.; Hosseini, S.H.; Blaabjerg, F.; Babaei, E.; Gharehpetian, G.B. A single-phase transformerless inverter with charge pump circuit concept for grid-tied PV applications. *IEEE Trans. Ind. Electron.* **2017**, *64*, 5403–5415. [CrossRef]

119. Wang, J.; Mu, X.; Li, Q.K. Study of passivity-based decoupling control of T-NPC PV grid-connected inverter. *IEEE Trans. Ind. Electron.* **2017**, *64*, 7542–7551. [CrossRef]

120. Guo, X.; Zhou, J.; He, R.; Jia, X.; Rojas, C.A. Leakage current attenuation of a three-phase cascaded inverter for transformerless grid-connected PV systems. *IEEE Trans. Ind. Electron.* **2018**, *65*, 676–686. [CrossRef]

121. Kakosimos, P.; Abu-Rub, H. Predictive control of a grid-tied cascaded full-bridge NPC inverter for reducing high-frequency common-mode voltage components. *IEEE Trans. Ind. Inform.* **2018**, *14*, 2385–2394. [CrossRef]

122. Guisso, R.A.; Andrade, A.M.S.S.; Hey, H.L.; Martins, M.L.d.S. Grid-tied single source quasi-Z-source cascaded multilevel inverter for PV applications. *Electron. Lett.* **2019**, *55*, 342–343. [CrossRef]

123. Ahmed, A.; Sundar Manoharan, M.; Park, J. An efficient single-sourced asymmetrical cascaded multilevel inverter with reduced leakage current suitable for single-stage PV systems. *IEEE Trans. Energy Convers.* **2019**, *34*, 211–220. [CrossRef]

124. Aly, M.; Rojas, C.A.; Ahmed, E.M.; Kouro, S. Leakage current elimination PWM method for fault-tolerant string H-NPC PV inverter. In Proceedings of the IECON 2019—45th Annual Conference of the IEEE Industrial Electronics Society, Lisbon, Portugal, 14–17 October 2019. [CrossRef]

125. Madasamy, P.; Kumar, V.S.; Sanjeevikumar, P.; Holm-Nielsen, J.B.; Hosain, E.; Bharatiraja, C. A three-phase transformerless T-Type- NPC-MLI for grid connected PV systems with common-mode leakage current mitigation. *Energies* **2019**, *12*, 2434. [CrossRef]

126. Taghvaie, A.; Haque, M.E.; Saha, S.; Mahmud, M.A. A new step-up switched-capacitor voltage balancing converter for NPC multilevel inverter-based solar PV system. *IEEE Access* **2020**, *8*, 83940–83952. [CrossRef]

127. Wang, K.; Zheng, Z.; Xu, L.; Li, Y. A generalized carrier-overlapped PWM method for neutral-point-clamped multilevel converters. *IEEE Trans. Power Electron.* **2020**, *35*, 9095–9106. [CrossRef]

128. Da Costa Bahia, F.A.; Jacobina, C.B.; Rocha, N.; de Sousa, R.P.R. Cascaded transformer multilevel inverters with asymmetrical turns ratios based on NPC. *IEEE Trans. Ind. Electron.* **2020**, *67*, 6387–6397. [CrossRef]

Energies **2020**, *13*, 3261

129. Zhang, J.; Wai, R.J. Design of new SVPWM mechanism for three-level NPC ZSI via line-voltage coordinate system. *IEEE Trans. Power Electron.* **2020**, *35*, 8593–8606. [CrossRef]
130. Yuan, Q.; Li, A.; Qian, J.; Xia, K. Neutral-point potential control for the NPC three-level inverter with a newly MPC-based virtual vector modulation. *IET Power Electron.* **2019**. [CrossRef]

 © 2020 by the authors. Licensee MDPI, Basel, Switzerland. This article is an open access article distributed under the terms and conditions of the Creative Commons Attribution (CC BY) license (http://creativecommons.org/licenses/by/4.0/).

Article

Adaptive Fuzzy Approximation Control of PV Grid-Connected Inverters

Myada Shadoul [1], Hassan Yousef [1,*], Rashid Al Abri [1] and Amer Al-Hinai [1,2]

[1] Department of Electrical and Computer Engineering, Sultan Qaboos University, Muscat-123, Oman; myada.shadoul@gmail.com (M.S.); arashid@squ.edu.om (R.A.A.); hinai@squ.edu.om (A.A.-H.)
[2] Sustainable Energy Research Center, Sultan Qaboos University, Muscat-123, Oman
* Correspondence: hyousef@squ.edu.om

Abstract: Three-phase inverters are widely used in grid-connected renewable energy systems. This paper presents a new control methodology for grid-connected inverters using an adaptive fuzzy control (AFC) technique. The implementation of the proposed controller does not need prior knowledge of the system mathematical model. The capabilities of the fuzzy system in approximating the nonlinear functions of the grid-connected inverter system are exploited to design the controller. The proposed controller is capable to achieve the control objectives in the presence of both parametric and modelling uncertainties. The control objectives are to regulate the grid power factor and the dc output voltage of the photovoltaic systems. The closed-loop system stability and the updating laws of the controller parameters are determined via Lyapunov analysis. The proposed controller is simulated under different system disturbances, parameters, and modelling uncertainties to validate the effectiveness of the designed controller. For evaluation, the proposed controller is compared with conventional proportional-integral (PI) controller and Takagi–Sugeno–Kang-type probabilistic fuzzy neural network controller (TSKPFNN). The results demonstrated that the proposed AFC showed better performance in terms of response and reduced fluctuations compared to conventional PI controllers and TSKPFNN controllers.

Keywords: adaptive; fuzzy; feedback linearization; photovoltaic (PV) grid inverter; voltage source inverter (VSI)

Citation: Shadoul, M.; Yousef, H.; Al Abri, R.; Al-Hinai, A. Adaptive Fuzzy Approximation Control of PV Grid-Connected Inverters. *Energies* **2021**, *14*, 942. https://doi.org/10.3390/en14040942

Academic Editor: Oscar Barambones

Received: 18 January 2021
Accepted: 5 February 2021
Published: 11 February 2021

Publisher's Note: MDPI stays neutral with regard to jurisdictional claims in published maps and institutional affiliations.

Copyright: © 2021 by the authors. Licensee MDPI, Basel, Switzerland. This article is an open access article distributed under the terms and conditions of the Creative Commons Attribution (CC BY) license (https://creativecommons.org/licenses/by/4.0/).

1. Introduction

Nowadays, renewable energy sources (RES) such as photovoltaic (PV) solar systems, wind turbines, and others are integrated into conventional power systems to avoid the high cost of constructing new or expanded facilities [1]. The final stage of the integration of PV systems consists of DC-AC inverters. Special consideration for inverter topologies and controls is required to preserve the network stability and to achieve acceptable dynamic performance of the voltage and frequency [2]. Different controllers for micro-grid inverters during grid-connected and islanded operation modes have been investigated in previous studies [3,4]. Corresponding to behavior and operating conditions of the electrical grid, the controllers of the inverter system can be classified as linear, non-linear, robust, adaptive, predictive, and intelligent controllers [5]. Various types of linear controllers for micro-grid inverters including classical controllers, Proportional Resonant (PR) controllers, and Linear Quadratic Gaussian (LQG) controllers were reported [6–8]. Non-linear controllers for grid-connected inverter systems (GCIS) such as sliding mode, feedback linearization, and hysteresis controllers have been proposed in [9–13]. In [14], a current-control is proposed for voltage-source inverters using the H_∞ robust control technique. Additionally, adaptive control techniques and model predictive controllers for grid-connected and standalone inverters were reported in [15–18]. In all studies referred to, the proposed non-linear controllers showed better performance when compared with linear controllers' performance. The main drawback of nonlinear control methods is that they rely on the system mathematical model and system parameters availability.

Intelligent control systems including neural network controllers, repetitive controllers, fuzzy logic controllers (FLCs), and autonomous controllers are introduced for nonlinear systems control. The advantage of intelligent controllers is that they do not rely on the system mathematical model and they can handle many nonlinear and uncertain systems. In [19], a discrete-time repetitive controller (RC) was proposed to improve the output current and to overcome the drawback of using a linear PI controller in the presence of non-linearities in the system components. Type-1 and Type-2 FLCs have been widely used in various applications and have achieved remarkable success in managing higher levels of uncertainty [20–25]. FLC applications also include intelligent control for marine applications, traction diesel engines, robotic control, internet bandwidth control, industrial system controllers, power management and electrical control, aircraft control evolutionary computing, and DC-DC converters. Moreover, type-2 fuzzy logic has also proven successful in clinical diagnosis, differential diagnosis, and nursing evaluation in the health field [26].

For GCIS, different FLCs were presented in [27–29]. In [28], the real time testing for FLC for three phase grid-connected inverters to control the voltage and the current was presented. The results demonstrated FLC ability to generate high-quality PV power while maintaining the power factor of unity. A grid side inverter system control using a simple FLC that works well for grid interconnected variable speed wind generators was proposed in [29]. In another work, type-2 FLC (T2FLC) was implemented to control a DC-DC buck converter [30]. For PV systems, interval T2FLC (IT2FLC) based on maximum power point tracking (MPPT) method was proposed in [31]. In addition, the work in [32] implemented a T2FLC as a MPPT to handle the rules' uncertainties during high weather condition variations. The proposed MPPT based on the FLC showed a faster response in the transient response and a stable steady state. A further IT2FLC was developed for single phase grid connected PV systems in [33], where IT2FLC was used as MPPT algorithm. Simulation results showed that the proposed IT2FLC-based MPPT controller has a fast transient response.

In [34], an FLC for inverters in PV applications was presented; the work discussed several factors and challenges and provided guidelines for developing capable and effective inverter control systems.

Moreover, a fuzzy neural network controller based on the Takagi–Sugeno–Kang type approach presented to control the active and reactive power of three-phase grid-connected PV systems during grid faults was reported in [35,36].

Furthermore, to overcome the disadvantages of conventional controllers for uncertain nonlinear systems, adaptive fuzzy control (AFC) techniques were proposed to control uncertain nonlinear systems [37,38]. Due to its advantage in handling complex uncertain nonlinear systems, researchers have used the AFC techniques in different applications [39–43]. The AFC technique was applied for induction motor control in [39], the optimal power conversion control for standalone wind energy conversion systems in [40], permanent magnet synchronous motor control, and fuzzy fault-tolerant switched systems in [41,42]. To the best of the authors' knowledge, there is no reported work available that describe the application of AFC to the GCIS. This motivates the authors to propose an AFC scheme that exploits the concept of the multi-input multi-output (MIMO) feedback linearization and the approximation capability of fuzzy systems. PV GCIS are highly nonlinear and uncertain systems due to the intermittent nature of the PVand the inverter pulse width-modulation (PWM) technique. Without fast-acting inverter control, these nonlinearities and uncertainties lead to power quality, output harmonics, voltage regulation, losses, and system implementation issues. The proposed AFC, based on the method of feedback linearization, is developed to solve these nonlinearities and uncertainties due to the method's ability to manage complex nonlinear control systems without the need for a mathematical model. The fuzzy system's capability to approximate unknown parameters of the GCIS for different operation cases will be used to design the controller. The objectives of the proposed AFC for GCIS are to control both the power factor and the dc voltage. The quality of the designed controller will be tested to validate its effectiveness in achieving the control objectives for different simulation cases. Moreover, to evaluate the efficiency of the proposed controller, a comparison between its performance, the PI controller, and a Takagi–Sugeno–Kang-type probabilistic fuzzy neural network controller (TSKPFNN) performances was conducted. The main contribution of the paper can be summarized in the following:

- The paper proposes an adaptive fuzzy approximation control scheme for GCIS.
- Excellent tracking performance of the proposed controller is obtained under different operating conditions such as power factor, parameter, and modelling uncertainties.

The rest of the paper is organized in five sections. The MIMO of GCIS and the feedback linearization are presented in Section 2. Section 3 gives the design of an AFC for a general MIMO. Based on the analysis presented in Section 3, the proposed AFC for GCIS is explained in Section 4. Simulation results are presented in Section 5 and conclusions are drawn in Section 6.

2. Grid-Connected Inverter System (GCIS)

2.1. GCIS Description

A GCIS is shown in Figure 1. The system contains a PV array, a DC link capacitor, a three-phase voltage source inverter (VSI), and a three-phase grid. The VSI facilitates the MPPT through regulation of the dc-link voltage, along with power transfer to the utility grid.

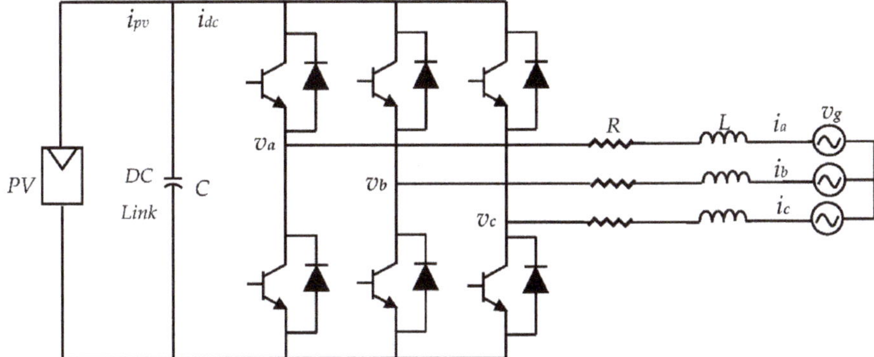

Figure 1. Three-phase grid-connected inverter.

The output power of the PV system is a highly nonlinear and uncertain system. The PV output voltage corresponding to the maximum output power of the PV array varies with cell temperature and solar irradiation. The PV system should always be designed to operate at its maximum output power level. The MPPT technique is usually incorporated with the PV system to adjust the PV array output voltage to obtain the maximum available power at any change in solar irradiation or temperature of the cells. In addition, the MPPT scheme has the capability to release the dc-link voltage reference command [44]. Many MPPT techniques have been reported for PV systems, however, in practice, the most commonly used methods are perturb and observe (P&O) and the incremental conductance (IC) techniques [45].

2.2. MIMO Model of GCIS

The model of the GCIS shown in Figure 1 can be represented by

$$v_a = Ri_a + L\frac{di_a}{dt} + v_{ga} \tag{1}$$

$$v_b = Ri_b + L\frac{di_b}{dt} + v_{gb} \tag{2}$$

$$v_c = Ri_c + L\frac{di_c}{dt} + v_{gc} \tag{3}$$

where v_{ga}, v_{gb}, v_{gc} are the grid voltage components, i_a, i_b, i_c are the grid current components, and v_a, v_b, v_c are the inverter output voltage [46]. Park's transformation is used to represent Equations (1)–(3) in the rotating dq frame as

$$v_d = v_{gd} + Ri_d + L\frac{di_d}{dt} + \omega L i_q \tag{4}$$

$$v_q = v_{gq} + Ri_q + L\frac{di_q}{dt} + \omega L i_d \tag{5}$$

where v_{gd}, v_{gq} are the dq grid voltage components, i_d, i_q are the dq grid current components, and v_d, v_q are the dq inverter output voltage components. Upon neglecting the power loss in the inverter switches [46], the dc-input side connection with the ac-output side are given by

$$v_{gd}i_d + v_{gq}i_q = v_{dc}i_{dc} \tag{6}$$

$$C\frac{dv_{dc}}{dt} = i_{pv} - i_{dc} = i_{pv} - \frac{v_{gd}i_d + v_{gq}i_q}{v_{dc}} \tag{7}$$

where v_{dc} and i_{pv} are the PV output voltage and current respectively and i_{dc} is the input current to the inverter.

Defining the state vector x and the control input u as

$$x = \begin{bmatrix} x_1 \\ x_2 \\ x_3 \end{bmatrix} = \begin{bmatrix} i_d \\ i_q \\ v_{dc} \end{bmatrix} \tag{8}$$

$$u = \begin{bmatrix} u_1 \\ u_2 \end{bmatrix} = \begin{bmatrix} v_d \\ v_q \end{bmatrix} \tag{9}$$

Then, the state model of the GCIS can be formed as in Equation (10).

$$\dot{x} = \begin{bmatrix} -\frac{R}{L}x_1 + \omega x_2 - \frac{v_{gd}}{L} \\ -\frac{R}{L}x_2 - \omega x_1 - \frac{v_{gq}}{L} \\ \frac{i_{pv}}{C} - \frac{v_{gd}x_1 + v_{gq}x_2}{Cx_3} \end{bmatrix} + \begin{bmatrix} \frac{1}{L} & 0 \\ 0 & \frac{1}{L} \\ 0 & 0 \end{bmatrix} u \tag{10}$$

The control objective is to regulate the power factor of the grid, through the q-component of the grid current, and the dc-input voltage v_{dc}. Therefore, the output vector of the system is considered as

$$y = \begin{bmatrix} y_1 \\ y_2 \end{bmatrix} = \begin{bmatrix} x_2 \\ x_3 \end{bmatrix} = \begin{bmatrix} i_q \\ v_{dc} \end{bmatrix} \tag{11}$$

The Equations (10) and (11) can be written in the following general expression of the MIMO system

$$\dot{x} = f(x) + g(x)u, \; y = h(x) \tag{12}$$

where x is a 3×1 state vector, u is a 2×1 control input vector, y is a 2×1 output vector, and $f(x)$ and $g(x)$ are defined by

$$f = \begin{bmatrix} f_1 \\ f_2 \\ f_3 \end{bmatrix} = \begin{bmatrix} -\frac{R}{L}x_1 + \omega x_2 - \frac{v_{gd}}{L} \\ -\frac{R}{L}x_2 - \omega x_1 - \frac{v_{gq}}{L} \\ \frac{i_{pv}}{C} - \frac{v_{gd}x_1 + v_{gq}x_2}{Cx_3} \end{bmatrix}, \; g = \begin{bmatrix} \frac{1}{L} & 0 \\ 0 & \frac{1}{L} \\ 0 & 0 \end{bmatrix} \tag{13}$$

The MIMO model of the GCIS in Equations (12) and (13) can be converted to a feedback linearizable form by using the input-output feedback linearization approach. In this approach, a nonlinear control signal is designed and used to convert the nonlinear system dynamics Equation (12) into decoupled linear subsystems. The feedback linearization for GCIS is presented next.

2.3. Input-Output Feedback Linearization of GCIS

In order to design a feedback linearization control, we used the notion of relative degree where each output is differentiated successively until one input u_1 or u_2 appears [47]. It can be shown that the relative degree r_1 for the first output y_1 is $r_1 = 1$ and the relative degree r_2 for the second output y_2 is $r_2 = 2$. The first derivative of y_1 and the second derivative of y_2 are given by Equations (14) and (15).

$$\dot{y}_1 = f_2 + \frac{1}{L} u_2 \tag{14}$$

$$\ddot{y}_2 = \dot{f}_3 = \frac{1}{C}\frac{di_{pv}}{dt} - \frac{1}{C.x_3}\left[v_{gd}\left(f_1 + \frac{1}{L}u_1\right) + v_{gq}\left(f_2 + \frac{1}{L}u_2\right)\right] + \frac{\left(v_{gd}x_1 + v_{gq}x_2\right)}{Cx_3^2} f_3 \tag{15}$$

Equations (14) and (15) can be cast in the following matrix form

$$\begin{bmatrix} \dot{y}_1 \\ \ddot{y}_2 \end{bmatrix} = \alpha(x) + \beta(x)\begin{bmatrix} u_1 \\ u_2 \end{bmatrix} \tag{16}$$

where

$$\alpha(x) = \begin{bmatrix} f_2 \\ m - \frac{1}{Cx_3}\left(v_{gd}f_1 + v_{gq}f_2\right) + \frac{\left(v_{gd}x_1 + v_{gq}x_2\right)}{Cx_3^2} f_3 \end{bmatrix} \tag{17}$$

$$\beta(x) = \begin{bmatrix} 0 & \frac{1}{L} \\ -\frac{v_{gd}}{LCx_3} & -\frac{v_{gq}}{LCx_3} \end{bmatrix} \tag{18}$$

and $m = \frac{1}{C}\frac{di_{pv}}{dt}$.

The control law in Equation (19) when used in Equation (16) yields to the linear input-output relation in Equation (20)

$$\begin{bmatrix} u_1 \\ u_2 \end{bmatrix} = \beta^{-1}(x)\begin{bmatrix} v_1 - \alpha_1 \\ v_2 - \alpha_2 \end{bmatrix} \tag{19}$$

$$\begin{bmatrix} \dot{y}_1 \\ \ddot{y}_2 \end{bmatrix} = \begin{bmatrix} v_1 \\ v_2 \end{bmatrix} \tag{20}$$

where v_1 and v_2 are external signals that can be chosen in a way to ensure asymptotic tracking of the outputs y_1 and y_2 to their references $y_{ref1} = i_{qref}$ and $y_{ref2} = v_{dcref}$. Defining the tracking errors $e_1 = \left(y_{ref1} - y_1\right)$ and $e_2 = \left(y_{ref2} - y_2\right)$, the signals v_1 and v_2 can be selected as

$$v_1 = k_{01}e_1 + \dot{y}_{ref1} \tag{21}$$

$$v_2 = k_{02}e_2 + k_{12}\dot{e}_2 + \ddot{y}_{ref2} \tag{22}$$

Now, substituting Equations (21) and (22) into Equation (20), we obtain the following tracking errors dynamics:

$$\dot{e}_1 + k_{01}e_1 = 0 \tag{23}$$

$$\ddot{e}_2 + k_{12}\dot{e}_2 + k_{02}e_2 = 0 \tag{24}$$

The coefficients k_{01}, k_{02}, and k_{12} are design parameters selected such that the characteristic polynomials of Equations (23) and (24) are Hurwitz and hence ensuring that the tracking errors e_1 and e_2 converge to zero asymptotically [47]. It is worth mentioning that the control law given in Equation (19) is implementable since the matrix $\beta(x)$ is non-singular, provided that $e_d \neq 0$ (which is the case for GCIS).

A main disadvantage of the control law in Equation (19) is that the exact values of the system parameters involved in $\alpha(x)$ and $\beta(x)$ should be known and any change in the parameters affects the output of the controller. In practice, the parameters of the GCIS may be unknown or imprecise and the uncertainty in these parameters is inevitable. To overcome this drawback, the universal approximation capability of the fuzzy systems is used to approximate the nonlinear functions $\alpha(x)$ and $\beta(x)$. In the next section, the proposed adaptive fuzzy controller for GCIS is presented.

3. The Proposed Controller

3.1. Adaptive Fuzzy Approximation Controller for GCIS

In this section, the proposed controller is developed using the Equation (12) which is a square MIMO nonlinear system. The input-output feedback linearization given in Equation (16) with $r_1 = 1$ and $r_2 = 2$ can be written in the form

$$y^{(r)} = \alpha(x) + \beta(x)u \tag{25}$$

where $y^{(r)} = \begin{bmatrix} y_1^{(r_1)} \\ y_2^{(r_2)} \end{bmatrix} = \begin{bmatrix} \dot{y}_1 \\ \ddot{y}_2 \end{bmatrix}$, $u = \begin{bmatrix} u_1 \\ u_2 \end{bmatrix}$, $\alpha(x)$ and $\beta(x)$ are as given in Equations (17) and (18), and their entries are in general nonlinear functions with imprecise parameters.

Approximations of the nonlinear functions $\alpha(x)$ and $\beta(x)$ were generated using a fuzzy logic system with singleton fuzzifier, product inference rule, and weighted average defuzzifier. To construct these estimates, the notion of the fuzzy basis function (FBF) expansion $\xi(x)$ was used [37]. The fuzzy estimates $\hat{a}_i(x)$ and $\hat{\beta}_{ij}(x)$ of the nonlinear functions $\alpha_i(x)$ and $\beta_{ij}(x)$, $i = 1, 2$ and $j = 1, 2$ were determined as

$$\hat{a}_i(x) = \theta_i^T \xi(x) \tag{26}$$

$$\hat{\beta}_{ij}(x) = \theta_{ij}^T \xi(x) \tag{27}$$

where $\theta_i \in R^{M \times 1}$ and $\theta_{ij} \in R^{M \times 1}$ represent vectors of adjustable parameters and $\xi(x) \in R^{M \times 1}$ represents the vector of FBFs. The FBF was generated using the weighted-average defuzzifier [48].

$$\xi_i(x) = \frac{\prod_{i=1}^{n} x_i \, \mu_{il}(x_i)}{\sum_{l=1}^{M} \left(\prod_{i=1}^{n} \mu_{il}(x_i) \right)} \tag{28}$$

where n is the number of states, $\mu_{il}(x_i)$ is the membership function of the ith state x_i in the lth rule, and M is the number of If-Then rules.

Upon replacing $\alpha(x)$ and $\beta(x)$ in Equation (25) by their corresponding fuzzy estimates Equations (26) and (27), we get

$$y^{(r)} = \hat{\alpha}(x) + \hat{\beta}(x)u \tag{29}$$

where $\hat{\alpha}(x) = \begin{bmatrix} \hat{\alpha}_1 \\ \hat{\alpha}_2 \end{bmatrix}$ and $\hat{\beta}(x) = \begin{bmatrix} \hat{\beta}_{11} & \hat{\beta}_{12} \\ \hat{\beta}_{21} & \hat{\beta}_{22} \end{bmatrix}$.

Therefore, the AFC can be written in terms of the fuzzy estimates Equations (26) and (27) as

$$u = \hat{\beta}^{-1}(x)\left(y_{ref}^{(r)} + Ke - \hat{\alpha}(x) \right) \tag{30}$$

where $y_{ref}^{(r)} = \begin{bmatrix} y_{ref1}^{(r_1)} & y_{ref2}^{(r_2)} \end{bmatrix}^T$, $K = diag[k_1 \ k_2]$, $e = \begin{bmatrix} e_1 & e_2 \end{bmatrix}^T$, $k_1 = k_{01}$, $k_2 = [k_{02} \ k_{12}]$, $e_1 = e_1$,

$e_2 = \begin{bmatrix} e_2 & \dot{e}_2 \end{bmatrix}$, and $e_i = y_{refi} - y_i$, $i = 1, 2$. It is worth mentioning that the implementation of the proposed AFC given in Equation (30) needs only the fuzzy estimates $\hat{\alpha}(x)$, $\hat{\beta}(x)$, the derivatives of the reference signal $y_{ref}^{(r)}$, and the tracking error e.

3.2. Closed-Loop Stability

In this section, we show the boundedness of both the tracking error and the adjustable parameters via the Lyapunov function analysis. Equation (30) can be rewritten as

$$\hat{\beta}(x)u = \left(y_{ref}^{(r)} - y^{(r)}\right) + K\underline{e} + y^{(r)} - \hat{\alpha}(x) \tag{31}$$

Using Equation (25) in Equation (31) to obtain the following error equation in terms of the fuzzy approximation errors, it becomes

$$\begin{bmatrix} e_1^{(r_1)} \\ e_2^{(r_2)} \end{bmatrix} = \left(y_{ref}^{(r)} - y^{(r)}\right) = -Ke + (\hat{\alpha}(x) - \alpha(x)) + (\hat{\beta}(x) - \beta(x))u \tag{32}$$

From Equation (32), the error equation for the ith output becomes

$$e_i^{r_i} = -k_i\underline{e}_i + \Delta\alpha_i(x) + \sum_{j=1}^{p} \Delta\beta_{ij}(x)u_j \tag{33}$$

where $\Delta\alpha_i(x) = \hat{\alpha}_i(x) - \alpha_i(x)$ and $\Delta\beta_{ij}(x) = \hat{\beta}_{ij}(x) - \beta_{ij}(x)$ are the fuzzy approximation errors.

In state-variable form, the error equation of the ith output Equation (33) takes the form

$$\dot{e}_i = A_i e_i + [\Delta\alpha_i(x) + \sum_{j=1}^{p=2} \Delta\beta_{ij}(x)u_j]b_i \tag{34}$$

where A_i and b_i are given by

$$\begin{cases} A_1 = -k_{01}, \; b_1 = 1 \\ A_2 = \begin{bmatrix} 0 & 1 \\ -k_{12} & -k_{02} \end{bmatrix}, \; b_2 = \begin{bmatrix} 0 \\ 1 \end{bmatrix} \end{cases} \tag{35}$$

Theorem 1. *The closed-loop tracking error* $e = \begin{bmatrix} e_1 & e_2 \end{bmatrix}^T$ *is globally ultimately bounded if the updating laws of the parameter vectors* $\theta_i \in R^{M\times1}$ *and* $\theta_{ij} \in R^{M\times1}$ *are chosen as in Equations (36) and (37):*

$$\dot{\theta}_i = -\gamma_i e_i^T P_i b_i \xi(x) \tag{36}$$

$$\dot{\theta}_{ij} = -\gamma_{ij} e_i^T P_i b_i \xi(x) u_j \tag{37}$$

where γ_i *and* γ_{ij} *are design parameters and* P_i *is a unique positive definite matrix solution of the Lyapunov Equation (38) with arbitrary positive definite matrix* Q_i

$$A_i^T P_i + P_i A_i = -Q_i \tag{38}$$

Proof. Define the minimum fuzzy approximation error w_i in terms of the optimal values of adjustable parameters θ_i^* and θ_{ij}^* as

$$w_i = [\hat{\alpha}_i(x \mid \theta_i^*) - \alpha_i(x)] + \sum_{j=1}^{p=2} \left[\hat{\beta}_{ij}\left(x \mid \theta_{ij}^*\right) - \beta_{ij}(x)\right]u_j \tag{39}$$

Add and subtract the terms $\hat{\alpha}_i\left(x \mid \theta_i^*\right)$ and $\hat{\beta}_{ij}\left(x \mid \theta_{ij}^*\right)$ to Equation (34) and then use the definition given in Equation (39) to obtain the following error equation

$$\dot{e}_i = A_i e_i + b_i\left[w_i + \varphi_{\alpha_i}^T \xi(x) + \sum_{j=1}^{p=2} \varphi_{\beta_{ij}}^T \xi(x)u_j\right] \tag{40}$$

where $\varphi_{\alpha_i} = \left(\theta_i - \theta_i^*\right)$ and $\varphi_{\beta_{ij}} = \left(\theta_{ij} - \theta_{ij}^*\right)$ are the parameter errors. Note that the derivatives of these parameter errors are given by:

$$\dot{\varphi}_{\alpha_i} = \dot{\theta}_i \tag{41}$$

$$\dot{\varphi}_{\beta_{ij}} = \dot{\theta}_{ij} \tag{42}$$

The following positive Lyapunov function is formulated as a quadratic function of the error involved, namely the tracking error (34) and the parameter error (41) and (42):

$$V_i = \frac{1}{2}e_i^T P_i e_i + \frac{1}{2\gamma_i}\varphi_{\alpha_i}^T \varphi_{\alpha_i} + \sum_{j=1}^{p=2} \frac{1}{2\gamma_{ij}}\varphi_{\beta_{ij}}^T \varphi_{\beta_{ij}} \tag{43}$$

The time derivative of Equation (43) along the trajectories Equations (40)–(42) is found as:

$$\dot{V}_i = -\frac{1}{2}e_i^T Q_i e_i + \frac{1}{\gamma_i}\varphi_{\alpha_i}^T\left(\dot{\theta}_i + \gamma_i e_i^T P_i b_i \xi(x)\right) + \left(\frac{1}{\gamma_{ij}}\sum_{j=1}^{p=2}\varphi_{\beta_{ij}}^T\dot{\theta}_{ij} + e_i^T P_i b_i \sum_{j=1}^{p=2}\varphi_{\beta_{ij}}^T \xi(x)u_j\right) + e_i^T P_i b_i w_i \tag{44}$$

Now, substituting the parameters' updating laws in Equations (36) and (37) in Equation (44) to get:

$$\dot{V}_i = -\frac{1}{2}e_i^T Q_i e_i + e_i^T P_i b_i w_i \tag{45}$$

Provided that $\left\|e_i\right\| \geq \frac{4\sigma_i \lambda_{max}(P_i)}{\beta_i \lambda_{min}(Q_i)} = r_i$, it is straightforward to write Equation (45) in the form

$$\dot{V}_i \leq -\frac{1}{2}(1 - \beta_i)\lambda_{min}(Q_i)\left\|e_i\right\|^2 \tag{46}$$

where $0 < \beta_i < 1, \sigma_i > 0$, such that $\left\|w_i\right\| \leq \sigma_i$, $\lambda_{min}(Q_i)$ and $\lambda_{max}(P_i)$ are the minimum and maximum eigenvalues of the indicated matrices and $\|.\|$ stands for the Euclidean norm. From the positive definiteness of Equation (43) and the negative definiteness of Equation (46), we conclude that the tracking error is globally ultimately bounded with bound $\mu_{bi} = r_i\sqrt{\frac{\lambda_{max}(P_i)}{\lambda_{min}(P_i)}}$ [47]. In Equation (43), the Lyapunov function is quadratic; the non-quadratic Lyapunov function can also be used in adaptive schemes for better performance [49]. $\square$

4. Implementation of the Proposed Adaptive Fuzzy Controller for GCIS

In order to implement the proposed AFC based on feedback linearization given by Equations (26), (27), (32), (36), and (37), fuzzy sets F_k^i have to be selected where $i = 1, 2, \ldots N$, N is the number of the fuzzy sets and $k = 1, 2, 3$. The fuzzy sets are utilized to determine the vector of FBFs given in Equation (28). To this end, three Gaussian fuzzy sets, namely Negative (N), Zero (Z), and Positive (P) are used to generate the FBFs for each state of the system. These fuzzy sets are characterized by the membership functions. The general form of the membership functions of Gaussian type is given by

$$\mu_{F_k^i}(x_k) = exp\left(-\frac{\left(x_k - \overline{x}_k^i\right)^2}{\sigma_k^i}\right) \tag{47}$$

where $\overline{x}_k^i$ and σ_k^i are the center and the width of the ith fuzzy set F_k^i.

The block diagram of the proposed controller is shown in Figure 2. It can be seen in the block diagram that the grid voltage and current are transformed into a dq frame from an abc frame. The control laws in Equations (36) and (37) were used to estimate the unknown parameters of GCIS, where the calculation initially started from chosen initial values of θ_i and θ_{ij}. Then, AFC low in Equation (30) was applied to generate the control signals. The PWM was generated by applying space vector pulse width modulation (SVPWM) to drive the inverter. Note that the signal V_{dcref} is released from the MPPT algorithm.

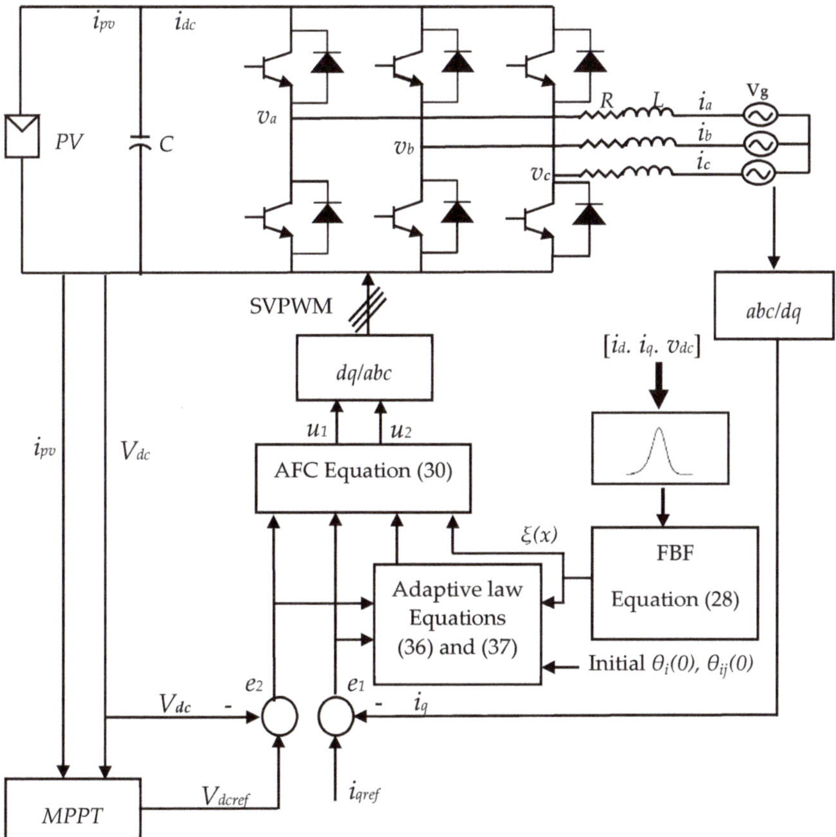

Figure 2. The proposed adaptive fuzzy control (AFC) technique for the grid connected inverter system (GCIS).

5. Simulation Cases and Results

To examine the effectiveness of the proposed controller performance, the proposed AFC was implemented and tested in the MATLAB/SIMULINK [50] environment for a GCIS having the parameters as listed in Table 1. The other design parameters were selected as

$k_{01} = 10$, $k_{02} = k_{12} = 10,000$, $\gamma_1 = 40$, $\gamma_2 = 0.01$, $\gamma_{11} = 0.01$, $\gamma_{12} = 0.1$, $\gamma_{21} = 0.1$, and $\gamma_{22} = 1$. The selected positive definite matrices Q_i and the unique positive definite matrix solution P_i, $i = 1, 2$ that appeared in Lyapunov Equation (38) are given by

$$Q_1 = 100, \ Q_2 = \begin{bmatrix} 2000 & 0 \\ 0 & 1 \end{bmatrix}, \ P_1 = 5, \ P_2 = \begin{bmatrix} 1000.6 & 0.1 \\ 0.1 & 0.00006 \end{bmatrix}$$

Table 1. System parameters.

Parameter	Value
Grid Voltage rms	120 V
Inductance L	2 mH
Resistor R	0.1 Ω
Grid Frequency	50 Hz
DC link capacitor	2200 μF
PV array voltage Vdc	540 V
PV array current Ipv	3.46 A

For each state of the system, the parameters of the Gaussian membership functions given in Equation (47) are listed in Table 2. The membership functions for the state x_1 are shown in Figure 3, as an example for states membership functions.

Table 2. Parameters of the Gaussian membership functions.

State$\downarrow$	Fuzzy Set$\rightarrow$	N	Z	P
x_1		$\bar{x}_1^N = -5$ $\sigma_1^N = 6$	$\bar{x}_1^Z = 0$ $\sigma_1^Z = 6$	$\bar{x}_1^P = 5$ $\sigma_1^P = 6$
x_2		$\bar{x}_2^N = -0.1$ $\sigma_2^N = 0.005$	$\bar{x}_2^Z = 0$ $\sigma_2^Z = 0.005$	$\bar{x}_2^P = 0.1$ $\sigma_2^P = 0.005$
x_3		$\bar{x}_3^N = 525$ $\sigma_3^N = 100$	$\bar{x}_3^Z = 550$ $\sigma_3^Z = 100$	$\bar{x}_3^P = 575$ $\sigma_3^P = 100$

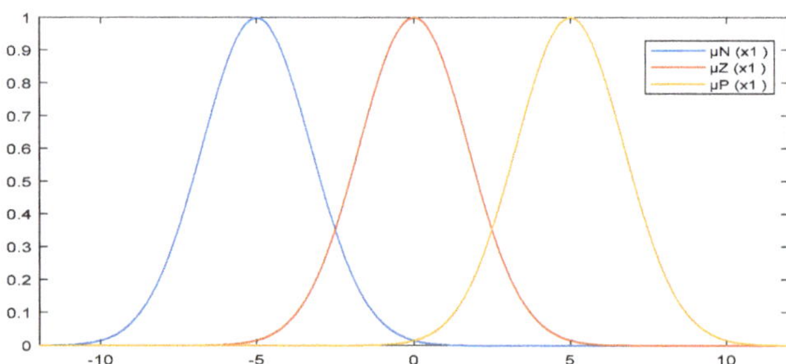

Figure 3. Membership functions for x_1.

The proposed AFC was studied under different operating cases as unity power factor tracking, tracking of power factor changes, and robust tracking. Smooth reference values were used for all simulation cases.

5.1. Case I: Unity Power Factor Tracking

In this case, simulation was carried out by selecting the reference grid current components as $i_{qref} = 0.0$ A, which corresponds to unity power factor. The output voltage and current are shown in Figure 4. The figure clearly shows that the grid current is in phase with grid voltage, which indicates unity PF operation. Figure 5a,b depicts the reactive i_q and active i_d current tracking output of the proposed controller. The DC voltage and reference voltage are shown in Figure 6. The control signal u_1 and u_2 are shown in Figure 7a,b, from which it can be noticed that they are bounded. From the obtained result in case of the unity power factor, it can be stated that the proposed controller provides excellent tracking performance with bounded tracking error and bounded control signals.

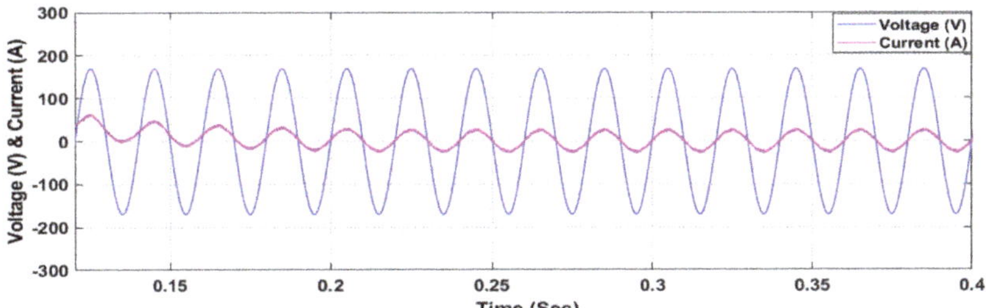

Figure 4. Output voltage current.

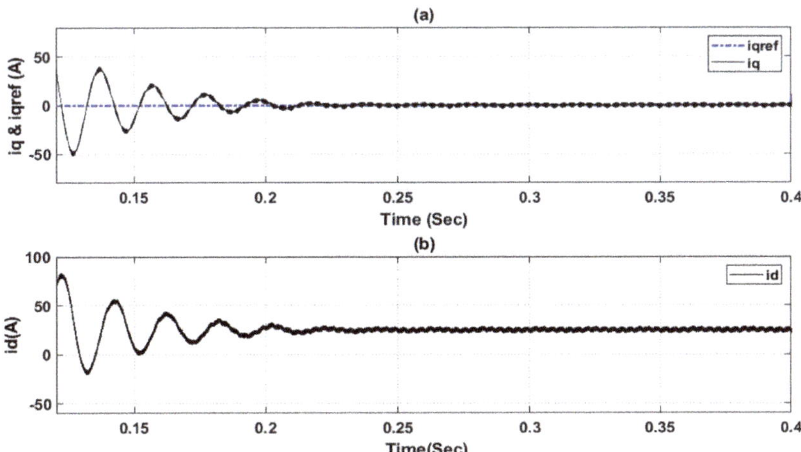

Figure 5. Grid current components: (**a**) i_q, i_{qref}; (**b**) i_d.

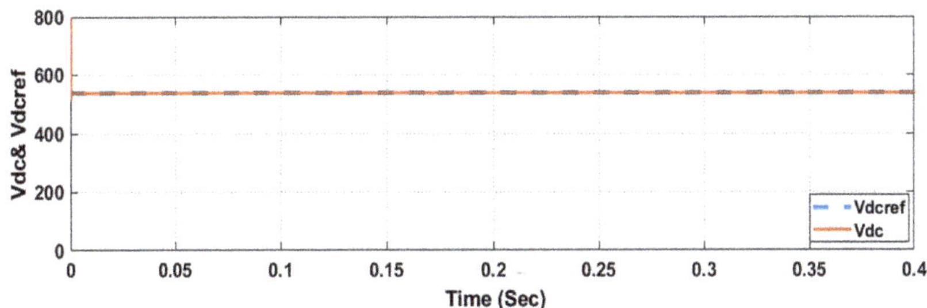

Figure 6. DC voltage and reference voltage.

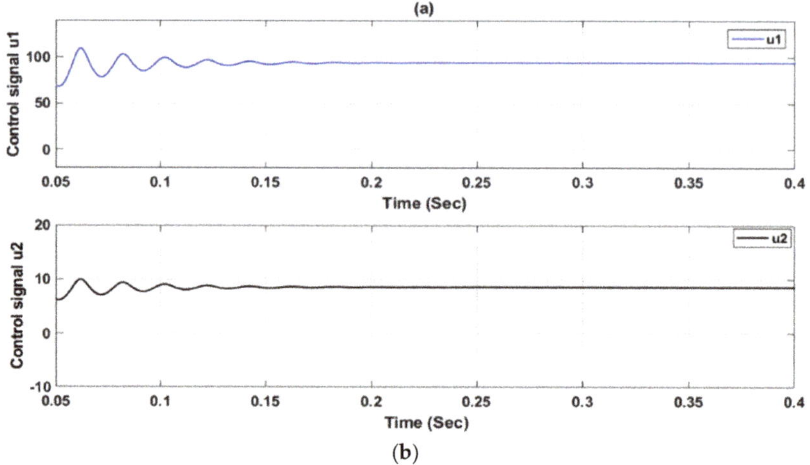

Figure 7. Control signals: (a) $u_1 = v_d$; (b) $u_2 = v_q$.

5.2. Case II: Tracking of Power Factor Changes

In this case study, the performance of the proposed AFC was tested for power factor tracking. At the start, the system was assumed to operate at unity power factor with $i_{qref} = 0$, then a step change of 10 A in i_{qref} at 0.4 s was applied. This change in i_{qref} corresponded to a change in the power factor to 0.937. Figure 8a,b display the effect of changes in i_{qref}, i_q, and i_d. The results demonstrate that i_q reaches its new reference quickly. Hence, the obtained results clearly prove that the proposed AFC has the ability to track power factor changes. The output voltage and current are shown in Figure 9. A phase shift can be noticed between current and voltage after t = 0.4 s, confirming the tracking of the desired power factor. The active and reactive power delivered by the inverter to the grid are shown in Figure 10a,b, confirming the proposed controller tracking ability. The bounded control signals $u_1 = v_d$ and $u_2 = v_q$ are shown in Figure 11a,b.

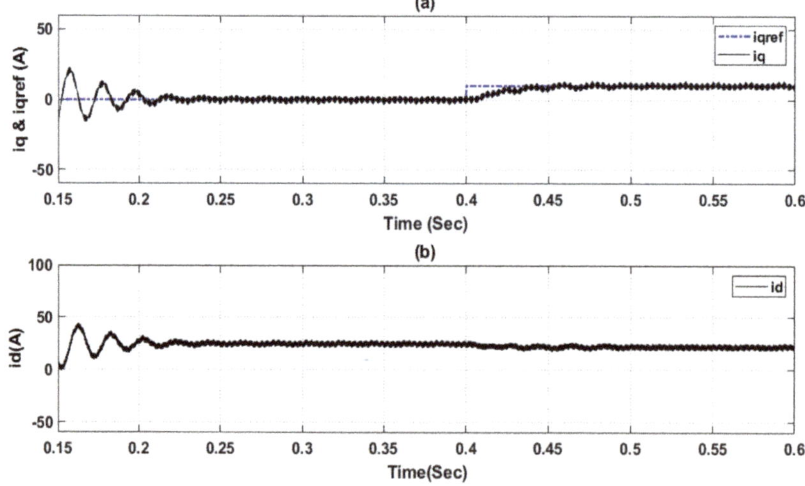

Figure 8. Grid current components: (a) i_q, i_{qref}; (b) i_d.

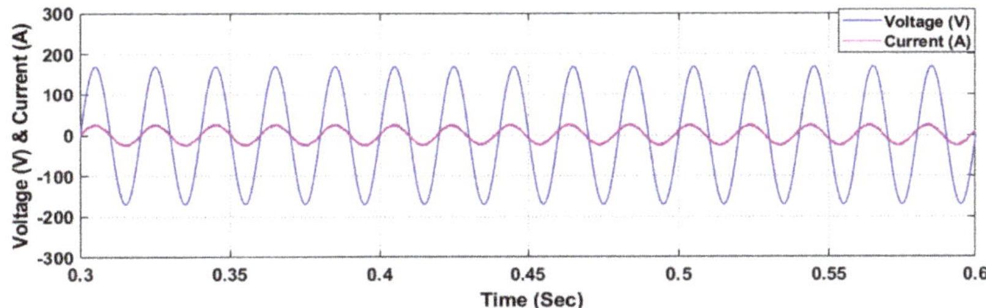

Figure 9. Voltage and current.

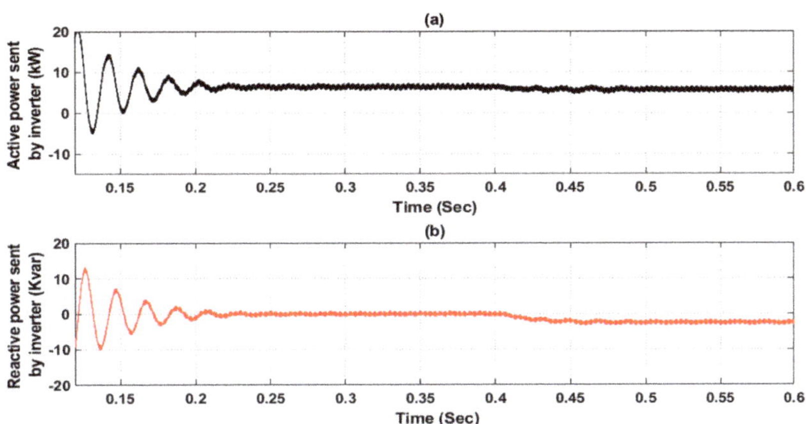

Figure 10. Power sent by inverter: (**a**) Active power; (**b**) Reactive power.

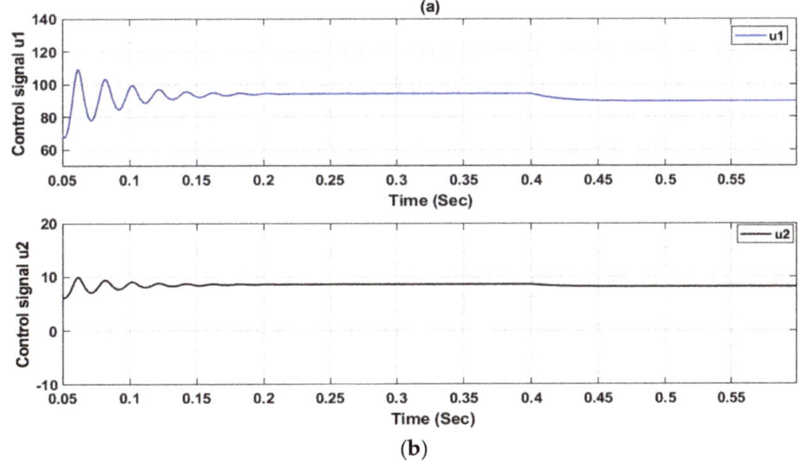

Figure 11. Control signals: (**a**) $u_1 = v_d$; (**b**) $u_2 = v_q$.

To evaluate the effectiveness of AFC, the performance of the proposed controller was compared with the PI controller as in [51]. The comparison was conducted for power factor change tracking case by applying a step change of 10 A in i_{qref} at 0.4 s. Figure 12 demonstrates the performance of the proposed AFC and PI controller. The result illustrates that the tracking between i_q and i_{qref} after the step change occurs has less fluctuations and overshooting in case of proposed AFC in comparison to the PI controller. Moreover, a comparison between the performance of the proposed controller, the PI controller, and the TSKPFNN controller presented in [35] is shown in Table 3. From the illustrated results in Table 3, it can be said that the performance of proposed AFC is better and exceeds the PI controller and TSKPFNN controller performances.

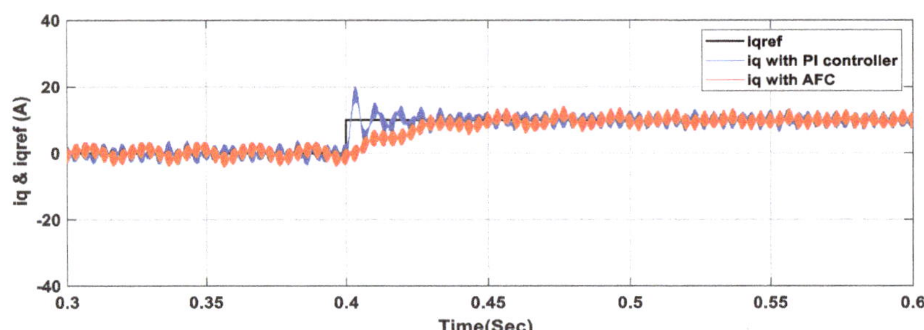

Figure 12. Comparison between PI controller and AFC with power factor tracking.

Table 3. Comparison between the performance of the proposed controller, PI, and TSKPFNN controllers.

Controller	Max Overshoot %	Settling Time (S)
PI controller	75	0.04
TSKPFNN controller	12.24	0.3
Proposed AFC	0.0	0.035

5.3. Case III: Robust Tracking

In certain cases, the parameters used in the GCIS are either time-varying or not precisely defined, so there are often parametric uncertainties where the filters connected to the grid inductance value change over time, affected by the impedance value of the grid which varies depending on the grid structure and conditions leading to resonance and instability problems. In addition, due to changes in ambient operating temperature or changes in applied voltage and frequency, DC Link capacitance values can change.

In this simulation case, the robustness of the proposed AFC was tested for GCIS parameter variations. Simulations were carried out for different percentages of variations in filter inductor L and dc-link capacitor C. Grid reactive current component i_q, with 10% variations in filter inductor L, is shown in Figure 13. Figure 14 illustrates the bounded control signals of the proposed controller with the same variation in L. The obtained results illustrate the robustness of the AFC with filter inductor increase.

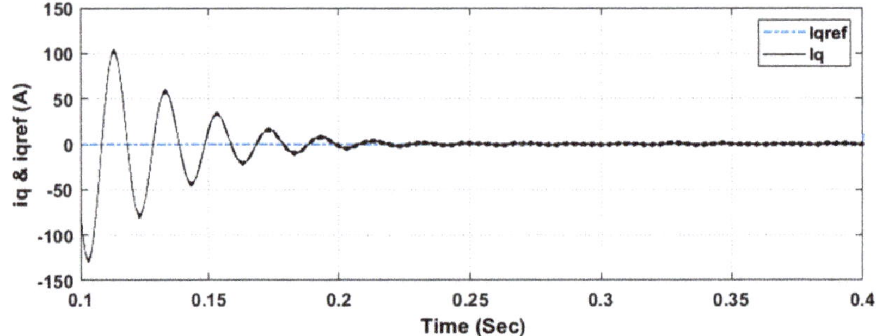

Figure 13. i_q & i_{qref} with 10% increase in L.

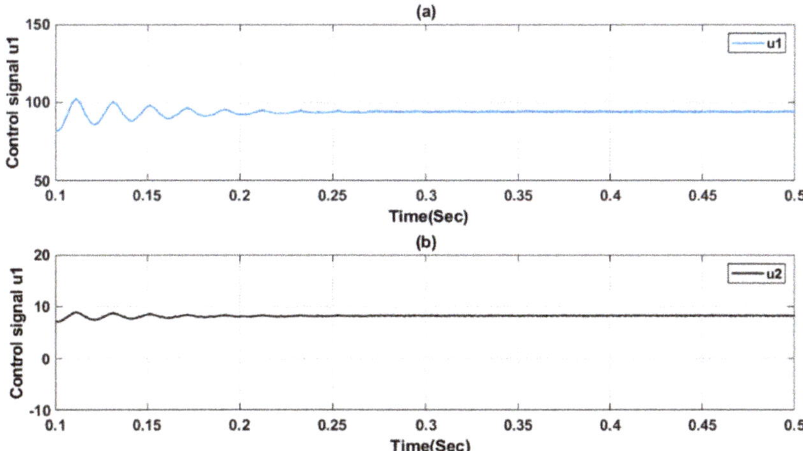

Figure 14. Control signals with 10% increase in L: (**a**) $u_1 = v_d$ (**b**) $u_2 = v_q$.

To study the robustness of the proposed AFC for variations dc-link capacitor C, simulations were carried out for 30% increase and carried again for 20% decrease in C. The performance of the GCIS with proposed AFC with applied variations is shown in Figures 15–18, where Figure 15a,b displays i_q and i_d with 30% increase in C. Grid voltage and current with the same increase in C are shown in Figure 16. For the case of the 20% decrease in C, Figure 17 illustrates the bounded control signals; tracking between i_q and i_{qref}, is shown in Figure 18. The obtained simulation results with variations in C prove the robustness of the controller. Furthermore, Figure 19 displays the performance of the proposed controller for variation in the inductor and capacitor at the same time with 10% increase in L and 30% increase in C.

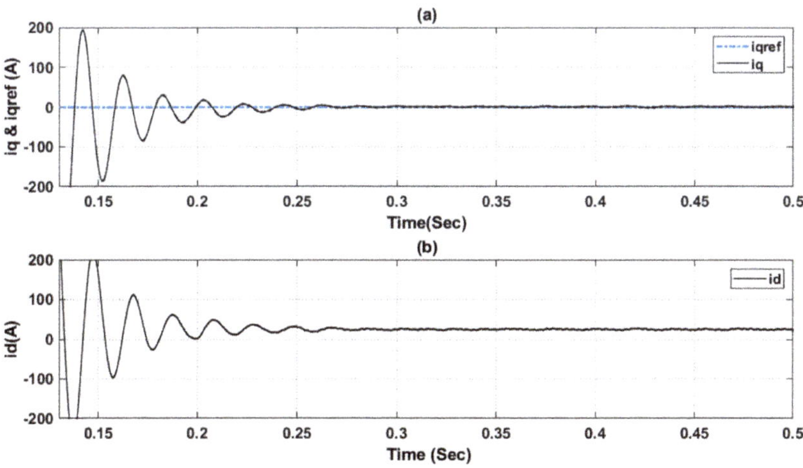

Figure 15. GCIS performance with 30% increase in C: (**a**) i_q, i_{qref}, (**b**) i_d.

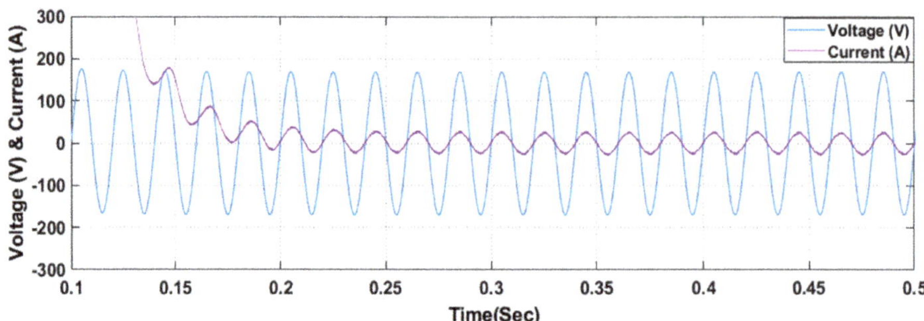

Figure 16. Grid voltage and current with 30% increase in C.

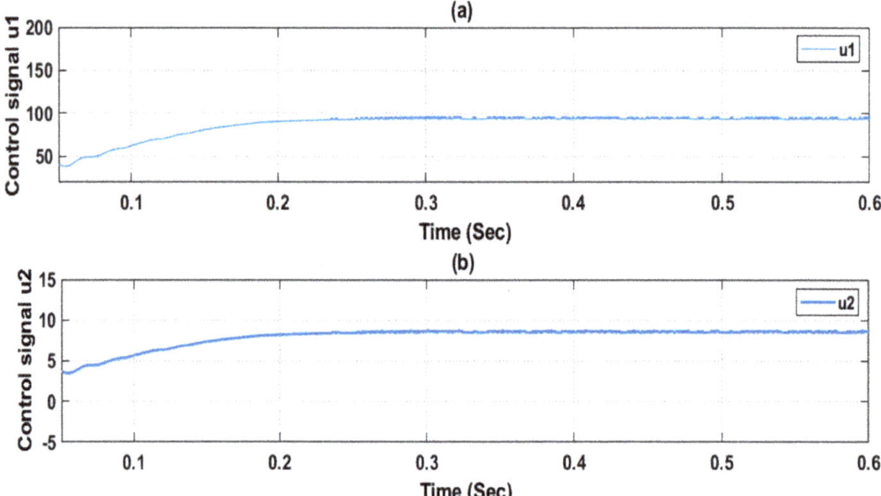

Figure 17. Control signals with 20% decrease in C: (**a**) $u_1 = v_d$ (**b**) $u_2 = v_q$.

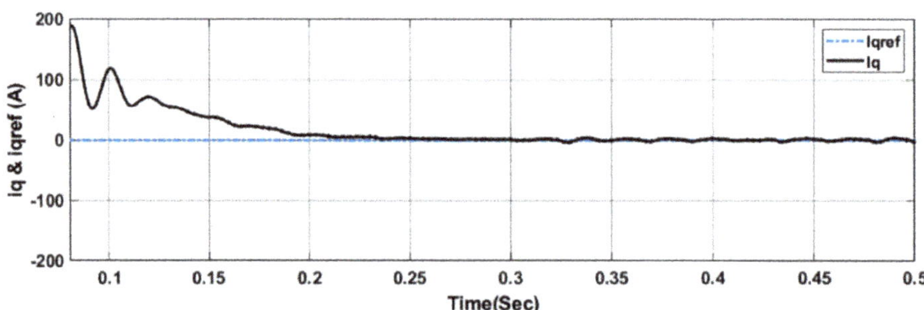

Figure 18. i_q & i_{qref} with 20% decrease in C.

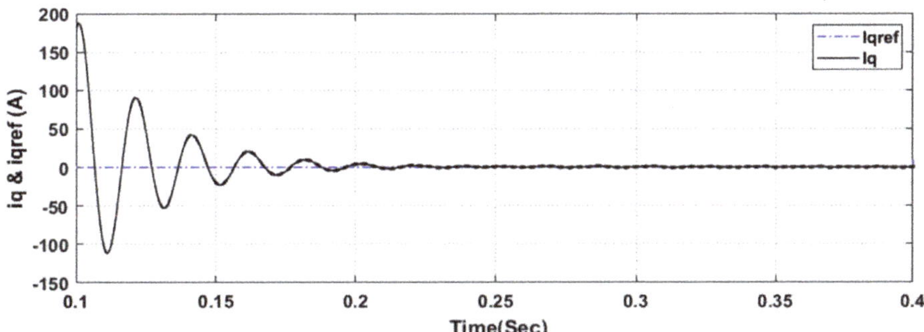

Figure 19. i_q & i_{qref} with simultaneous 10% increase in L and 30% increase in C.

From all conducted simulation results for parameter uncertainties, the performance of the GCIS proves that the proposed AFC is capable to cope with the uncertainty of the GCIS parameters and achieve the desired tracking performance.

5.4. Case IV: Tracking in the Presence of Model Uncertainity

The proposed AFC given in (30) can achieve tracking in presence of modelling uncertainties that are inherent in the nature of the GCIS. The presence of the PV in the GCIS model given by (10) is the main reason for the modelling uncertainties and to account for these uncertainties, Equation (10) can be rewritten as follows:

$$\dot{x} = (f(x) + \Delta f(x)) + g(x)u \tag{48}$$

where $\Delta f(x)$ is the uncertainty associated with $f(x)$ and given by

$$\Delta f = \begin{bmatrix} \Delta f_1 \\ \Delta f_2 \\ \Delta f_3 \end{bmatrix} \tag{49}$$

In this case, the function $\alpha(x)$ given in Equation (17) that results from feedback linearization will be perturbed by $\Delta\alpha(x)$ given by

$$\Delta\alpha(x) = \begin{bmatrix} \Delta f_2 \\ \Delta\alpha_2 \end{bmatrix} \tag{50}$$

In (50), $\Delta\alpha_2$ is given by

$$\Delta\alpha_2 = -\frac{1}{Cx_3}\left(v_{gd}\Delta f_1 + v_{gq}\Delta f_2\right) + n\,\Delta f_3 \tag{51}$$

where $n = \dfrac{(v_{gd}x_1 + v_{gq}x_2)}{Cx_3^2}$.

To test the tracking performance of the proposed AFC against modeling uncertainty, we assumed there is an uncertainty $\Delta f_3 = 5\%$ which is mainly due to the presence of the PV current. The other uncertainties $\Delta f_1 = \Delta f_2$ were assumed zero. The simulation result for tracking i_{qref} is shown in Figure 20. It can be seen that even with modeling uncertainty, the proposed AFC controller is able to track the reference reactive current i_{qref} and keep the GCIS operating at the unity power factor.

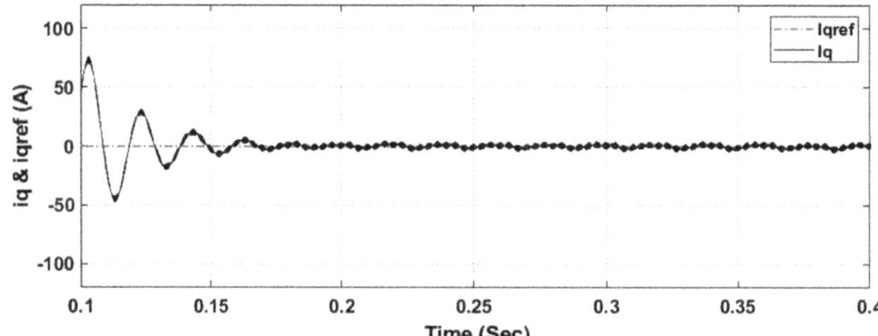

Figure 20. i_q & i_{qref} with modeling uncertainty $\Delta f_3 = 5\%$.

6. Conclusions

The grid-connected inverter control is studied in this paper. To solve the nonlinearity and uncertainty issues of GCIS, an AFC approach for GCIS is proposed. The developed controller relies on two principles: namely, the input-output feedback linearization principle and the approximation capability of the fuzzy system. The GCIS is modeled as a nonlinear MIMO system. A fuzzy system with weighted average defuzzifier, singleton fuzzifier, and product inference rule is utilized to develop the AFC law through approximation of the unknown nonlinear functions that appear in the input–output linearizing model. Due to the ability of the proposed controller to estimate unknown parameters for different operation cases, the controller is robust against parametric uncertainties. The closed-loop stability using the Lyapunov function analysis is established to show that the output tracking error is globally ultimately bounded. To test the effectiveness of the proposed approach, the proposed AFC was implemented and tested in the MATLAB/SIMULINK environment for a GCIS for different operating cases as unity power factor tracking, tracking of power factor changes and robust tracking. The obtained simulation results showed that the proposed AFC provides excellent tracking performance with bounded tracking error and bounded control signals. In comparison to the PI control and TSKPFNN controller, the proposed AFC showed superiority in terms of response and reduced fluctuations in case of power factor change tracking. Moreover, the results showed that the proposed AFC is very robust against parametric and model uncertainties.

Author Contributions: Methodology, H.Y. and M.S.; Software, M.S.; Validation, H.Y., R.A.A. and A.A.-H.; Writing—original draft preparation, M.S.; Supervision, H.Y.; Writing—review and editing, H.Y., R.A.A. and A.A.-H. All authors have read and agreed to the published version of the manuscript.

Funding: This research was funded by the Sustainable Energy Research Center (SERC) at Sultan Qaboos University (SQU) under grant number IG/DVC/SERC/18/01.

Acknowledgments: The authors acknowledge the financial support provided by Sustainable Energy Research Center (SERC) at Sultan Qaboos University (SQU) under grant number IG/DVC/SERC/18/01.

Conflicts of Interest: The authors declare no conflict of interest.

Abbreviations

The following abbreviations are used in this manuscript:

AFC	Adaptive fuzzy control
DG	Distributed generation
FLC	Fuzzy logic controllers
GCIS	Grid-connected inverter systems
IC	Incremental conductance
IT2FLC	Interval Type-2 fuzzy logic controller
LQG	Linear Quadratic Gaussian
MIMO	Multi-input multi-output
PI	Proportional-integral
PR	Proportional resonant
P&O	Perturb and observe
PV	Photovoltaic
PWM	Pulse width-modulation
RC	Repetitive controller
RES	Renewable energy sources
SVPWM	Space vector pulse width modulation
T2FLC	Type-2 fuzzy logic controller
THD	Total harmonic distortion
TSKPFNN	Takagi–Sugeno–Kang-type probabilistic fuzzy neural network
VSI	Voltage source inverter

References

1. Zahedi, A. A review of drivers, benefits, and challenges in integrating renewable energy sources into electricity grid. *Renew. Sustain. Energy Rev.* **2011**, *15*, 4775–4779. [CrossRef]
2. Dragicevic, T.; Lu, X.; Vasquez, J.C.; Guerrero, J.M. DC Microgrids—Part II: A Review of Power Architectures, Applications, and Standardization Issues. *IEEE Trans. Power Electron.* **2016**, *31*, 3528–3549. [CrossRef]
3. Abbasi, A.K.; Mustafa, M.W. Mathematical Model and Stability Analysis of Inverter-Based Distributed Generator. *Math. Probl. Eng.* **2013**, *2013*, 1–7. [CrossRef]
4. Hadisupadmo, S.; Hadiputro, A.N.; Widyotriatmo, A. A small signal state space model of inverter-based microgrid control on single phase AC power network. *Internetwork. Indones. J.* **2016**, *8*, 71–76.
5. Monica, P.; Kowsalya, M. Control strategies of parallel operated inverters in renewable energy application: A review. *Renew. Sustain. Energy Rev.* **2016**, *65*, 885–901. [CrossRef]
6. Castilla, M.; Miret, J.; Camacho, A.; Matas, J.; De Vicuna, L.G. Reduction of Current Harmonic Distortion in Three-Phase Grid-Connected Photovoltaic Inverters via Resonant Current Control. *IEEE Trans. Ind. Electron.* **2011**, *60*, 1464–1472. [CrossRef]
7. Shen, G.; Zhu, X.; Zhang, J.; Xu, D. A New Feedback Method for PR Current Control of LCL-Filter-Based Grid-Connected Inverter. *IEEE Trans. Ind. Electron.* **2010**, *57*, 2033–2041. [CrossRef]
8. Huerta, F.; Pizarro, D.; Cobreces, S.; Rodriguez, F.J.; Giron, C.; Rodriguez, A. LQG Servo Controller for the Current Control of LCLLCL Grid-Connected Voltage-Source Converters. *IEEE Trans. Ind. Electron.* **2011**, *59*, 4272–4284. [CrossRef]
9. Merabet, A.; Labib, L.; Ghias, A.M.Y.M.; Ghenai, C.; Salameh, T. Robust Feedback Linearizing Control with Sliding Mode Compensation for a Grid-Connected Photovoltaic Inverter System Under Unbalanced Grid Voltages. *IEEE J. Photovolt.* **2017**, *7*, 828–838. [CrossRef]
10. Mahmud, M.A.; Hossain, M.J.; Pota, H.R.; Roy, N.K. Robust Nonlinear Controller Design for Three-Phase Grid-Connected Photovoltaic Systems Under Structured Uncertainties. *IEEE Trans. Power Deliv.* **2014**, *29*, 1221–1230. [CrossRef]
11. Mohiuddin, S.; Mahmud, A.; Haruni, A.; Pota, H.R. Design and implementation of partial feedback linearizing controller for grid-connected fuel cell systems. *Int. J. Electr. Power Energy Syst.* **2017**, *93*, 414–425. [CrossRef]
12. Zhang, X.; Wang, Y.; Yu, C.; Guo, L.; Cao, R. Hysteresis Model Predictive Control for High-Power Grid-Connected Inverters With Output LCL Filter. *IEEE Trans. Ind. Electron.* **2016**, *63*, 246–256. [CrossRef]
13. López-Estrada, F.-R.; Rotondo, D.; Valencia-Palomo, G. A Review of Convex Approaches for Control, Observation and Safety of Linear Parameter Varying and Takagi-Sugeno Systems. *Process* **2019**, *7*, 814. [CrossRef]
14. Hornik, T.; Zhong, Q.-C. A Current-Control Strategy for Voltage-Source Inverters in Microgrids Based on H∞H∞ and Repetitive Control. *IEEE Trans. Power Electron.* **2010**, *26*, 943–952. [CrossRef]
15. Hasanien, H.M. An Adaptive Control Strategy for Low Voltage Ride Through Capability Enhancement of Grid-Connected Photovoltaic Power Plants. *IEEE Trans. Power Syst.* **2016**, *31*, 3230–3237. [CrossRef]
16. Jorge, S.G.; Busada, C.A.; Solsona, J.A. Frequency-Adaptive Current Controller for Three-Phase Grid-Connected Converters. *IEEE Trans. Ind. Electron.* **2012**, *60*, 4169–4177. [CrossRef]

17. Li, X.; Zhang, H.; Shadmand, M.B.; Balog, R.S. Model Predictive Control of a Voltage-Source Inverter With Seamless Transition Between Islanded and Grid-Connected Operations. *IEEE Trans. Ind. Electron.* **2017**, *64*, 7906–7918. [CrossRef]
18. Errouissi, R.; Muyeen, S.M.; Al-Durra, A.; Leng, S. Experimental Validation of a Robust Continuous Nonlinear Model Predictive Control Based Grid-Interlinked Photovoltaic Inverter. *IEEE Trans. Ind. Electron.* **2015**, *63*, 4495–4505. [CrossRef]
19. Almeida, P.M.; Duarte, J.L.; Ribeiro, P.F.; Barbosa, P.G. Repetitive controller for improving grid-connected photovoltaic systems. *IET Power Electron.* **2014**, *7*, 1466–1474. [CrossRef]
20. Harirchian, E.; Lahmer, T. Improved Rapid Visual Earthquake Hazard Safety Evaluation of Existing Buildings Using a Type-2 Fuzzy Logic Model. *Appl. Sci.* **2020**, *10*, 2375. [CrossRef]
21. Tarbosh, Q.A.; Aydogdu, O.; Farah, N.; Talib, H.N.; Salh, A.; Cankaya, N.; Omar, F.A.; Durdu, A. Review and Investigation of Simplified Rules Fuzzy Logic Speed Controller of High Performance Induction Motor Drives. *IEEE Access* **2020**, *8*, 49377–49394. [CrossRef]
22. Pushpavalli, M.; Swaroopan, N.M.J. KY converter with fuzzy logic controller for hybrid renewable photovoltaic/wind power system. *Trans. Emerg. Telecommun. Technol.* **2020**, *31*, 3989. [CrossRef]
23. Mumtaz, S.; Ahmad, S.; Khan, L.; Ali, S.; Kamal, T.; Hassan, S.Z. Adaptive Feedback Linearization Based NeuroFuzzy Maximum Power Point Tracking for a Photovoltaic System. *Energies* **2018**, *11*, 606. [CrossRef]
24. Hosseinzadeh, M.; Salmasi, F.R. Power management of an isolated hybrid AC/DC micro-grid with fuzzy control of battery banks. *IET Renew. Power Gener.* **2015**, *9*, 484–493. [CrossRef]
25. Harirchian, E.; Lahmer, T. Developing a hierarchical type-2 fuzzy logic model to improve rapid evaluation of earthquake hazard safety of existing buildings. *Structures* **2020**, *28*, 1384–1399. [CrossRef]
26. Mittal, K.; Jain, A.; Vaisla, K.S.; Castillo, O.; Kacprzyk, J. A comprehensive review on type 2 fuzzy logic applications: Past, present and future. *Eng. Appl. Artif. Intell.* **2020**, *95*, 103916. [CrossRef]
27. Chen, G.; Pham, T.T.; Boustany, N. Introduction to Fuzzy Sets, Fuzzy Logic, and Fuzzy Control Systems. *Appl. Mech. Rev.* **2001**, *54*, B102–B103. [CrossRef]
28. Hannan, M.; Ghani, Z.A.; Mohamed, A.; Uddin, M.N. Real-time testing of a fuzzy-logic-controller-based grid-connected photovoltaic inverter system. *IEEE Trans. Ind. Appl.* **2015**, *51*, 4775–4784. [CrossRef]
29. Muyeen, S.M.; Al-Durra, A. Modeling and Control Strategies of Fuzzy Logic Controlled Inverter System for Grid Interconnected Variable Speed Wind Generator. *IEEE Syst. J.* **2013**, *7*, 817–824. [CrossRef]
30. Lin, P.-Z.; Hsu, C.-F.; Lee, T.-T. Type-2 Fuzzy Logic Controller Design for Buck DC-DC Converters. In Proceedings of the 14th IEEE International Conference on Fuzzy Systems, FUZZ '05, Reno, NV, USA, 25–25 May 2005. [CrossRef]
31. Altın, N. Interval Type-2 Fuzzy Logic Controller Based Maximum Power Point Tracking in Photovoltaic Systems. *Adv. Electr. Comput. Eng.* **2013**, *13*, 65–70. [CrossRef]
32. El Khateb, A.H.; Rahim, N.A.; Selvaraj, J. Type-2 Fuzzy Logic Approach of a Maximum Power Point Tracking Employing SEPIC Converter forPhotovoltaic System. *J. Clean Energy Technol.* **2013**, *1*, 41–44. [CrossRef]
33. Altin, N. Single phase grid interactive PV system with MPPT capability based on type-2 fuzzy logic systems. In Proceedings of the 2012 International Conference on Renewable Energy Research and Applications (ICRERA), Nagasaki, Japan, 11–14 November 2012; pp. 1–6.
34. Hannan, M.A.; Ghani, Z.A.; Hoque, M.; Ker, P.J.; Hussain, A.; Mohamed, A. Fuzzy Logic Inverter Controller in Photovoltaic Applications: Issues and Recommendations. *IEEE Access* **2019**, *7*, 24934–24955. [CrossRef]
35. Lin, F.-J.; Lu, K.-C.; Ke, T.-H.; Yang, B.-H.; Chang, Y.-R. Reactive Power Control of Three-Phase Grid-Connected PV System During Grid Faults Using Takagi–Sugeno–Kang Probabilistic Fuzzy Neural Network Control. *IEEE Trans. Ind. Electron.* **2015**, *62*, 5516–5528. [CrossRef]
36. Lin, F.-J.; Lu, K.-C.; Yang, B.-H. Recurrent Fuzzy Cerebellar Model Articulation Neural Network Based Power Control of a Single-Stage Three-Phase Grid-Connected Photovoltaic System During Grid Faults. *IEEE Trans. Ind. Electron.* **2016**, *64*, 1258–1268. [CrossRef]
37. Wang, L.-X. *A Course in Fuzzy Systems and Control*; Prentice-Hall, Inc.: Upper Saddle River, NJ, USA, 1996.
38. Wang, L.X.; Ying, H. Adaptive fuzzy systems and control: Design and stability analysis. *J. Intell. Fuzzy Syst. Appl. Eng. Technol.* **1995**, *3*, 187.
39. Yousef, H.A.; Wahba, M.A. Adaptive fuzzy mimo control of induction motors. *Expert Syst. Appl.* **2009**, *36*, 4171–4175. [CrossRef]
40. Nguyen, H.M. Advanced Control Strategies for Wind Energy Conversion Systems. Ph.D. Thesis, Idaho State University, Pocatello, ID, USA, 2013.
41. Zhou, C.; Quach, D.-C.; Xiong, N.; Huang, S.; Zhang, Q.; Yin, Q.; Vasilakos, A.V. An Improved Direct Adaptive Fuzzy Controller of an Uncertain PMSM for Web-Based E-Service Systems. *IEEE Trans. Fuzzy Syst.* **2014**, *23*, 58–71. [CrossRef]
42. Liu, X.; Zhai, D.; Dong, J.; Zhang, Q. Adaptive fault-tolerant control with prescribed performance for switched nonlinear pure-feedback systems. *J. Frankl. Inst.* **2018**, *355*, 273–290. [CrossRef]
43. Li, Y.-X.; Yang, G.-H. Adaptive fuzzy fault tolerant tracking control for a class of uncertain switched nonlinear systems with output constraints. *J. Frankl. Inst.* **2016**, *353*, 2999–3020. [CrossRef]
44. Yazdani, A.; Di Fazio, A.R.; Ghoddami, H.; Russo, M.; Kazerani, M.; Jatskevich, J.; Strunz, K.; Leva, S.; Martinez, J.A. Modeling Guidelines and a Benchmark for Power System Simulation Studies of Three-Phase Single-Stage Photovoltaic Systems. *IEEE Trans. Power Deliv.* **2010**, *26*, 1247–1264. [CrossRef]

45. Chen, P.-C.; Liu, Y.-H.; Chen, J.-H.; Luo, Y.-F. A comparative study on maximum power point tracking techniques for photovoltaic generation systems operating under fast changing environments. *Sol. Energy* **2015**, *119*, 261–276. [CrossRef]
46. Lalili, D.; Mellit, A.; Lourci, N.; Medjahed, B.; Boubakir, C. State feedback control and variable step size MPPT algorithm of three-level grid-connected photovoltaic inverter. *Sol. Energy* **2013**, *98*, 561–571. [CrossRef]
47. Khalil, H.K.; Grizzle, J.W. *Nonlinear Systems*; Prentice Hall: Upper Saddle River, NJ, USA, 2002; Volume 3.
48. Ross, T.J. *Fuzzy Logic with Engineering Applications*, 4th ed.; Wiley: Hoboken, NJ, USA, 2010.
49. Hosseinzadeh, M.; Yazdanpanah, M.J. Performance enhanced model reference adaptive control through switching non-quadratic Lyapunov functions. *Syst. Control. Lett.* **2015**, *76*, 47–55. [CrossRef]
50. MATLAB. *MATLAB R2018b*; The MathWorks: Natick, MA, USA, 2018.
51. Arzani, A.; Arunagirinathan, P.; Venayagamoorthy, G.K.; Ali, A.; Paranietharan, A.; Kumar, V.G. Development of Optimal PI Controllers for a Grid-Tied Photovoltaic Inverter. In Proceedings of the 2015 IEEE Symposium Series on Computational Intelligence, Institute of Electrical and Electronics Engineers. Cape Town, South Africa, 8–10 December 2015; pp. 1272–1279.

Article

Conceptual Study of Vernier Generator and Rectifier Association for Low Power Wind Energy Systems

Philippe Enrici [1,*], Ivan Meny [2] and Daniel Matt [1]

[1] Institut d'Electroniques et des Systèmes (IES)-CNRS UMR 5214, Université de Montpellier, 34095 Montpellier, France; daniel.matt@umontpellier.fr

[2] Département Génie Electrique Informatique Industrielle, Université de Rouen-Normandie, 76130 Mont-Saint-Aignan, France; ivan.meny@univ-rouen.fr

* Correspondence: Philippe.enrici@umontpellier.fr

Abstract: In this paper, we study a wind energy conversion system designed for domestic use in urban or agricultural areas. We first present the turbine, which was specifically designed to be installed on the buildings that it supplies. Based on turbine characteristics, we perform analytical sizing of a Permanent Magnet Vernier Machine (PMVM), which will be used as a generator in our energy conversion system. We show the influence of this generator on system operation by studying its association with a PWM rectifier and with a diode bridge rectifier. We then seek to improve generator design so that the turbine operates closely to maximum power points, while using a simple and robust energy conversion system. We use simulation to show the improvements achieved by taking into account the entire energy conversion system during machine design.

Keywords: converter–machine association; direct drive machine; Permanent Magnet Vernier Machine; synchronous generator; wind energy system for domestic applications; renewable energy

Citation: Enrici, P.; Meny, I.; Matt, D. Conceptual Study of Vernier Generator and Rectifier Association for Low Power Wind Energy Systems. *Energies* **2021**, *14*, 666. https://doi.org/10.3390/en14030666

Academic Editor: Adolfo Dannier
Received: 14 December 2020
Accepted: 25 January 2021
Published: 28 January 2021

Publisher's Note: MDPI stays neutral with regard to jurisdictional claims in published maps and institutional affiliations.

Copyright: © 2021 by the authors. Licensee MDPI, Basel, Switzerland. This article is an open access article distributed under the terms and conditions of the Creative Commons Attribution (CC BY) license (https://creativecommons.org/licenses/by/4.0/).

1. Introduction

Concurrently with the increasing existence of wind farms offering power of several megawatts [1], the expansion of small wind turbines can also be observed, with power ranges from 1 to 50 kW [2–4]. When associated with other energy sources, these turbines can provide a self-sufficient supply of power for a home or remote location. As such, they avoid costly, or even impractical, connections to the grid, which can actually be a significant source of energy loss with regard to the power transmitted over the line.

In electrical installations already connected to the grid, small turbines help reduce the relatively low efficiency of centralized production, while increasing the share of renewable energies in electric power production. In addition, their smaller footprints, and the fact that they are not grouped in wind farms, reduce visual impact, which is one of the greatest reasons for opposing the development of this energy source. However, if the risk of nuisance is rather low in rural areas, turbines must be designed specifically for use in urban areas, so as to be both acoustically and visually discreet.

In this article, we study the entire energy conversion system associated with a wind turbine designed specifically for use in urban areas. This system includes a Vernier machine, which is considered as a suitable alternative for direct-drive applications. The performance of this machine has already been the subject of several studies, but its integration in an electromechanical conversion system has not been investigated thoroughly. Our objective in this paper is to design a Vernier machine to extract a maximum amount of energy from the turbine by implementing a simple and robust energy conversion system. To reach this goal, we are particularly interested in the association of the Vernier machine with a diode bridge rectifier.

2. The Vertical Axis Wind Turbine

In order to meet market requirements for small wind turbines designed to be used in urban areas, the Gual Industrie Company has developed a vertical axis turbine [5], the StatoEolian, shown in Figure 1. This wind turbine is located in Occitanie region in the south of France. This turbine is comprised of an external stator surrounding a paddle rotor. The stator channels the wind at its optimal force onto the rotor. Performance improvement requires a thorough study of the interactions between the different geometric parameters of the turbine. An important quality of this device is that it even remains operational in severe storms.

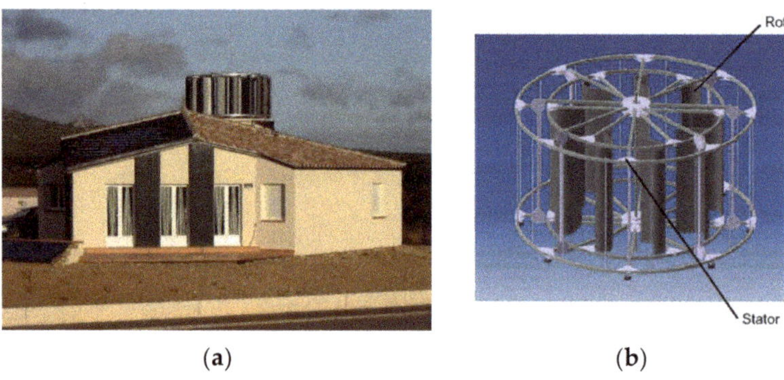

(**a**)   (**b**)

Figure 1. (**a**) The vertical axis turbine installed on a house; (**b**) the wind turbine.

2.1. Characterization of the Turbine

A measurement campaign over several weeks made by the laboratory enabled us to model the aerodynamic torque T_T (in Nm) developed by the turbine as a function of wind speed S_w (in m/s), and of rotation speed N_T (in rpm):

$$T_T = 1.5 \cdot S_W^2 - 0.275 \cdot N_T \cdot S_W \tag{1}$$

This model was used to deduce torque/speed (Figure 2) and power/speed (Figure 3) characteristics for wind speeds ranging from 6 m/s to 22 m/s.

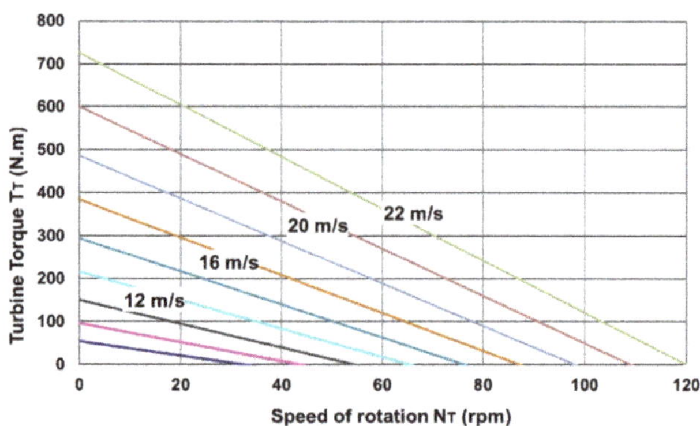

Figure 2. Turbine torque for wind speeds ranging from 6 m/s to 22 m/s.

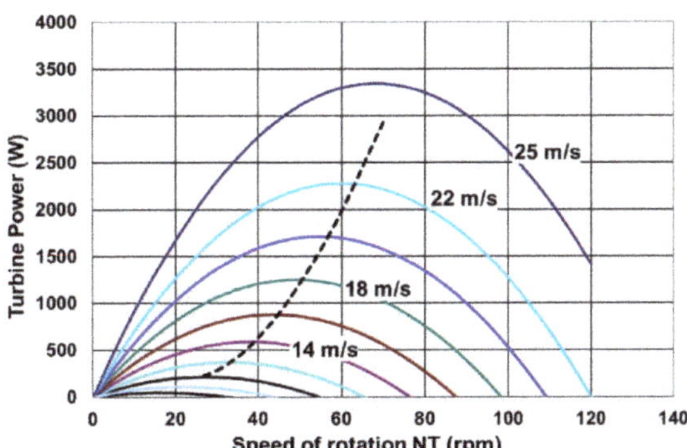

Figure 3. Turbine power for wind speeds ranging from 6 m/s to 25 m/s.

Unlike horizontal axis turbines, the torque value is particularly high when the turbine starts, and it decreases when the rotation speed rises. A more classical bell-shaped curve is observed for power characteristic. One of the main purposes of the energy conversion system, associated with the turbine, will therefore be to keep its working point as close as possible to the maximum power points (dotted curve in Figure 3).

2.2. Generator for the Middle Wind Turbine Characterization of the Turbine

The gearbox is a significant cause of breakdown in a wind turbine, and therefore requires maintenance operations to prevent or correct these failures. This last point is particularly problematic in the case of domestic installations, where users are not likely to possess the skills needed for repairs. The installation could thus be out-of-order for a relatively long time, particularly in a remote area.

The entire energy conversion system must be sturdy. It can face extreme conditions, such as violent winds, possibly without any people being present to perform a safety shut-down. To meet these constraints, and to ensure operation of the turbine without failure for as long as possible, the gearbox is generally oversized in domestic wind turbines. As a result, the gearbox price rises and its integration into the wind energy system becomes more difficult. Another solution is to eliminate the gearbox by using a direct-drive generator.

To make a generator with a high torque-to-weight ratio, essential for implementing a direct-drive, we propose to use a Surface Permanent Magnet Vernier Machine (SPMVM). In this machine, the high torque feature is brought about by the so-called "magnetic gearing effect": a small movement of the rotor induces a large change in flux, which results in high torque.

By taking into account the high value of mean wind speeds recorded at the turbine installation site, as well as the dimensioning variables of the generator, we calculated its rated values for a wind speed of 19 m/s.

The choice of this wind speed was made based on the wind speed readings at the site where the vertical axis wind turbine is installed. This location is one of the windiest French regions with 300 to 350 days of wind per year. The wind is gusty, with large wind variations and up to maximum wind speeds of 24 m/s. The energetic study allowed the determination of the most interesting peak wind speed for the dimensioning. The choice of this wind speed could have been lower but the aim was also in the case of this turbine to show its ability to operate under high wind speeds. For other sites the choice of this peak speed must be predetermined.

We thus obtained:

- rated torque T_{Tn} = 270 Nm;
- rated speed N_{Tn} = 54 rpm;
- rated power P_{Tn} = 1.5 kW.

3. The Permanent Magnet Vernier Machine

3.1. Principle of Permanent Magnet Vernier Machine

The Permanent Magnet Vernier Machine (PMVM) is an interesting alternative to a conventional Permanent Magnet Synchronous Machine (PMSM). It is less well known despite being the subject of many studies [6–10].

PMVMs allow attainment of high mass torques of interest to obtain direct drive generators suitable for low-speed vertical axis turbines used in proximity to wind turbines. They are also efficient for horizontal axis wind turbines compared to synchronous machines with a large number of poles. The PMVMs have sinusoidal electromotive force (e.m.f.) and the torque ripple is almost zero.

The manufacturing cost and reliability of a PMVM is identical to that of a conventional synchronous machine. The objective of the rest of the article is to show the interest of associating a PMVM with a diode rectifier whose association has been optimized. Thus, the economic cost and reliability of the unit are interesting for an urban or rural environment for small powers.

The Permanent Magnet Vernier Machine we present in this article (PMVM) is an evolution of the Vernier Reluctance Machine (VRM).

The polar coupling in a permanent magnet machine is defined when the fundamental interaction of currents and magnets takes place at the pole pitch level, which is the repeating pattern of the stator winding. The toothed coupling is defined when the fundamental interaction of currents and magnets takes place at the tooth pitch scale, which is the length between two slots. When the winding is distributed, the actuator has two forms of coupling: a toothed type and a polar type. This machine is also called a Vernier effect machine [11].

As shown in Figure 4, the rotor teeth row of the VRM has been replaced by an alternate magnets row to obtain the VRMM.

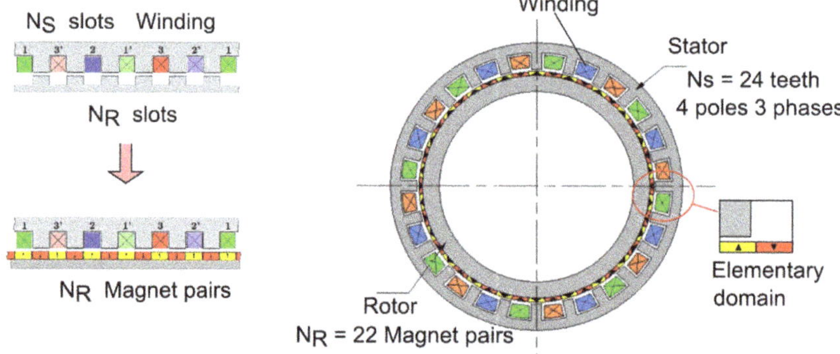

Figure 4. Principle of the Permanent Magnet Vernier Machine.

In a motor, the torque is produced by the interaction of stator and rotor magnetic fields. For its two machines the winding is identical, we have a polyphase winding with pairs of p poles distributed with NS number of slots.

In the PMVM, the tooth pitch is nearly equal to mechanical pitch, defined as the angle covered during an electric period. The number of rotor magnet pairs N_R is different from the number of stator teeth N_S. The condition to be met is:

$$|N_S - N_R| = p \tag{2}$$

Moreover, the electrical frequency f of the PMVM is uniquely linked to the number of pairs of magnets N_R, as seen in the following formula:

$$f = \frac{1}{T} = \frac{N_R \cdot \Omega_R}{2\pi} \tag{3}$$

with Ω_R is the rotor speed in rd/s.

The PMVM is a tooth coupling machine. However, the study of this type of machine is identical to that of a conventional synchronous machine. From the Maxwell tensor we can write for the expression of the torque:

$$T_{em} = K_V \cdot R_e^2 \cdot L \cdot \int_{2\pi} b_{1an} \cdot \lambda_1 \cdot d\theta \tag{4}$$

The dimensions R_e and L of expression (4) represent the air gap radius and iron length. The coefficient K_v, the speed ratio, is called the Vernier Ratio. This coefficient is difference between the stator field speed and the rotor speed. The stator field speed Ω_S is:

$$\Omega_S = \frac{\omega}{p} \tag{5}$$

The rotor speed Ω_R is:

$$\Omega_R = \frac{\omega}{N_R} \tag{6}$$

We obtain:

$$\frac{\Omega_S}{\Omega_R} = K_V = \frac{N_R}{P} \tag{7}$$

We use for our study the linear density of current, λ_1, equivalent to stator winding. The periodicity for λ_1 is equal to $2\pi/p$, which we express as:

$$\lambda_1 = A_1 \cdot \cos(p \cdot \theta) \tag{8}$$

The amplitude A_1 of the linear density depends on the current in the slot, the winding coefficient and the shape of the slot. The stator air gap permeance has equal to $2\pi/N_S$.

The air gap magnetomotive force created by the magnet has a periodicity equal to $2\pi/|N_S - N_R|$. The fundamental field component b_{1an} can be expressed as follows:

$$b_{1an} = k_1 \cdot M \cdot \cos((N_S - N_R) \cdot \theta) \tag{9}$$

with M as the Remanent flux density of magnet. The coefficient k_1, which defines the field amplitude $b_{1an}(\theta)$, is deduced using the finite element [10]. Its value, which depends on the ratios of the dimensional proportion parameters, is generally between 0.1 and 0.2.

In the PMVM, the increase in operating frequency allows a gain on the mass–power ratio at very low speed compared to a PMSM.

3.2. Example of Permanent Magnet Vernier Machine Prototype

The Figure 5 shows an example of a prototype designed to simulate a Vernier effect generator for an energy conversion system.

This prototype has the following characteristics:

- $p = 2$;
- $Ns = 24$ stator teeth;
- $Nr = 22$ pairs of magnets;
- Xs synchronous inductance: 61.2 mH;
- The electromotrice force (e.m.f.) amplitude varies linearly with the rotating speed (emf coefficient: 0.825 V/rad/s).

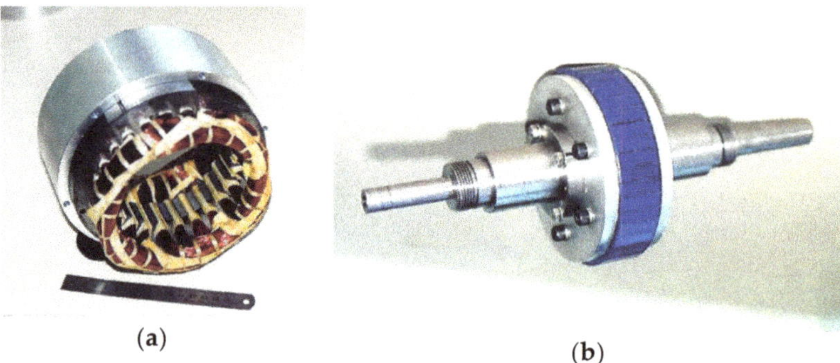

(a)   (b)

Figure 5. (**a**) Stator of the Vernier machine; (**b**) rotor of the Vernier machine.

A particularity of the Vernier machine is that its fem is sinusoidal. There is an e.m.f. of the prototype on Figure 6.

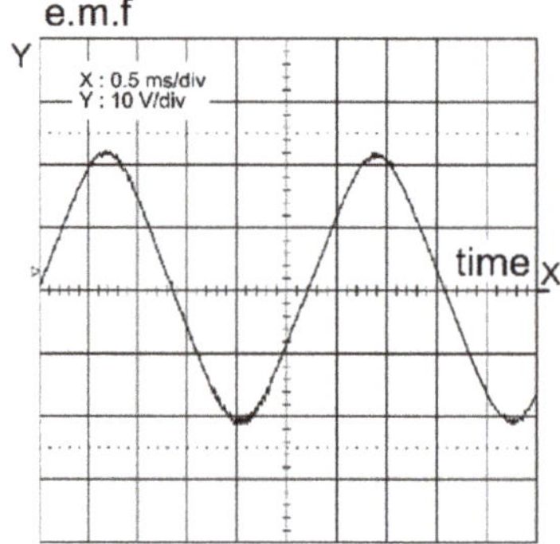

Figure 6. Electromotive force (e.m.f.) induction of Vernier machine at 1000 rpm and f = 367 Hz.

4. Sizing of Generator

4.1. Introduction

Precise calculation of the working of the generator can only be achieved with the finite elements method. However, here we use analytic relations with the aim of making many calculations and making some comparisons between the Vernier structure and synchronous machines with many poles. Most of these analytic relations can be found in [10]. In particular, for the calculation of the torque, we use the relation:

$$T_{em} = 4 \cdot \pi \cdot R_e^3 \cdot K_f \cdot F_S \tag{10}$$

R_e: bore radius (m) and K_f, form-factor, defined as the ratio: $L/(2R_e)$, L being the laminations' length.

For the synchronous machine, if the machine is driven so that electromagnetic force F_S is in phase with current:

$$F_S = \frac{\sqrt{2}}{2} \cdot M \cdot \lambda_0 \tag{11}$$

M: remanence of rotor magnets (T); λ_0: length density of current (A/m), connected to the armature magnetic field.

For the Vernier structure:

$$F_S = M \cdot \frac{a}{a+e} \cdot H \cdot K_S \tag{12}$$

a: thickness of magnets (m); *e*: air gap (m); *H*: magnetic field density of armature reaction (A/m); K_S: coupling coefficient, function of the waveform of feed currents and of the geometric proportions of the machine structure.

We can define an elementary domain, as in Figure 7. This last one is a repetitive cell of the structure. It contains a stator slot and one alternate magnet couple [11].

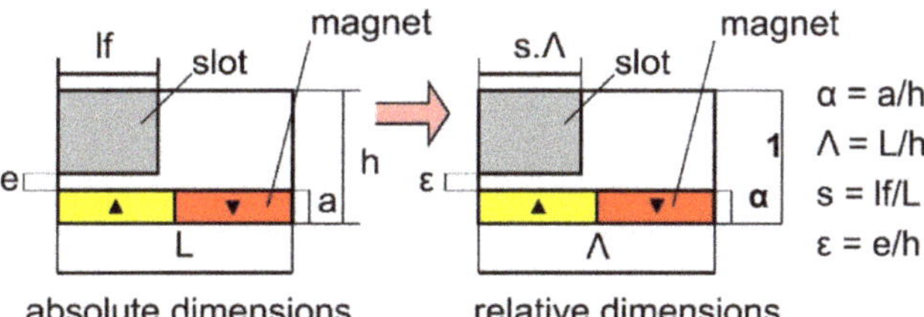

Figure 7. Magnet–slot interaction in the elementary domain.

This elementary domain is defined by five dimensional parameters which have L = domain length (from slot to slot), a = magnet thickness, lf = slot width, e = air gap thickness, and h = domain height. We have the normalized domain with α, ε, Λ and s. It has been demonstrated that the average tangential force for an elementary domain can be written as:

$$F_S = M \cdot \frac{\alpha}{\alpha + \varepsilon} \cdot H \cdot K_S \tag{13}$$

We have the characteristics on Figure 8.

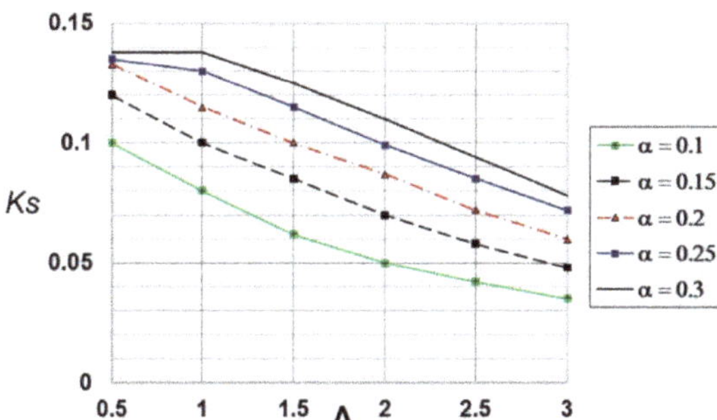

Figure 8. Coupling coefficient K_S.

4.2. Estimate of Losses and Temperature Rise Product

To design the generator driven by the vertical axis wind turbine, we performed analytical sizing for a rapid overview of the different feasible PMVM with the rated values T_{Tn}, N_{Tn} and P_{Tn}. The calculations are made by taking into account thrusts deduced from geometric, electromagnetic, and thermal constraints.

Copper losses are calculated with the relation (14)

$$P_{copper} = \rho \cdot \frac{m_{copper}}{\rho_{copper}} \cdot J^2 \tag{14}$$

ρ: copper resistivity (2×10^{-8} Ω·m); m_{copper}: copper mass in the machine (kg); ρ_{copper}: copper density (8.9×10^3 kg/m^3); J: surface density of current in stator winding (A/m^2).

Core losses are calculated on the basis of data given for laminations designed to work at 400 Hz, by using the relation given in [12], valid with sinusoidal waveforms:

$$P_{iron} = \left(4 \cdot k \cdot B_{max}^2 \cdot f + 2 \cdot \pi^2 \cdot \alpha \cdot B_{max}^2 \cdot f^2\right) \cdot \frac{m_{iron}}{\rho_{iron}} \tag{15}$$

m_{iron}: laminations mass; ρ_{iron}: alloy density (7.6×10^3 kg/m^3); f: rated frequency (Hz); B_{max}: maximum flux density in the laminations (1.5 T).

As the rated frequency is often next to 200 Hz, we will use 0.2 mm thick laminations. With this thickness: $\alpha = 6.7 \times 10^{-3}$ and $k = 58$.

4.3. Results of Analitical Sizing

In terms of the permanent magnet, Neodynium–Iron–Boron Magnets were chosen because of their costs, lower specific weight and a mechanical strength much higher than samarium cobalt.

The PMVM has straight teeth while the PMSM has slot isthmus, so to compare the two machines we estimated the equivalent air gap for a PMSM without slot isthmus at a value of 0.6 mm.

From the rated values of the generator, we calculate the possible solutions, by varying:

1. the form factor K_f;
2. the number of slots per pole and per phase;
3. the number of stator poles;
4. λ_0 (for the classical synchronous machine) or H (for the Vernier structure);
5. the thickness of magnets and by taking into account mechanic and electromagnetic limiting factors;
6. demagnetization constraint of magnets;
7. slot pitch > 5 mm;
8. maximum frequency: e.g., 400 Hz;
9. thickness of the yoke;
10. current density in armature winding $J < 5$ A/mm^2.

The evolution of the torque/weight ratio as a function of the rated torque for a rated power of 1.5 kW confirms that the performances of the Vernier machine stands out of those of the synchronous machine all the more because the rated speed of the turbine is low (Figure 9). The same dimensioning can be done for a horizontal axis wind turbine or a different power output.

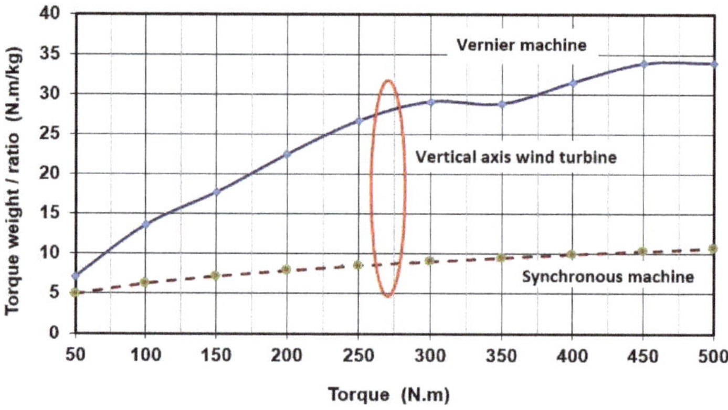

Figure 9. Torque/weight ratio versus rated torque for a 1.5 kW-rated power.

To analyze the results thus obtained, we plot the efficiency of the feasible solutions as a function of their torque-to-weight ratio (Figure 10) for the PMVM, these two parameters being of prime importance for a direct drive.

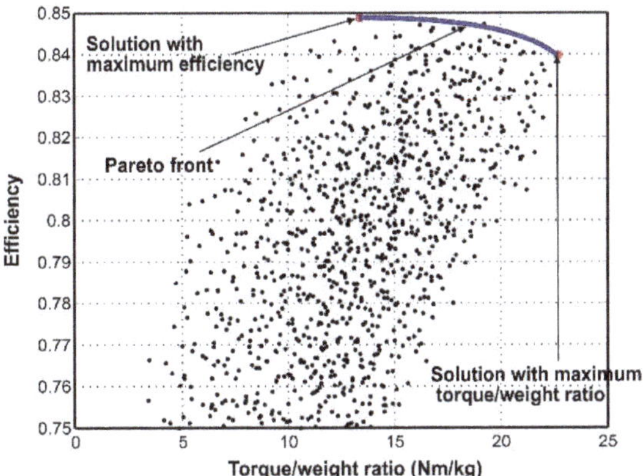

Figure 10. Feasible solutions with T_{Tn} = 270 Nm and N_{Tn} = 54 rpm (analytical sizing).

With the chosen two objectives, the Pareto front links the solution with the highest efficiency and the solution with the highest torque/weight ratio. As seen in Figure 7, there is a slight difference of efficiency between the solution presenting the highest torque/weight ratio and the solution with the highest efficiency. On the other hand, a small improvement in efficiency leads to a significant decrease in the torque/weight ratio. As a consequence, if the same weighting is given to both objectives, the solution with the highest torque/weight ratio is the more interesting one to implement in our wind energy system. The main characteristics of the resulting machine are listed in Table 1.

Table 1. Characteristics of the solutions with the highest torque-to-weight ratio (analytical sizing).

Designation	Vernier Machine	Synchronous Machine
Rated efficiency (%)	84	86
Torque/weight ratio (Nm·kg)	**22.7**	**11.8**
Outer diameter (mm)	441	528
Inner diameter (mm)	410	484
Total length (mm)	109	107
Air gap (mm)	0.5	0.6
Total mass (kg)	**11.9**	**22.8**
Number of phases	3	3
Number of stator poles	26	102
Power factor	**0.52**	**1**
Rated frequency (Hz)	**199**	**46**
Force density (N/cm^2)	**1.17**	**0.71**

For comparison, we performed sizing on a conventional synchronous machine with the same rated values, also maximizing the torque/weight ratio. The torque/weight ratio is twice as high with the SPMVM. However, we can note the low power factor of the Vernier machine, when it is close to one for the conventional machine.

As a reminder, the calculations are made for a torque T_{TN} = 270 N.m and a rotation speed N_{TN} = 5 rpm.

A characteristic of Vernier machines is that they have a lower power factor than conventional synchronous machines. During motor operation it allows us to dimension the power system. The over-sizing of the power system is largely compensated by the high torque of the actuator.

With the SPMVM behaving externally like a synchronous machine without saliency. The electrical model for this machine is the same as that of a classical synchronous magnet machine. We have E electromotive force (e.m.f.), Rs stator resistance, Xs synchronous reactance and V single voltage at the stator terminals.

The Fresnel diagrams obtained with a diode rectifier and with a Power Wave Modulation (PWM) rectifier are presented in Figure 11. The phasors shown in the diagrams are fundamental quantities.

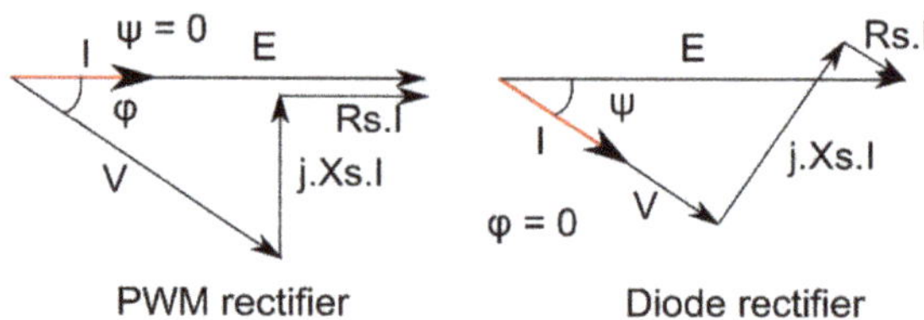

Figure 11. Fresnel diagram of the generator when associated with a rectifier.

These diagrams show that a low power factor will require oversizing of the PWM rectifier to achieve the same transmitted power. Concerning the diode rectifier, the torque/weight ratios presented above were obtained by assuming that the e.m.f. was in phase with the stator current, so the expected performance will not be reached at the rated current.

As the low power factor of the generator is a drawback with the two usual types of rectifiers, we want to increase its value. With the efficiency and torque/weight ratio, we then have three optimization objectives. To potentially identify others, we will now consider the energy conversion system as a whole.

5. Study of the Vernier Generator and Rectifier Association

5.1. Introduction

To investigate the association of the Vernier machine with the two types of rectifiers, we model the complete energy conversion system with Simulink. The turbine is simulated with (1). The PMVM is modeled using the classical relations of the synchronous machine, with some adaptations to take its particularities into account. The converters are represented by average models. With experiments on a test bed [13], we have shown that the overall model offers a good estimate of the energy produced by the turbine, but that it is less precise for converter and machine losses.

As a model of wind, we used a profile with a mean speed of 8 m/s and including sharp variations, so as to study the dynamic behavior of the conversion system when submitted to uneven wind gusts that commonly arise in urban areas (Figure 12).

We used the parameters given in Table 1 for the SPMVM. With an additional condition imposing a 120-V phase voltage under rated speed, we calculated the parameters of the machine's electrical model:

- Back-emf coefficient: Ke = 11.3 V/rad/s
- Stator inductance: Ls = 10.4 mH
- Stator resistance: Rs = 1.12 Ω.

The rectifier associated with the generator can feed a DC link with a fixed or variable voltage (see Figure 12). The load can be an accumulator, possibly fed by a chopper, or an inverter connected to a local grid [14–17].

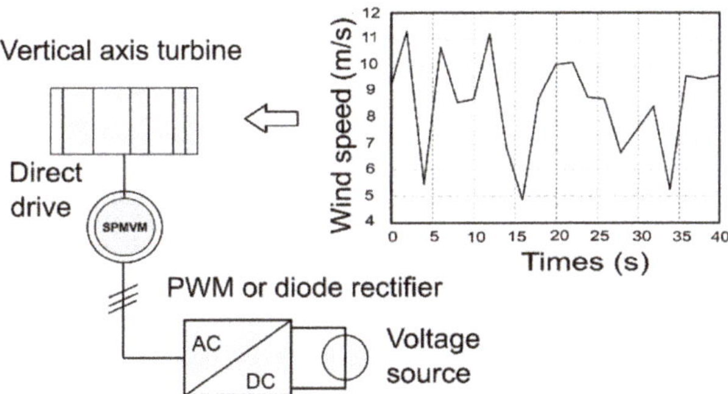

Figure 12. Wind profile used for simulation and the wind energy conversion system.

Considering the high synchronous reactance of the PMVM, it is better to use a voltage source at the output of the diode bridge rectifier. We thus obtain phase currents that are naturally sinusoidal in the machine and the total harmonic distortion of the stator voltage is lower than with a smoothing inductor.

5.2. Vernier Machine PWM Rectifier Association

With a PWM rectifier, we can drive the generator to maintain the operating point of the turbine close to the maximum power point locus (Figure 3) [18–22]. The torque vs. speed characteristic of the turbine being known, we can use the driving method shown in Figure 13.

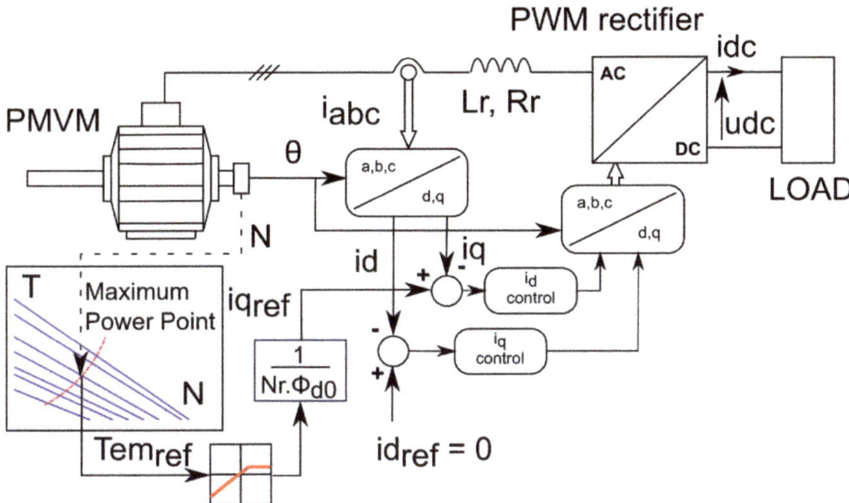

Figure 13. Driving method with a PWM rectifier.

From a speed measurement, the torque set point necessary to keep the working point of the turbine on the maximum power point locus is deduced. This value is used to calculate the stator current set point in a rotating frame, and the resulting control action is transformed back to the stationary frame for execution. The working point trajectory of the turbine resulting from this control method is shown in Figure 14.

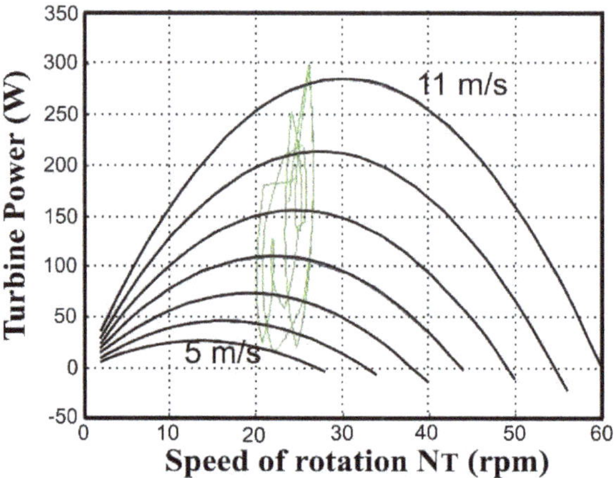

Figure 14. Working point trajectory of the turbine.

5.3. Vernier Machine Diode Rectifier Association

The diode bridge rectifier is an easy-to-use, low-cost, and sturdy converter, which is what makes it a very interesting choice for a domestic installation. On the other hand, as it is not a driven converter, it obviously does not make it possible to impose a working point on the turbine. Consequently, the energy conversion system must be designed as a whole to obtain a working point trajectory naturally, which will come close to the one shown in Figure 14.

5.3.1. Constant DC-Link Voltage

In the case of constant DC-link voltage, the voltage value must be chosen to reach a compromise. For a high voltage value, the diode bridge will conduct only for high-speed, low-power working points of the turbine. With a low voltage value, the working point of the turbine will settle at low speeds, which also correspond to low power.

Therefore, there is an optimal voltage value with which the working point will settle close to the maximum power point locus (Figure 15), thus leading to maximum energy recuperation (Figure 16).

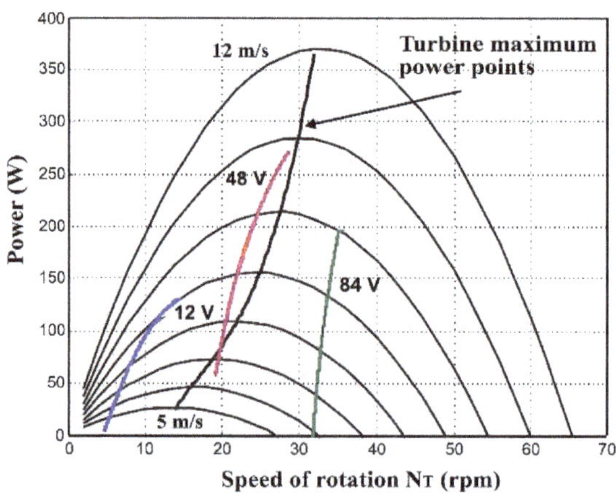

Figure 15. Working points for generator power with several constant DC-link voltages.

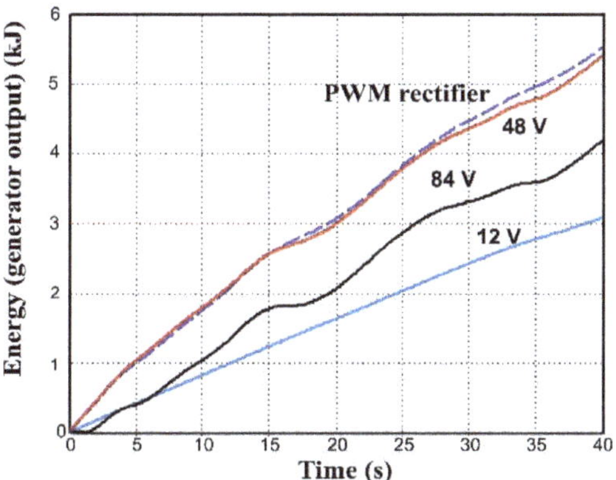

Figure 16. Working points for generator energy with several constant DC-link voltages.

If we pay attention to the energy produced by this conversion system, depending on whether a diode bridge or a PWM rectifier is used, the difference is much more distinct for high wind speeds: with the diode rectifier, when the wind speed rises, the derivative of the generator torque with respect to the speed of rotation $\delta Tg/\delta\Omega$ tends towards negative values. In that case, the working point becomes unstable and the turbine speed increases rapidly. The working point moves away from maximum power points and, if we refer to what is produced with a PWM rectifier, the turbine is clearly under-exploited. As a consequence, we cannot take advantage of the turbine's ability to work with strong winds.

To illustrate what happens under high wind speed conditions, we use a wind profile with the same shape as that in Figure 8, but with a mean value of 19 m/s, that is, the wind speed with which we determined the rated values. With a DC-link voltage of 48 V, which is the optimal value for an average wind speed of 8 m/s, the working point trajectory of the resulting turbine is shown in Figure 17, and the energy collected on the DC-link is divided by more than four compared to the energy produced with a PWM rectifier (Figure 18).

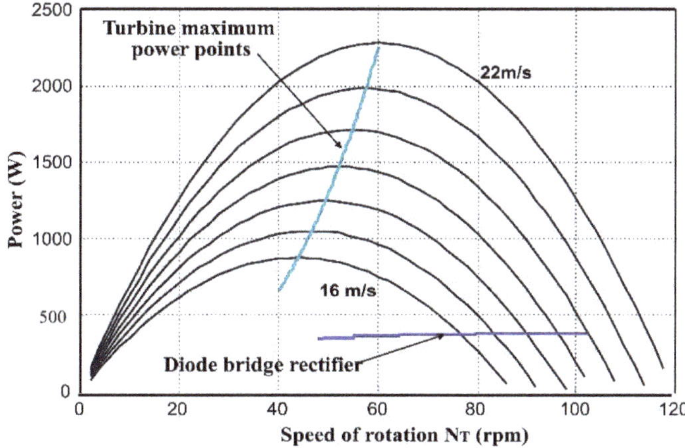

Figure 17. Working points for generator power with a diode bridge rectifier for an average wind speed of 19 m/s (DC-link voltage = 48 V).

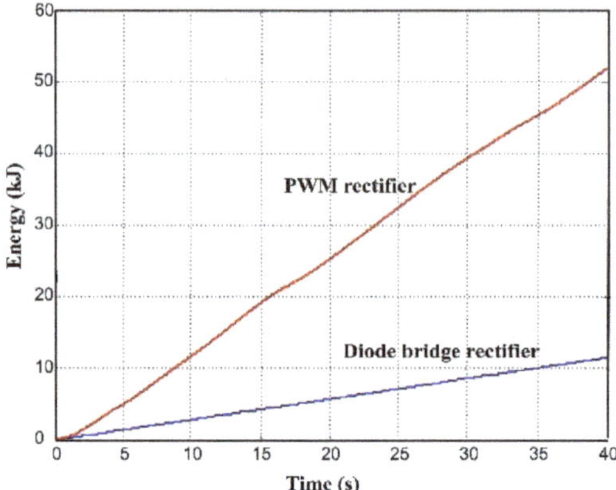

Figure 18. Working points for generator energy with a diode bridge rectifier for an average wind speed of 19 m/s (DC-link voltage = 48 V).

5.3.2. Variable DC-Link Voltage

As a starting point, we use the power expression on the generator output. From there, we can obtain the expression of the DC bus voltage uDCM for which the generator associated with a diode bridge supplies maximum power to the DC bus. Using the notation from Figure 11 and with a unit power factor, we obtain:

$$P = 3 \cdot V \cdot I = 3 \cdot \left(\frac{\sqrt{2}}{\pi} \cdot U_{DC} \right) \cdot I \tag{16}$$

In addition, from Figure 8, diode rectifier, we can deduce the stator current (I):

$$I = \frac{-V \cdot R_S}{R_S^2 + X_S^2} + \frac{\sqrt{\left(\frac{K_e \cdot \omega}{N_R} \right)^2 \cdot (R_S^2 + X_S^2) - (X_S \cdot V)^2}}{R_S^2 + X_S^2} \tag{17}$$

K_e being the e.m.f. coefficient of the PMVM.

Expression (16) thus becomes:

$$P = 3 \cdot \frac{V}{R_S^2 + X_S^2} \cdot \left(-V \cdot R_S + \sqrt{\left(\frac{K_e \cdot \omega}{N_R} \right)^2 \cdot (R_S^2 + X_S^2) - (X_S \cdot V)^2} \right) \tag{18}$$

Differentiating this expression with respect to the DC bus voltage UDC, we deduce that power P is maximum for U_{DCM}:

$$U_{DCM}^2 = \frac{\left(\frac{K_e \cdot \omega}{N_R} \right)^2 \cdot \sqrt{R_S^2 + X_S^2} \cdot \left(\sqrt{R_S^2 + X_S^2} - R_S \right)}{\left(\frac{2}{\pi} \right)^2 \cdot X_S^2} \tag{19}$$

If we note stator angular frequency ω, as $X_S^2 = (L_S \cdot \omega)^2 >> R_S^2$, this expression becomes:

$$U_{DCM}^2 = \frac{\left(\frac{K_e \cdot \omega}{N_R} \right)^2 \cdot (L_S \cdot \omega - R_S)}{\left(\frac{2}{\pi} \right)^2 \cdot L_S \cdot \omega} = \left(\frac{K_e \cdot \omega \cdot \pi}{2 \cdot N_R} \right)^2 \cdot \left(1 - \frac{R_S}{L_S \cdot \omega} \right) \tag{20}$$

If the wind speed is 8 m/s, maximum turbine power is reached when N = 20 rpm. At this operating point, the stator frequency is 68 Hz and it can be found with (20) that the optimal DC bus voltage is about 48 V. If the wind speed reaches 19 m/s, the turbine's maximum power is obtained at nominal rotating speed, giving a stator frequency of 199 Hz. An amount of 48 V as the DC bus value is no longer optimal: the new optimal voltage UDCM is obtained from (20) and is about 96 V.

If we may vary the DC bus voltage, we can improve the operation of the energy conversion system with a diode bridge, by adapting this voltage to the rotating speed. However, this solution reduces simplicity, sturdiness, and does not bring significant improvement in power production. If we apply (18) with a voltage U_{DC} = 96 V, the power is about 470 W, which is much lower than the 1470 W that the turbine can potentially produce when the wind reaches 19 m/s.

5.3.3. Machine Optimization for a Diode Bridge Rectifier

Using (18) and (20), we can plot the evolution of the maximum power versus speed produced by the association of the generator and the diode bridge rectifier. Figure 19 is obtained by using this plot and the characteristic giving the maximum power produced by the turbine versus speed.

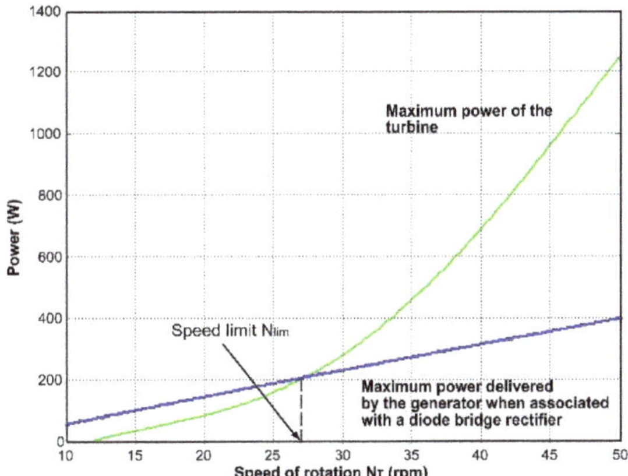

Figure 19. Evolution of the maximum power of the generator and the turbine.

There is a speed limit N_{lim}, beyond which the generator cannot transmit all the power produced by the turbine to the DC bus.

The generator's rated values are specified for a wind speed of 19 m/s. Thus, among the solutions plotted in Figure 10, we look for a generator giving a speed N_{lim} as near as possible to 54 rpm (speed of the turbine maximum power point for this wind speed in Figure 3).

We can deduce N_{lim} or here Ω_{lim} expression from the relation giving the maximum power of the turbine as a function of the rotating speed:

$$P_M = K_T \times \Omega^3 = \left(\frac{1}{2} \cdot C_{p_{opt}} \cdot S_T \cdot \rho \cdot \frac{R^3}{\lambda_{opt}^3} \right) \times \Omega^3 = 9.4 \times \Omega^3 \tag{21}$$

C_{Popt}: maximum power coefficient of the turbine: C_{Popt} = 7.62/100; S_T: turbine area "seen" by the wind: S_T = 4.5 m²; ρ: air density: ρ = 1.25 kg/m³; R: turbine rotor radius: R = 0.9 m; λ represents the ratio: $(R \cdot \Omega / S_w)$, Sw being the wind speed; λ_{opt} is the value of λ for which Cp = C_{Popt} either λ_{opt} = 0.26.

From (18) and (20), we can then deduce the expression of the maximum power at the generator output as a function of speed, and equate it to (22). The only coherent solution is:

$$\Omega_{\text{lim}}\left(rd.s^{-1}\right) = \frac{K_e \cdot \sqrt{\frac{3}{2}}}{\sqrt{K_T \cdot L_S \cdot N_R}} - \frac{R_S}{2 \cdot L_S \cdot N_R} = \Omega_1(1) - \Omega_1(2) \tag{22}$$

From this expression, we can make the following remarks:

- The second term of (22) is smaller than the first: using the simulation parameters listed in introduction of Section 5, we find that the second term of the expression Ω_1 (2) represents about 8% of the first one.
- K_e is proportional to the number of winding turns, and L_S are proportional to the number of turns squared. Thus, Ω_{lim} is independent of this number of turns: changing the winding cannot improve system operation. Thus, if we want to modify Ω_{lim}, we are interested at first by the first term Ω_1 (1) of (22).

To obtain an expression of K_e and L_S, we use the following relations:

$$K_e = \frac{U_r \cdot P_F}{\Omega_r} \tag{23}$$

with U_r: stator voltage at rated speed (U_r = 120 V), P_F: power factor, Ω_r: rated speed (rad·s^{-1}).

$$L_S = \frac{\sin\left(\cos^{-1}(P_F)\right) \cdot U_r}{2 \cdot \pi \cdot I_r \cdot f_r} \tag{24}$$

f_r rated frequency:

$$f_r = \frac{N_R.\Omega_r}{2 \cdot \pi} \tag{25}$$

I_r rated stator current:

$$I_r = \frac{P_r}{3 \cdot U \cdot P_F} \tag{26}$$

P_r: rated power.
Using relations (23) to (26), we find that the first term in (22) can be written:

$$\Omega_1(1) = \sqrt{\frac{P_r \cdot P_F}{2 \cdot K_T \cdot \Omega_r \cdot \sin(\cos^{-1}(P_F))}} \tag{27}$$

The application sets the rated values for power and rotating speed. Thus only the power factor remains as a variable.

The second part of Equation (22) can be written as:

$$\Omega_1(2) = \frac{P_r \cdot R_S \cdot \Omega_r}{6 \cdot U_r^2 \cdot P_F \cdot \Omega_r \cdot \sin(\cos^{-1}(P_F))} \tag{28}$$

$\Omega_1(2)$ being small in comparison with $\Omega_1(1)$, we can approximate that stator resistance does not change from one machine to another. The stator voltage at rated speed U_r will remain unchanged too. We can then plot Ω_{lim} as a function of the power factor (Figure 19).

Ω_{lim} increases far more quickly when the power factor is greater than 0.9. If we want to obtain Ω_{lim} = 54 rpm, we need a power factor equal to 0.928. However, the curve on Figure 20 is plotted without imposing any limit on the machine stator current.

Using (18), we calculate that a machine with a power factor of 0.928, a rated power of 1.5 kW, and a rated voltage of 120 V, would have a rated current of 4.5 A, but its stator current would reach 7.5 A at the working point. So as to not oversize the machine, among the solutions calculated in Table 1, we seek the solution that represents the best compromise between efficiency and power factor while maintaining a high torque to weight ratio (Figure 21).

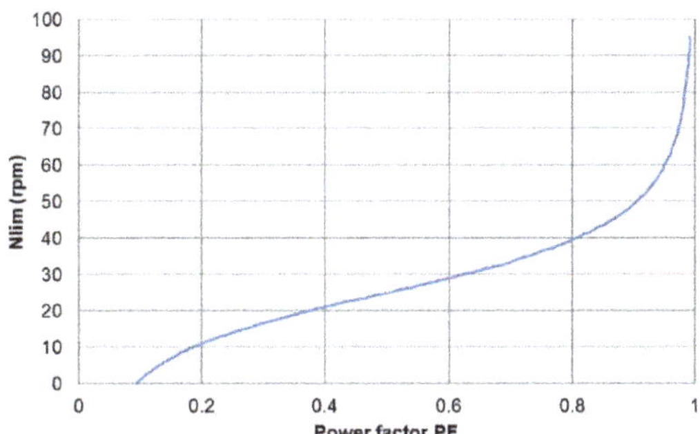

Figure 20. Speed limit (Nlim) versus power factor (PF).

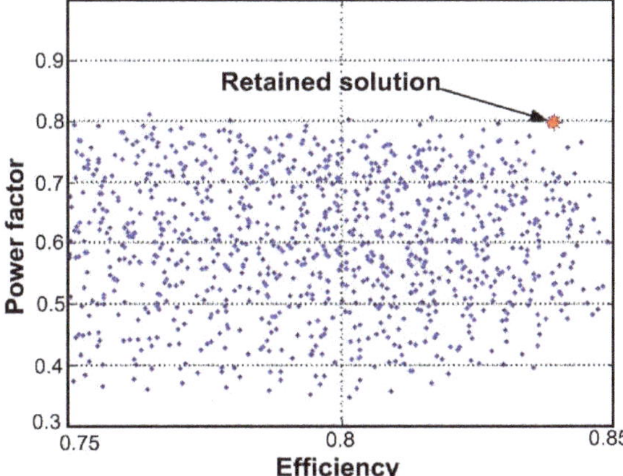

Figure 21. Power factor of feasible solutions versus efficiency.

In Table 2, the chosen solution is compared to the PMVM presented in Table 1.

Table 2. Characteristics of the solutions with the highest torque-to-weight ratio (analytical sizing).

Designation	PMVM Optimized for Torque/Weight Ratio	PMVM Optimized for Association with a Diode Bridge Rectifier
Rated efficiency (%)	84	83.9
Torque/weight ratio (Nm·kg)	**22.7**	**18.6**
Outer diameter (mm)	441	453
Inner diameter (mm)	410	420
Total length(mm)	109	121
Air gap (mm)	0.5	0.5
Total mass (kg)	**11.9**	**14.5**
Number of stator poles	26	20
Power factor	**0.52**	**0.8**
Rated frequency (Hz)	**199**	**99**
Force density (N/cm^2)	**1.17**	**1.06**

5.3.4. Improvement Brought by the Generator with a High Power Factor

To simplify the energy conversion system, we work with a constant DC voltage at the rectifier output and, by using simulations, we try to evaluate the improvements that can be expected in energy production with a machine optimized for association with a diode bridge rectifier. In these simulations, we use, for every generator, the DC voltage calculated using (20) for speed Ω_{lim}: this voltage seems to be the most appropriate for maximum energy production over the entire wind speed variation range. For the generator optimized for the torque-to-weight ratio: $U_{DC} = 48$ V, for the generator optimized for association with the diode rectifier: $U_{DC} = 96$ V.

With an average wind of 8 m/s, Figure 16 shows that the generator optimized by considering the torque/weight ratio can produce the power with a PWM rectifier. In comparison with this generator, we obtain 12% less power with the generator optimized for association with the diode bridge (Figure 22).

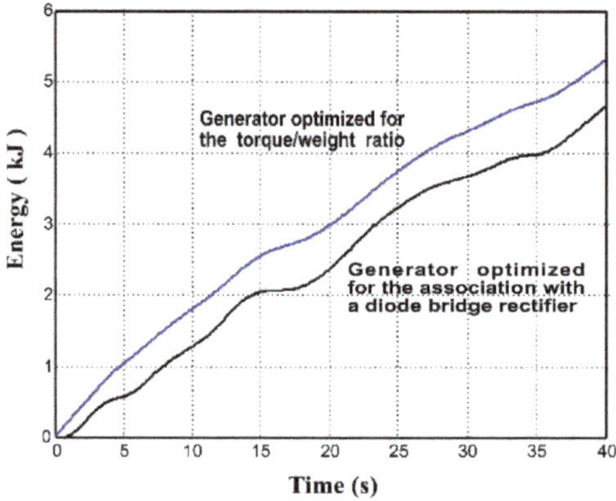

Figure 22. Energy at generator output for 8-m/s average wind speed.

With a wind profile centered on 19 m/s, the generator optimized for the diode bridge makes it possible to obtain working points nearer the maximal power points than with the machine with a high torque-to-weight ratio (Figure 23).

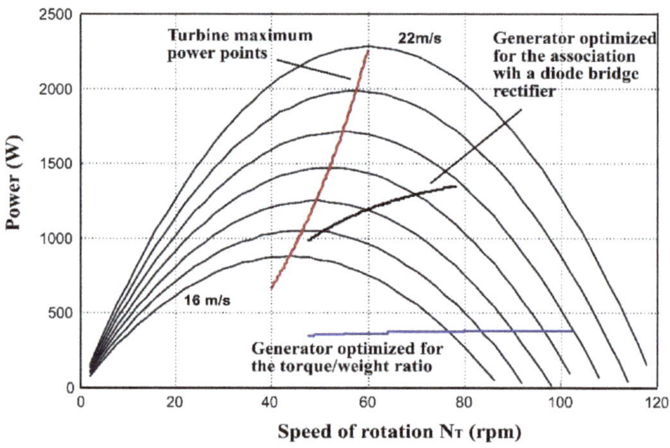

Figure 23. Power results obtained for 19-m/s average wind speed.

The energy at the generator output, despite being 30% lower than with a PWM rectifier, represents three times the energy we can obtain with a high torque/weight generator associated with a diode bridge (Figure 24).

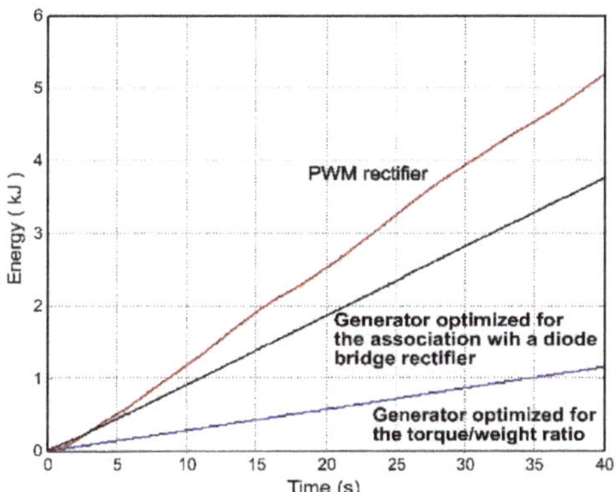

Figure 24. Results obtained for energy for 19-m/s average wind speed.

With a high power factor generator, we can then maintain high-performance over a wide wind speed range.

6. Conclusions

By using simulations, we showed that the association of a PMVM with a diode rectifier in a wind energy conversion system can result in good performance, requiring a specific design of the generator for the association.

In the application studied in this article, the system prioritizing generator weight and compactness seems to offer high performance for low wind speeds, but is far less efficient in case of strong winds. Therefore, to best exploit the wind turbine without oversizing the generator, a concession must be made on the torque-to-weight ratio in favor of the power factor. This concession, however, is justified by the simplicity and the sturdiness of the energy conversion system thereby obtained.

We are also studying other converter architectures associated with the Vernier generator effect, its converters already being studied for wind energy conversion.

The combination of an optimized Vernier generator with a diode rectifier can be an economical solution for small vertical or horizontal axis wind turbines. For this machine there are no mechanical feasibility constraints even for small yoke, tooth or magnet dimensions.

Author Contributions: Conceptualization, designed, analysis, writing—original draft preparation, P.E. and I.M.; supervision this project, P.E. and D.M.; writing—review and editing, P.E. All authors have read and agreed to the published version of the manuscript.

Funding: This research received no external funding.

Institutional Review Board Statement: Not applicable.

Informed Consent Statement: Not applicable.

Data Availability Statement: The study did not report any data.

Conflicts of Interest: The authors declare no conflict of interest.

References

1. Chen, Z.; Guerrero, J.M.; Blaabjerg, F. A review of the state of the art of power electronics for wind turbines. *IEEE Trans. Power Electron.* **2009**, *24*, 1859–1875. [CrossRef]
2. Milivojevic, N.; Stamenkovic, I.; Schofield, N. Power and Energy Analysis of Commercial Small Wind Turbine Systems. In Proceedings of the IEEE International Conference Industrial Technology (ICIT), Vina del Mar, Chile, 14–17 March 2010; pp. 1739–1744. [CrossRef]
3. Corbus, D.; Baring-Gould, I.; Drouilhet, S.; Jimenez, T.; Newcomb, C.; Flowers, L. Small wind turbine testing and develop-ment applications. In Proceedings of the WindPower'99, Burlington, VT, USA, 20–23 June 1999; AC36-99GO10337.

4. Ani, S.O.; Polinder, H.; Ferreira, J.A. Comparison of Energy Yield of Small Wind Turbines in Low Wind Speed Areas. *IEEE Trans. Sustain. Energy* **2012**, *4*, 42–49. [CrossRef]
5. Modi, V.J.; Fernando, U.K. On the performance of the Savonius wind turbine. *J. Sol. Energy Eng.* **1998**, *111*, 71–81. [CrossRef]
6. Llibre, J.F.; Matt, D. A Cylindrical Vernier Reluctance Permanent-Magnet Machine. *Electromotion J.* **1998**, *5*, 35–39.
7. Lipo, T.; Toba, A. Generic torque-maximizing design methodology of surface permanent-magnet vernier machine. *IEEE Trans. Ind. Appl.* **2000**, *36*, 1539–1546. [CrossRef]
8. Mény, I.; Enrici, P.; Huselstein, J.J.; Matt, D. Direct driven synchronous generator for wind turbines (Vernier reluctance magnet machine). In Proceedings of the ICEM'2004, Cracovie, Pologne, 5–8 September 2004.
9. Mény, I.; Enrici, P.; Didat, J.R.; Matt, D. Analytical dimensioning of a direct-driven wind generator. Use of a variable reluc-tance magnet machine with Vernier effect. In Proceedings of the Electromotion 05, Lausanne, Switzerland, 27–29 July 2005.
10. Llibre, J.F.; Matt, D. Performances comparées des machines à aimants et à réluctance variable. Maximisation du couple massique ou volumique. *J. Phys. III* **1995**, *5*, 1621–1641.
11. Enrici, P.; Dumas, F.; Ziegler, N.; Matt, D. Design of a High-Performance Multi-Air Gap Linear Actuator for Aeronautical Applications. *IEEE Trans. Energy Convers.* **2016**, *31*, 896–905. [CrossRef]
12. Hoang, E. Etude, modélisation et mesure des pertes magnétiques dans les moteurs à réluctance variable à double saillance. Ph.D. Thesis, Laboratoire d'Electricité, Signaux et Robotique (LESiR), Ecole normale Supérieure de Cachan, Cachan, France, 19 December 1995.
13. Meny, I. Chaîne de Conversion Éolienne de Petite Puissance: Optimisation D'une Génératrice Spécifique à Entraînement Di-rect, Développement de la Chaîne de Conversion. Ph.D. Thesis, Department of Electrical Engineering, Université Montpellier, Montpellier, France, 6 July 2005.
14. Matt, D.; Martirè, T.; Enrici, P.; Jac, J.; Ziegler, N. Passive Wind Energy Conversion System Association of a direct-driven synchronous motor with Vernier effect and a diode rectifier. In Proceedings of the Electrotechnical conference MELECON 2008, Ajaccio, France, 5–7 May 2008. [CrossRef]
15. Wang, J.; Xu, D.; Wu, B.; Luo, Z. A Low-Cost Rectifier Topology for Variable-Speed High-Power PMSG Wind Turbines. *IEEE Trans. Power Electron.* **2011**, *26*, 2192–2200. [CrossRef]
16. Kana, C.; Thamodharan, M.; Wolf, A. System management of a wind-energy converter. *IEEE Trans. Power Electron.* **2001**, *16*, 375–381. [CrossRef]
17. Di Gerlando, A.; Foglia, G.; Iacchetti, M.F.; Perini, R. Analysis and Test of Diode Rectifier Solutions in Grid-Connected Wind Energy Conversion Systems Employing Modular Permanent-Magnet Synchronous Generators. *IEEE Trans. Ind. Electron.* **2011**, *59*, 2135–2146. [CrossRef]
18. Baroudi, J.A.; Dinavahi, V.; Knight, A.M. A review of power converter topologies for wind generators. *Renew. Energy* **2007**, *32*, 2369–2385. [CrossRef]
19. Mirecki, A.; Roboam, X.; Richardeau, F. Architecture Complexity and Energy Efficiency of Small Wind Turbines. *IEEE Trans. Ind. Electron.* **2007**, *54*, 660–670. [CrossRef]
20. Wang, Q.; Chang, L. An Intelligent Maximum Power Extraction Algorithm for Inverter-Based Variable Speed Wind Turbine Systems. *IEEE Trans. Power Electron.* **2004**, *19*, 1242–1249. [CrossRef]
21. Lumbreras, C.; Guerrero, J.M.; Garcia, P.; Briz, F.; Reigosa, D.D. Control of a Small Wind Turbine in the High Wind Speed Region. *IEEE Trans. Power Electron.* **2016**, *31*, 6980–6991. [CrossRef]
22. Sunan, E.; Kucuk, F.; Goto, H.; Guo, H.-J.; Ichinokura, O. Three-Phase Full-Bridge Converter Controlled Permanent Magnet Reluctance Generator for Small-Scale Wind Energy Conversion Systems. *IEEE Trans. Energy Convers.* **2014**, *29*, 585–593. [CrossRef]

Article

Fractional-Order Control of Grid-Connected Photovoltaic System Based on Synergetic and Sliding Mode Controllers

Marcel Nicola [1,*] and Claudiu-Ionel Nicola [1,2,*]

[1] Research and Development Department, National Institute for Research, Development and Testing in Electrical Engineering—ICMET Craiova, 200746 Craiova, Romania
[2] Department of Automatic Control and Electronics, University of Craiova, 200585 Craiova, Romania
* Correspondence: marcel_nicola@yahoo.com or marcel_nicola@icmet.ro (M.N.);
claudiu@automation.ucv.ro (C.-I.N.)

Abstract: Starting with the problem of connecting the photovoltaic (PV) system to the main grid, this article presents the control of a grid-connected PV system using fractional-order (FO) sliding mode control (SMC) and FO-synergetic controllers. The article presents the mathematical model of a PV system connected to the main grid together with the chain of intermediate elements and their control systems. To obtain a control system with superior performance, the robustness and superior performance of an SMC-type controller for the control of the u_{dc} voltage in the DC intermediate circuit are combined with the advantages provided by the flexibility of using synergetic control for the control of currents i_d and i_q. In addition, these control techniques are suitable for the control of nonlinear systems, and it is not necessary to linearize the controlled system around a static operating point; thus, the control system achieved is robust to parametric variations and provides the required static and dynamic performance. Further, by approaching the synthesis of these controllers using the fractional calculus for integration operators and differentiation operators, this article proposes a control system based on an FO-SMC controller combined with FO-synergetic controllers. The validation of the synthesis of the proposed control system is achieved through numerical simulations performed in Matlab/Simulink and by comparing it with a benchmark for the control of a grid-connected PV system implemented in Matlab/Simulink. Superior results of the proposed control system are obtained compared to other types of control algorithms.

Keywords: photovoltaic system; grid; sliding mode control; synergetic control; fractional-order control

Citation: Nicola, M.; Nicola, C.-I. Fractional-Order Control of Grid-Connected Photovoltaic System Based on Synergetic and Sliding Mode Controllers. *Energies* **2021**, *14*, 510. https://doi.org/10.3390/en14020510

Received: 4 December 2020
Accepted: 14 January 2021
Published: 19 January 2021

Publisher's Note: MDPI stays neutral with regard to jurisdictional claims in published maps and institutional affiliations.

Copyright: © 2021 by the authors. Licensee MDPI, Basel, Switzerland. This article is an open access article distributed under the terms and conditions of the Creative Commons Attribution (CC BY) license (https://creativecommons.org/licenses/by/4.0/).

1. Introduction

It is a well-known fact that it is important to use, to an increasing extent, renewable energy, characterized by the fact that it is generated from easily renewable sources which can thus be considered unlimited energy. Among these types of renewable energy sources, we refer to solar energy, wind energy, water energy, geothermal energy, etc. There is also a strong upward trend in the study and use of the PV system technology. Obviously, the study of its control systems was also developed in parallel with it [1].

Moreover, among the general approaches to the study of microgrid systems control, we can mention the study of the transient stability of a hybrid microgrid [2,3], the use of algorithms to offset lagging in the microgrid system [4], the optimization of the charging process of microgrid batteries [5,6], the study of the topologies of the converters used in the microgrid [7], the parallel coupling of the inverters in a microgrid [8], problems related to fault tolerance in the microgrid [9], and also problems related to the multi-grid dispatching in view of obtaining an economic optimum [10–14].

An inherent problem that arises is the control of the process of the PV system connection to a main grid. The components which ensure the connection of the PV system to the main grid are the DC-DC boost converter, the DC intermediate circuit, the DC-AC converter, the filtering block, and the transformer for the connection to the main grid, together with their control systems. Further, a task of maximum importance in terms of the control of the grid-connected PV system is to maintain the DC voltage in the intermediate circuit—DC link voltage—u_{dc} as precisely as possible. The control of this voltage is performed by a cascade control system in which the outer control loop controls the level of u_{dc} voltage, and the inner control loops control currents i_d and i_q in the dq rotating reference frame. Usually, the synchronization with the main grid is performed through a phase-locked loop (PLL), and the controllers of the control loops are of the classic proportional integrator (PI) type [15].

In order to obtain a more precise control of the voltage u_{dc}, we can use more complex control algorithms of the adaptive [16], robust [17,18], and predictive [19] types, but also intelligent control algorithms, such as fuzzy [20], neuro-fuzzy [21], genetics [22], particle swarm optimization (PSO) [23], and reinforcement learning [24].

A type of special controller used for the control of linear and nonlinear systems is based on PI and passivity theory [25]. This type of control, known at the beginning as the hyperstability control theory, is based on the appropriate description of the system in the closed loop in the form of the Lagrangian or the port-controlled Hamiltonian, which defines the behavior of the system through energy functions. This description is generally expressed in the form of solutions to partial differential equations, which require an increased degree of difficulty in the implementation of these types of controllers.

An alternative for the synthesis of some controllers for nonlinear systems, which ensures the parametric robustness and maintains a low order of the synthesized control law, is the use of SMC [26]. Among the disadvantages of the SMC-type control, we mention the occurrence of the chattering phenomenon, which represents the occurrence of oscillations in the control input due to the process of synthesis of the controller. For this purpose, to reduce oscillations, transition functions smoother than the sgn function are used (between conventional thresholds +1 and -1) followed by a corresponding filtration. Further, a type of controller suitable for the control of nonlinear systems is the synergetic controller [27]. It is important to recall that, by using such a controller, it is not necessary to linearize the nonlinear model around a static operating point because this type of controller provides good performance for the entire operating range of the nonlinear model.

By adding the FO calculus and the fractional differentiation and integration operators [28,29], control laws can be obtained with a higher degree of refinement due to the additional control parameter which represents the order of the fractional differentiation and integration operator. Thus, a control structure in which classical PI controllers are replaced with FO-synergetic controllers is presented in [30]. Superior results are obtained by using the control structure proposed in this article, as well as the proposed macro-variables for the synthesis of control laws.

The main contributions of this article consist in replacing the classic PI-type controllers in the control loops of u_{dc} voltage and i_d and i_q currents with, respectively, the FO-SMC and FO-synergetic type controllers. Thus, the synthesis of the control laws related to these types of controllers is presented, as well as the results obtained by numerical simulations in Matlab/Simulink for the control of the grid-connected PV system, in which classic PI, synergetic, or FO-synergetic controllers are used for the inner control loops of currents i_d and i_q, and classic PI, SMC, or FO-SMC controllers are used for the outer control loop of u_{dc}. Owing to the levels of freedom and the refinement brought by the fractional calculus for the SMC and synergetic algorithms, the performances of the control system will be superior to those presented in the benchmark implementation in [15] but will also compare to the results obtained in other papers using the basic approach presented in [15].

The main contributions of this paper can be summarized as follows:

- We propose the cascade structure of the control system of the grid-connected PV system based on the robustness of an SMC-type controller for the control of u_{dc} voltage in the DC intermediate circuit, combined with the flexibility of using synergetic control for the control of currents i_d and i_q.

- The synthesis of the control laws by SMC- and synergetic-type controllers using the fractional calculus for integration operators and differentiation operators.
- We realized the numerical simulations in Matlab/Simulink, and by comparing them with a benchmark for the control of a grid-connected PV system implemented in Matlab/Simulink, superior results of the proposed control system compared to other types of control algorithms are presented.

The other sections of the paper are structured as follows: Section 2 presents a mathematical model of the grid-connected PV system. Section 3 presents the control of the grid-connected PV system using the FO calculus for the SMC controller and synergetic controllers. Section 4 presents the numerical simulations in Matlab/Simulink for the control of the grid-connected PV system using FO-SMC and FO-synergetic controllers and the analysis thereof, and some conclusions are presented in Section 5.

2. Mathematical Model of the Grid-Connected PV System

Following [15,27,31], which show the mathematical model of a grid-connected PV system, Figure 1 shows the general diagram of such a system. The PV array system is modeled in [15,27,31] and the input parameters are radiation and temperature. The list of component blocks additionally includes a DC boost converter along with the maximum power point tracking (MPPT) module and a three-phase DC-AC converter. The notations in Figure 1 are the usual ones.

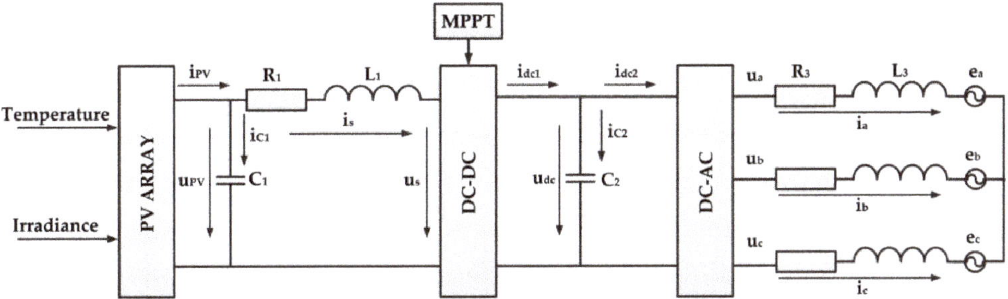

Figure 1. Block diagram of the main circuit diagram of the grid-connected PV system.

Thus, the following equations can be written to describe the operation of the grid-connected PV system:

$$C_1\frac{du_{PV}}{dt} = i_{PV} - i_s \tag{1}$$

$$u_{PV} = R_1 i_s + L_1\frac{di_s}{dt} + u_s \tag{2}$$

$$C_2\frac{du_{dc}}{dt} = i_{dc1} - i_{dc2} \tag{3}$$

$$u_{abc} - e_{abc} = R_3 i_{abc} + L_3\frac{di_{abc}}{dt} \tag{4}$$

where u_{abc} represents the output voltage of the DC-AC (voltage source converter—VSC) converter with $u_{abc} = \begin{bmatrix} u_a & u_b & u_c \end{bmatrix}^T$, e_{abc} represents the grid voltage with $e_{abc} = \begin{bmatrix} e_a & e_b & e_c \end{bmatrix}^T$, and i_{abc} represents the alternating current with $i_{abc} = \begin{bmatrix} i_a & i_b & i_c \end{bmatrix}^T$.

Park's transformation based on the P matrix is well known:

$$P = \begin{bmatrix} \sin(\omega t) & \sin\left(\omega t - \frac{2\pi}{3}\right) & \sin\left(\omega t + \frac{2\pi}{3}\right) \\ \cos(\omega t) & \cos\left(\omega t - \frac{2\pi}{3}\right) & \cos\left(\omega t + \frac{2\pi}{3}\right) \\ \frac{1}{2} & \frac{1}{2} & \frac{1}{2} \end{bmatrix} \tag{5}$$

By applying this transformation to the abc reference system, we obtain the usual quantities in Figure 1 in the dq reference system ($u_{dq0} = Pu_{abc}$, $e_{dq0} = Pe_{abc}$, $i_{dq0} = Pi_{abc}$). Thus, Equation (4) becomes

$$u_{dq0} - e_{dq0} = R_3 i_{dq0} + L_3 \frac{di_{dq0}}{dt} + L_3 \begin{bmatrix} -\omega i_q \\ \omega i_d \\ 0 \end{bmatrix} \tag{6}$$

By components, this equation can be rewritten in the form of Equations (7) and (8):

$$L_3 \frac{di_d}{dt} = -R_3 i_d + \omega L_3 i_q - e_d + u_d = u_{3d} + u_d \tag{7}$$

$$L_3 \frac{di_q}{dt} = -R_3 i_q - \omega L_3 i_d - e_d + u_q = u_{3q} + u_q \tag{8}$$

where u_{id} and u_{iq} represent the control variables for the command of the DC-AC converter (which is of voltage source converter—VSC-type). Equations (7) and (8) include the following notations: $u_{3d} = -R_3 i_d + \omega L_3 i_q - e_d$ and $u_{3q} = -R_3 i_q - \omega L_3 i_d - e_q$.

The MPPT algorithm is presented in [15,26,29]; it will provide the duty cycle signal (D) to control the DC boost converter. Thus, the following relations can be written:

$$i_{dc1} = (1 - D)i_s \tag{9}$$

$$u_s = (1 - D)u_{dc} \tag{10}$$

3. Control of the Grid-Connected PV System

The control of a grid-connected PV system is presented at length in [15,27,31], both in normal operation and in low-voltage ride through (LVRT). The controllers used for the inner control loops of currents i_d and i_q are of the classic PI type or synergetic type [27], while the controller for the outer control loop of u_{dc} is of the classic PI type [15,31]. Figure 2 shows the control scheme for the connection of a PV system to the power grid under normal operation. The control loops of currents i_d and i_q and of u_{dc} are presented schematically in Figure 3. In this section, we will present several basic elements of the FO calculus to synthesize the fractional-type control laws in the case of using the design and synthesis procedures of the SMC and synergetic controllers.

3.1. Notions and Notations for Fractional-Order Calculus

To achieve a refinement of the differential and integral calculus, the non-integer order operator is added as aD_t^{α}, where the FO is noted with α, and the limits of the use of the operator are denoted a and t [28,29].

$$aD_t^{\alpha} = \begin{cases} \frac{d^{\alpha}}{dt^{\alpha}} & \text{Re}(\alpha) > 0 \\ 1 & \text{Re}(\alpha) = 0 \\ \int_a^t (dt)^{-\alpha} & \text{Re}(\alpha) < 0 \end{cases} \tag{11}$$

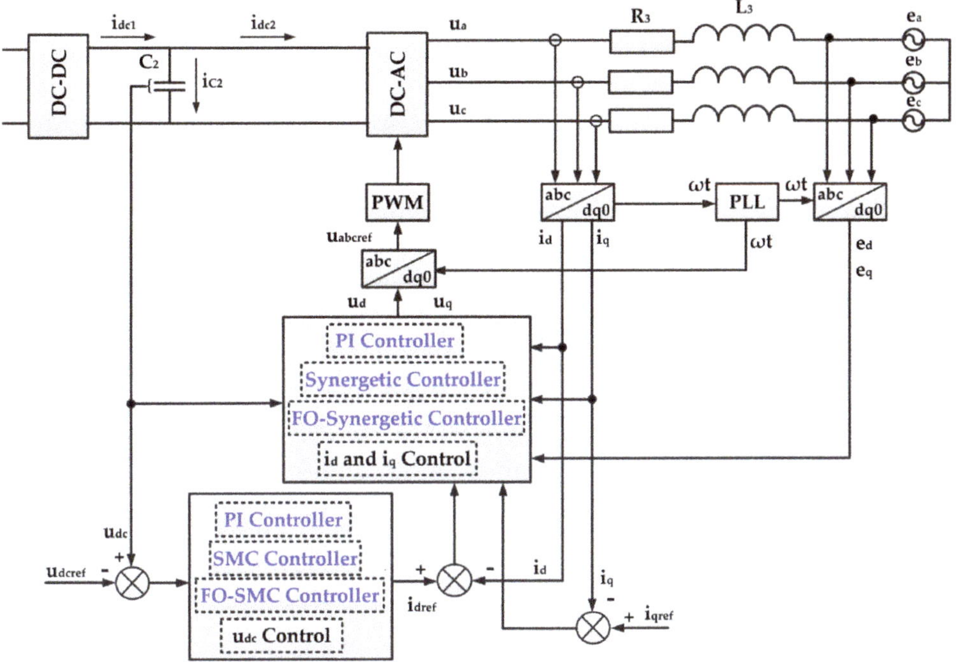

Figure 2. Block diagram of fractional-order sliding mode control (FO-SMC) and FO-synergetic control of the grid-connected PV system.

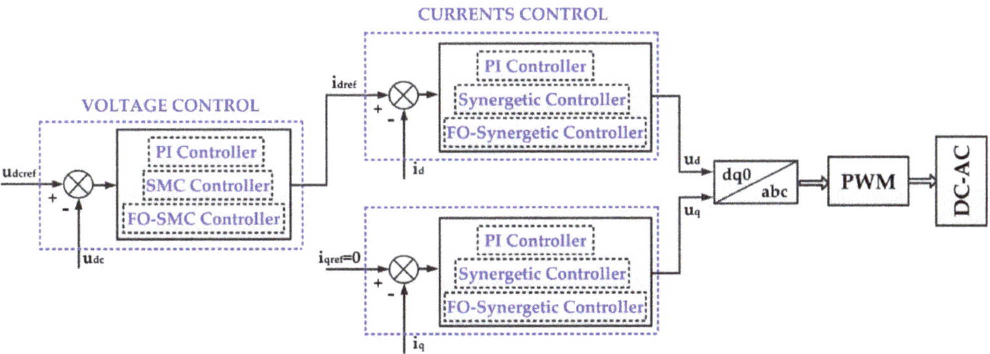

Figure 3. General scheme for cascade control of the grid-connected PV system.

Further, a common alternative definition is given by the Riemann–Liouville differintegral [28,29]:

$$aD_t^\alpha f(t) = \frac{1}{\Gamma(m-\alpha)} \left(\frac{d}{dt}\right)^m \int_\alpha^t \frac{f(\tau)}{(t-\tau)^{\alpha-m+1}} d\tau \tag{12}$$

where $m-1 < \alpha < m$, $m \in N$, and $\Gamma(\cdot)$ is Euler's gamma function.

For the practical implementation by numerical calculation, the Grünwald–Letnikov definition is presented as follows [28,29]:

$$aD_t^\alpha f(t) = \lim_{h \to 0} \frac{1}{h^\alpha} \sum_{j=0}^{\left(\frac{t-a}{h}\right)} (-1)^j \binom{\alpha}{j} f(t - jh) \tag{13}$$

where $(\cdot)$ is the integer part.

The Laplace transform can also be applied in the non-integer case similarly to the integer case (in terms of the power of the complex operator s). A special case is when the power α of operator s is of the commensurate order type q, $(q \in R^+, 0 < q < 1, \alpha_k = kq)$. For $\lambda = s^q$, the transfer function $H(\lambda)$ can be written as

$$H(\lambda) = \frac{\sum\limits_{k=0}^{m} b_k \lambda^k}{\sum\limits_{k=0}^{n} a_k \lambda^k} \tag{14}$$

For the numerical implementations in embedded systems in real time, it is important to emphasize that the results of the fractional calculus cannot be implemented directly, but an integer-order approximation of these calculi is used on a specified frequency range (ω_b, ω_h), by using Oustaloup recursive filters.

For s^γ with $0 < \gamma < 1$, an approximation can be used as follows [28,29]:

$$G_f(s) = K \prod_{k=-N}^{N} \frac{s + \omega_k'}{s + \omega_k} \tag{15}$$

where ω_k', ω_k, and K are given by

$$\omega_k' = \omega_b \left(\frac{\omega_h}{\omega_b}\right)^{\frac{k+N+\frac{1}{2}(1-\gamma)}{2N+1}}; \; \omega_k = \omega_b \left(\frac{\omega_h}{\omega_b}\right)^{\frac{k+N+\frac{1}{2}(1+\gamma)}{2N+1}}; \; k = \omega_h^\gamma \tag{16}$$

A refined form of Oustaloup-type filters is given by the following relations [28]:

$$s^\alpha \approx \left(\frac{d\omega_h}{b}\right)^\alpha \left(\frac{ds^2 + b\omega_h s}{d(1-\alpha)s^2 + b\omega_h s + d\alpha}\right) G_p \tag{17}$$

$$G_p = K \prod_{k=-N}^{N} \frac{s + \omega_k'}{s + \omega_k}; \; \omega_k = \left(\frac{b\omega_h}{d}\right)^{\frac{\alpha+2k}{2N+1}}; \; \omega_k' = \left(\frac{d\omega_b}{b}\right)^{\frac{\alpha-2k}{2N+1}} \tag{18}$$

where usually parameters $b = 10$ and $d = 9$.

3.2. Fractional-Order Sliding Mode Control

Starting from Equation (3) and using S_a, S_b, and S_c to denote the switching function for the DC-AC converter in Figure 1, the following equation is obtained in the *abc* frame:

$$C_2 \frac{du_{dc}}{dt} = i_{dc1} - (i_a S_a + i_b S_b + i_c S_c) \tag{19}$$

The switching functions Sd and S_q can be obtained by using transformation (5):

$$\begin{bmatrix} S_d & S_q & 0 \end{bmatrix}^T = P \begin{bmatrix} S_a & S_b & S_c \end{bmatrix}^T \tag{20}$$

Based on these, Equation (19) becomes

$$C_2 \frac{du_{dc}}{dt} = i_{dc1} - \frac{3}{2}\left(i_d S_d + i_q S_q\right) \tag{21}$$

In Equation (21), i_{dc1} is given by the relations (9) and (10) which depend on the DC boost converter and the MPPT algorithm, which we will consider as an optimized form given by [15,31], so we will consider that it is necessary for the other terms of the right member to be calculated by the sliding mode control technique to maintain u_{dc} at a prescribed value u_{dcref} (which is considered constant or has slow variations relative to the variation of the other quantities in the control system). Additionally, in [26], it is demonstrated that, when the three-phase grid system is symmetrical, i_d represents the direct current and reference $i_{qref} = 0$ is selected, and thus Equation (21) becomes:

$$C_2 \frac{du_{dc}}{dt} = i_{dc1} - \frac{3}{2} i_{dref} S_d \tag{22}$$

Following the sliding mode control design procedure, the reference i_{dref} for the inner control loop of currents i_d and i_q will be determined. Thus, we define the state variable x_1 as

$$x_1 = u_{dc} - u_{dcref} \tag{23}$$

We define the switching surface S:

$$\begin{cases} S = c_1 x_1 + x_2 \\ \dot{S} = c_1 x_2 + \dot{x}_2 \end{cases} \tag{24}$$

where the state variable x_2 is defined by:

$$x_2 = \dot{x}_1 = -\dot{u}_{dc} \tag{25}$$

To achieve convergence, the following is required:

$$\dot{S} = -\varepsilon \operatorname{sgn} S - kS \tag{26}$$

where ε and k are positive constants.
From the calculation, we obtain:

$$\ddot{x}_1 = \dot{x}_2 = -\ddot{u}_{dc} = \frac{3}{2} \frac{S_d}{C_2} \dot{i}_{dref} - \frac{\dot{i}_{dc1}}{C_2}, \tag{27}$$

and thus the following can be written:

$$-\varepsilon \operatorname{sgn} S - kS = c_1 x_2 + \frac{3}{2} \frac{1}{C_2} S_d \dot{i}_{dref} - \frac{\dot{i}_{dc1}}{C_2} \tag{28}$$

Following [32], to improve convergence and reduce high-frequency oscillations, the sgn function is replaced with the function below:

$$h(x) = \frac{2}{1 + e^{-a(x-b)}} - 1 \tag{29}$$

For $a = 4$ and $b = 0$, $h \in [-1 \ 1]$, and a smoothed transition is achieved between -1 and 1. From this, the output of the SMC-type controller can be inferred:

$$i_{dref} = \frac{2}{3} \frac{C_2}{S_d} \int_0^t \left[-(c_1 x_2 + kS - \varepsilon h(S)) + \frac{\dot{i}_{dc1}}{C_2} \right] dt \tag{30}$$

For the fractional case, the switching surface is defined as:

$$S = c_1 x_1 + c_2 D^{\mu} x_1 = c_1 x_1 + c_2 D^{\mu-1} x_2, \tag{31}$$

where the fractional differential operator D is defined in relation (11).

By calculating $\dot{S}$, we obtain:

$$\dot{S} = c_1 \dot{x}_1 + c_2 D^{\mu+1} x_1 = c_1 x_2 + c_2 D^{\mu-1} \dot{x}_2, \tag{32}$$

which can be rewritten using Equation (27):

$$\dot{S} = c_1 x_2 + c_2 D^{\mu-1} \left(\frac{3}{2} \frac{1}{C_2} S_d \dot{i}_{dref} - \frac{\dot{i}_{dc1}}{C_2} \right) \tag{33}$$

Using Equation (26), we obtain:

$$- \varepsilon h(S) - kS - c_1 x_2 = c_2 D^{\mu-1} \left(\frac{3}{2} \frac{1}{C_2} S_d \dot{i}_{dref} - \frac{\dot{i}_{dc1}}{C_2} \right) \tag{34}$$

By applying operator $D^{1-\mu}$ to relation (34), we obtain:

$$D^{1-\mu}(-\varepsilon h(S) - kS - c_1 x_2) = c_2 \frac{3}{2} \frac{1}{C_2} S_d \dot{i}_{dref} - c_2 \frac{\dot{i}_{dc1}}{C_2} \tag{35}$$

The output of the FO-SMC type controller can thus be inferred from the following:

$$i_{dref} = \frac{2}{3c_2} \frac{C_2}{S_d} \int_0^t \left[c_2 \frac{\dot{i}_{dc1}}{C_2} + D^{1-\mu}(-\varepsilon h(S) - kS - c_1 x_2) \right] dt \tag{36}$$

Both in Equation (30) and in Equation (36), in order to avoid the uncontrolled increase in i_{dref} due to the zero-crossings of the signal S_d, it will be replaced in the practical implementation with $S'_d = S_d + c_3$, where $c_3 > 0$ becomes a new level of freedom in the design of the FO-SMC controller.

3.3. Fractional-Order Synergetic Control

It is well known that synergetic control can be considered as a generalization of sliding mode control. Thus, synergetic control can be applied to nonlinear systems described by the general form [27,32]

$$\dot{x} = f(x, u, t) \tag{37}$$

where x represents the state vector, $x \in \Re^n$; $f(.)$ represents the continuous nonlinear function; u represents the control vector, $u \in \Re^m$, $(m < n)$.

The synergetic control procedure includes the selection of a macrovariable $\psi(x, t)$ which depends on the states of the system, for each control input. The system will be forced to evolve to manifolds $\psi = 0$, according to the following equation:

$$T\dot{\psi} + \psi = 0 \tag{38}$$

where $T > 0$ is selected to obtain the desired convergence rate.

By differentiating the macrovariable Ψ, we obtain:

$$\dot{\psi} = \frac{\partial \psi}{\partial x} \dot{x}, \tag{39}$$

and by inserting the relation (39) into Equation (38), we obtain:

$$T \frac{\partial \psi}{\partial x} \dot{x} + \psi = 0 \tag{40}$$

The explicit forms of $\dot{x}$ states in the mathematical model of the controlled system are inserted into Equation (40). This results in the control law as follows:

$$u = u(x, \psi(x,t), T, t) \tag{41}$$

Next, we will apply the integer-order and fractional-order synergetic control procedures to replace the classic PI-type control loops of currents i_d and i_q. The outputs of the synergetic controller will be u_d and u_q (see Figure 2).

For the d-axis, for $k_d > 0$, we choose the macrovariable Ψ_d as follows:

$$\psi_d = \left(u_{dcref} - u_{dc} \right) + k_d \left(i_{dref} - i_d \right) \tag{42}$$

We define another state variable x_2 (in addition to the state variable x_1 in Equation (23)):

$$\begin{cases} x_1 = u_{dcref} - u_{dc} \\ x_2 = i_{dref} - i_d \end{cases} \tag{43}$$

Based on the relation (43) for slow variations of the reference quantities or quasi-stationary regime, the following relation can be written:

$$\begin{cases} \dot{x}_1 = -\dot{u}_{dc} \\ \dot{x}_2 = -\dot{i}_d \end{cases} \tag{44}$$

Using these, we obtain the macrovariable derivative Ψ_d defined in relation (42), of the following form:

$$\dot{\psi}_d = \dot{x}_1 + k_d \dot{x}_2 = -\dot{u}_{dc} - k_d \dot{i}_d \tag{45}$$

Based on these, for $T = T_1$, Equation (40) becomes:

$$T_1 \left(-\dot{u}_{dc} - k_d \dot{i}_d \right) + \left(u_{dcref} - u_{dc} \right) + k_d \left(i_{dref} - i_d \right) = 0 \tag{46}$$

Using Equation (7), Equation (46) becomes:

$$-T_1 \dot{u}_{dc} - T_1 k_d \frac{1}{L_3}(u_{3d} - u_d) + \left(u_{dcref} - u_{dc} \right) + k_d \left(i_{dref} - i_d \right) = 0 \tag{47}$$

After rearranging the terms in Equation (47), we can write:

$$T_1 k_d \frac{1}{L_3} u_d = -T_1 \dot{u}_{dc} - T_1 k_d \frac{1}{L_3} u_{3d} + \left(u_{dcref} - u_{dc} \right) + k_d \left(i_{dref} - i_d \right) \tag{48}$$

Based on this, we obtain the control law u_d as follows:

$$u_d = \frac{L_3}{T_1 k_d} \left[-T_1 \dot{u}_{dc} - T_1 k_d \frac{1}{L_3} u_{3d} + \left(u_{dcref} - u_{dc} \right) + k_d \left(i_{dref} - i_d \right) \right] \tag{49}$$

For axis d in the fractional case, the macrovariable is chosen:

$$\psi_d = D^\mu x_1 + k_d x_2 \tag{50}$$

By deriving Equation (50), we obtain:

$$\dot{\psi}_d = D^\mu \dot{x}_1 + k_d \dot{x}_2 = -D^\mu \dot{u}_{dc} - k_d \dot{i}_d \tag{51}$$

Based on these, Equation (40) becomes

$$T_1 \left(-D^\mu \dot{u}_{dc} - k_d \dot{i}_d \right) + D^\mu \left(u_{dcref} - u_{dc} \right) + k_d \left(i_{dref} - i_d \right) = 0 \tag{52}$$

Using Equation (7), Equation (52) becomes:

$$- T_1 D^{\mu+1} u_{dc} - T_1 k_d \frac{1}{L_3}(u_{3d} + u_d) + D^{\mu}\left(u_{dcref} - u_{dc}\right) + k_d\left(i_{dref} - i_d\right) = 0 \tag{53}$$

After rearranging the terms in Equation (53), we can write:

$$T_1 k_d \frac{1}{L_3} u_d = -T_1 D^{\mu+1} u_{dc} - T_1 k_d \frac{1}{L_3} u_{3d} + D^{\mu}\left(u_{dcref} - u_{dc}\right) + k_d\left(i_{dref} - i_d\right) \tag{54}$$

Based on this, we obtain the control law u_d as follows:

$$u_d = \frac{L_3}{T_1 k_d}\left[-T_d D^{\mu+1} u_{dc} - T_1 k_d \frac{1}{L_3} u_{3d} + D^{\mu}\left(u_{dcref} - u_{dc}\right) + k_d\left(i_{dref} - i_d\right)\right] \tag{55}$$

For the q-axis, for $k_q > 0$, we choose the macrovariable Ψ_q of the following form:

$$\psi_q = i_{qref} - i_q \tag{56}$$

We define another state variable x_3 (in addition to the state variables x_1 and x_2 in Equation (43)):

$$\begin{cases} x_1 = u_{dcref} - u_{dc} \\ x_2 = i_{dref} - i_d \\ x_3 = i_{qref} - i_q \end{cases} \tag{57}$$

Based on relation (57), because i_{qref} is set to zero, we can write:

$$\begin{cases} \dot{x}_1 = -\dot{u}_{dc} \\ \dot{x}_2 = -\dot{i}_d \\ \dot{x}_3 = -\dot{i}_q \end{cases} \tag{58}$$

Using these, we obtain the macrovariable derivative Ψ_q defined in relation (56), of the following form:

$$\dot{\psi}_q = \dot{x}_3 \tag{59}$$

Based on these, for $T = T_2$, Equation (40) becomes:

$$- T_2 \dot{i}_q + \left(i_{qref} - i_q\right) = 0 \tag{60}$$

Using Equation (8), Equation (60) becomes:

$$- T_2 \frac{1}{L_3}(u_{3q} + u_q) + i_{qref} - i_q = 0 \tag{61}$$

After rearranging the terms in Equation (61), we can write:

$$(u_{3q} + u_q) = \frac{L_3}{T_2}\left(i_{qref} - i_q\right) \tag{62}$$

Based on this, we obtain the control law u_q as follows:

$$u_q = \frac{L_3}{T_2}\left(i_{qref} - i_q\right) - u_{3q} \tag{63}$$

For the q-axis in the fractional case, the macrovariable is chosen:

$$\psi_q = D^{\mu} x_3 + k_q \int_0^t x_3(t)\,dt \tag{64}$$

By deriving Equation (64), we obtain:

$$\dot{\psi}_q = D^\mu \dot{x}_3 + k_q x_3 = -D^\mu \dot{i}_q + k_q \left(i_{qref} - i_q \right) \tag{65}$$

Based on these, Equation (40) becomes:

$$T_2 \left[D^\mu \left(-\frac{1}{L_3} \left(u_{3q} + u_q \right) \right) + k_q \left(i_{qref} - i_q \right) \right] + D^\mu \left(i_{qref} - i_q \right) + k_q \int_0^t \left(i_{qref} - i_q \right) dt = 0 \tag{66}$$

By using Equation (8) and applying the fractional operator defined in relation (11) to both members of Equation (66), and considering that $D^{-\mu}$ becomes I_μ, we can write:

$$-\frac{T_2}{L_3} \left(u_{3q} + u_q \right) + T_2 k_q I_\mu \left(i_{qref} - i_q \right) + \left(i_{qref} - i_q \right) + k_q I_{\mu+1} \left(i_{qref} - i_q \right) = 0 \tag{67}$$

After rearranging the terms in Equation (67), we can write:

$$\frac{T_2}{L_3} u_q = T_q k_q I_\mu \left(i_{qref} - i_q \right) + \left(i_{qref} - i_q \right) + k_q I_{\mu+1} \left(i_{qref} - i_q \right) - \frac{T_2}{L_3} u_{3q} \tag{68}$$

Based on this, we obtain the control law u_q as follows:

$$u_q = \frac{L_3}{T_2} \left[T_2 k_q I_\mu \left(i_{qref} - i_q \right) + \left(i_{qref} - i_q \right) + k_q I_{\mu+1} \left(i_{qref} - i_q \right) - \frac{T_2}{L_3} u_{3q} \right] \tag{69}$$

In the case of integer-order synergetic control, the control inputs u_d and u_q are provided by Equations (49) and (63), and in the case of the fractional synergetic control, the control inputs u_d and u_q are provided by Equations (55) and (69). By applying the inverse Park transform, the actual control inputs u_{abc} (see Figure 2) are obtained as follows:

$$u_{abc} = P^{-1} u_{dq0} \tag{70}$$

4. Numerical Simulations and Analysis for the Control of the Grid-Connected PV System Using FO-SMC and FO-Synergetic Controllers

In this section, starting from the controllers synthesized in the previous section, we will present the obtained results of the global system for the control of the grid-connected PV system, in which classic PI, synergetic, or FO-synergetic controllers are used for the inner control loops of currents i_d and i_q and classic PI, SMC, or FO-SMC controllers are used for the outer control loop of u_{dc}. Owing to the levels of freedom and the refinement brought by the fractional calculus for the SMC and synergetic algorithms, it will be demonstrated, through numerical simulations, using the Matlab/Simulink environment, that superior performance is achieved using the proposed control system. The system described in Figures 2 and 3 is implemented in Matlab/Simulink and the block diagram is presented in Figure 4.

The presented implementation starts from an example in Matlab/Simulink [15], which presents the control system and the performances for a 100 kW model of the grid-connected PV array. The reference value of the voltage in the DC intermediate circuit is set to 500 V and the rated AC voltage supplied by the DC-AC converter is of 260 V. Further, the load is connected to the main grid through a distribution transformer with a rated voltage of 25 kV/260 V. The MPPT algorithm and its performances are presented and implemented in [15,31] and are used in the implementation presented in this article as a block function.

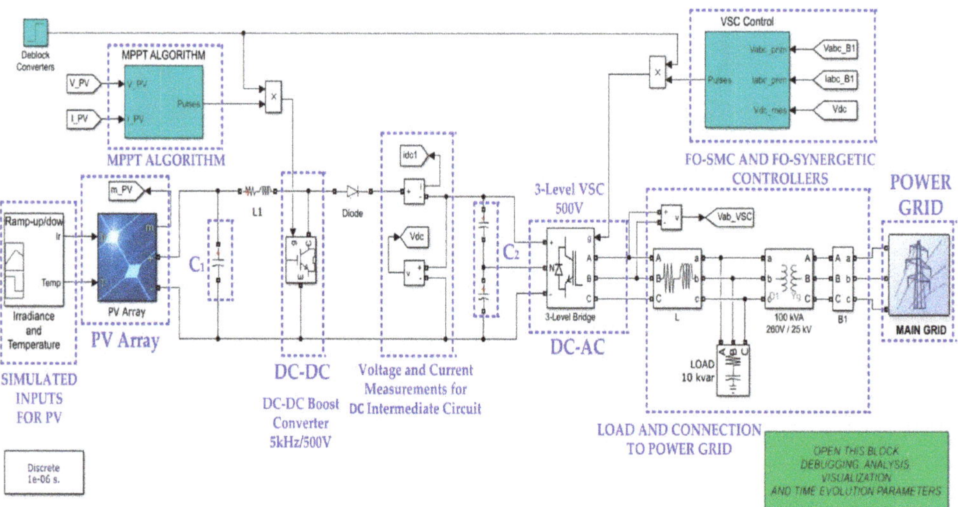

Figure 4. Matlab/Simulink implementation block diagram for control of the grid-connected PV system using FO-SMC and FO-synergetic controllers.

In order to have the smallest possible fluctuations when the DC-AC converter supplies a variable load, the importance of the precise control of the voltage level in the DC intermediate circuit—u_{dc}—is well known. For this, two cascade control loops are used, an outer one for the control of u_{dc} and two inner loops for the control of currents i_d and i_q. The current reference and i_{dref} are supplied by the output of the outer loop controller, and i_{qref} is set to zero [26].

Figures 5 and 6 show the Matlab/Simulink implementations of the control laws synthesized in Section 3 for the most complex case, where the controller of the outer control loop of voltage u_{dc} is of the FO-SMC type, and the inner control loops of currents i_d and i_q are of the FO-synergetic type. Moreover, in Figure 4, the signal filtering at the DC-AC converter output is achieved using a 10 kvar bank capacitor, which can be considered as the load for the control system.

The operation of the PV array is also implemented in [15], and the time variation of the irradiance and temperature input signals is shown in Figure 7. The PV array includes 330 SunPower-type modules which can supply a maximum of 100.7 kW (305.2 W/modules), and each module is characterized by an open-circuit voltage of $V_{oc} = 64.2$ V and a short-circuit current of $I_{sc} = 5.96$ A. In the Matlab/Simulink implementation, the sample time used is of one microsecond for the Pulse Width Modulation (PWM) generator command signals for the DC-DC and DC-AC converters. For the control system of the voltage and currents, but also for the PLL-type synchronization loop, the sampling period is of 0.1 ms.

In the Matlab/Simulink implementation in [15], for the first 50 ms, the operation of the converters is bypassed during the period when the control system operates in the open loop. After the first 50 ms, the controllers come into operation both for the DC-DC converter and for the MPPT algorithm provided by [31], but also for the control of the DC-AC converter whose improved FO-SMC and FO-synergetic controllers proposed in this article provide superior performance compared to the classical PI controllers proposed in [15], both in stationary mode and in dynamic mode (see Figures 8–11).

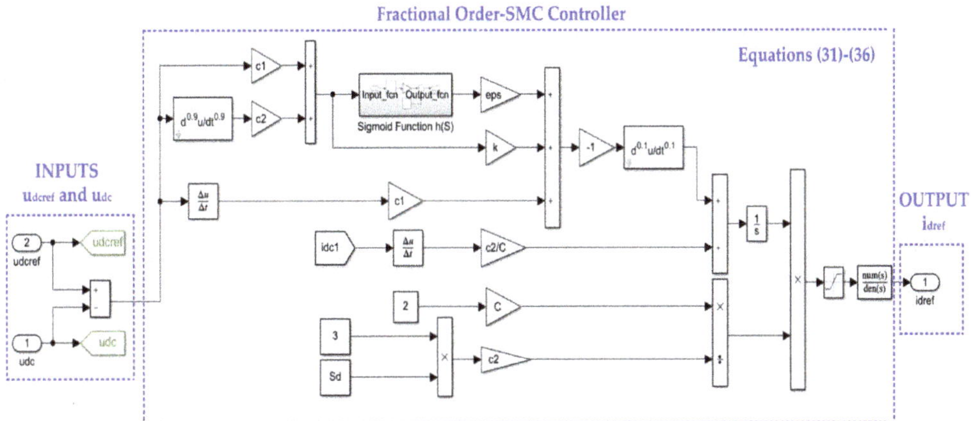

Figure 5. Matlab/Simulink implementation block diagram for the FO-SMC controller.

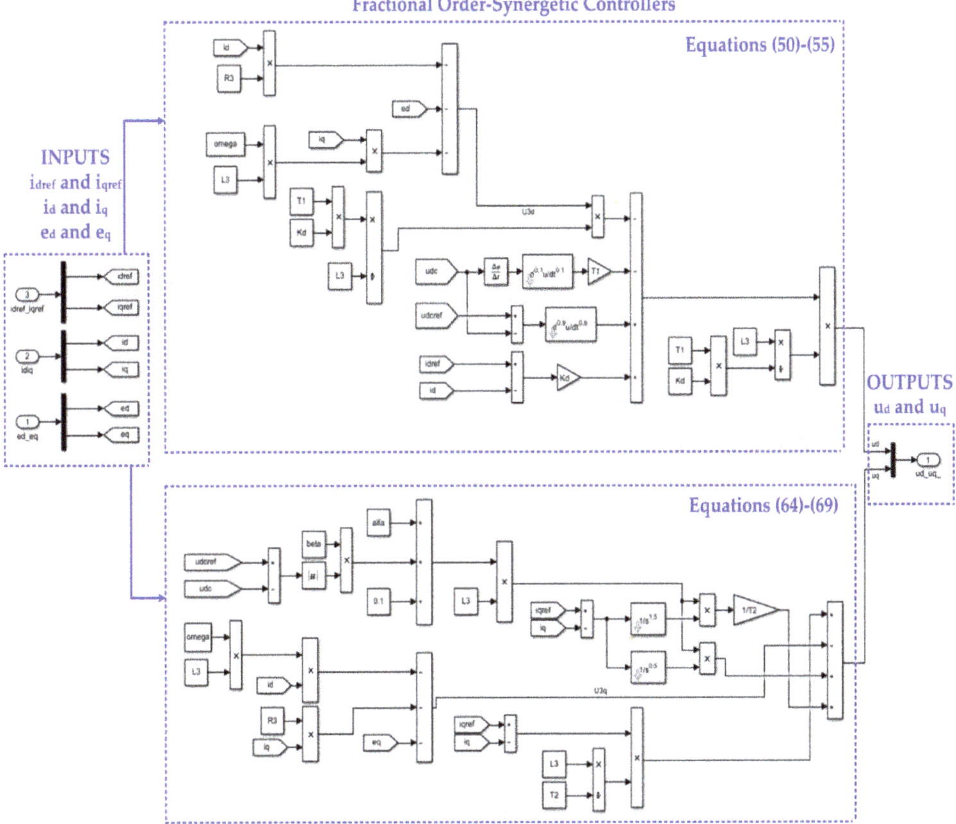

Figure 6. Matlab/Simulink implementation block diagram for FO-synergetic controllers.

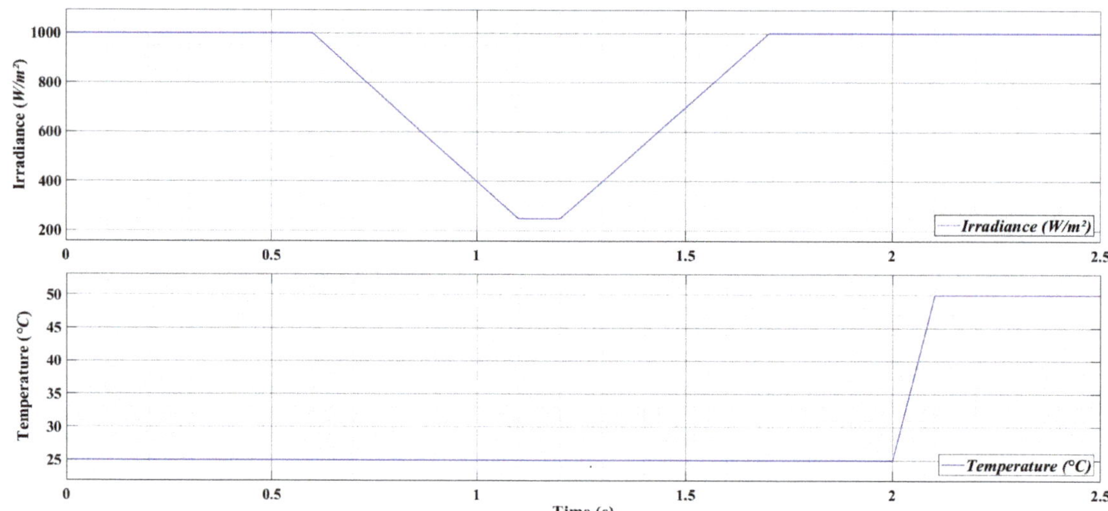

Figure 7. Time evolution of the irradiance and temperature signals type 1.

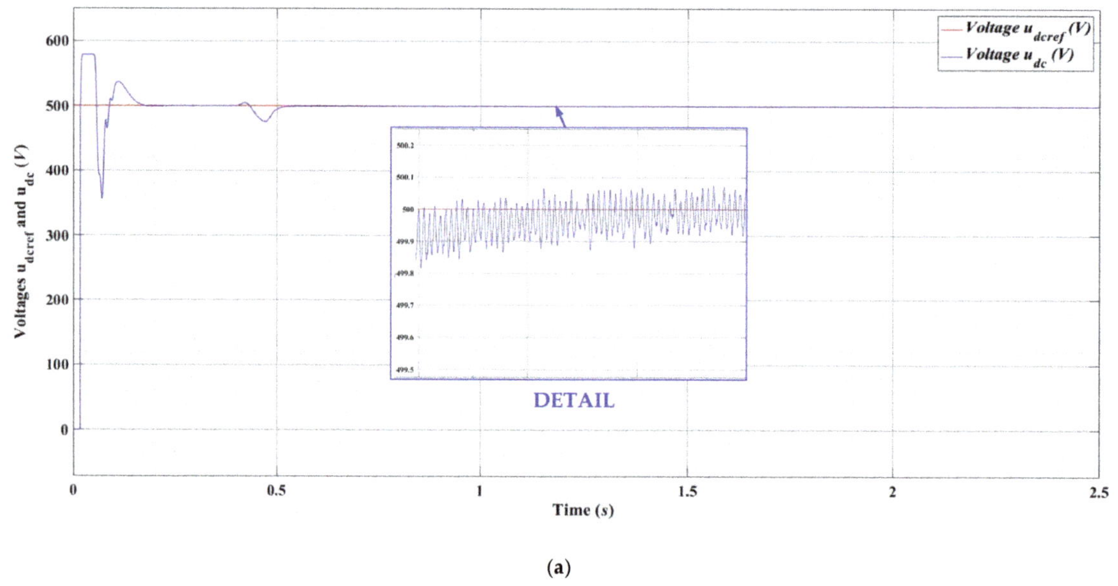

(**a**)

Figure 8. *Cont.*

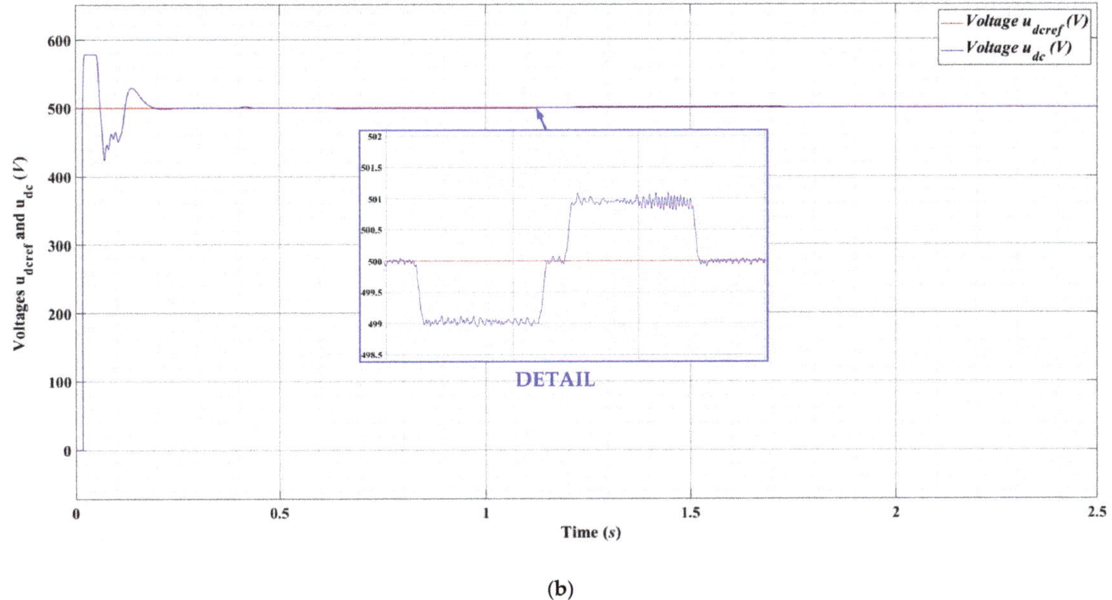

(b)

Figure 8. Time evolution of the u_{dc} for the irradiance and temperature signals type 1, at 10 kvar load: (**a**) FO-SMC/FO-SYN controllers; (**b**) classical PI controllers.

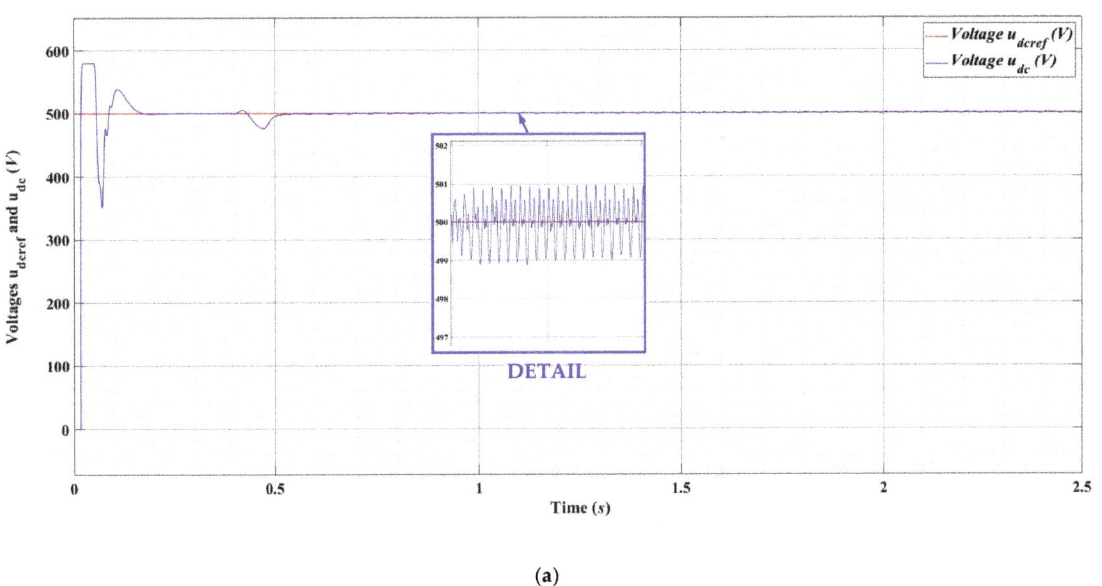

(a)

Figure 9. *Cont.*

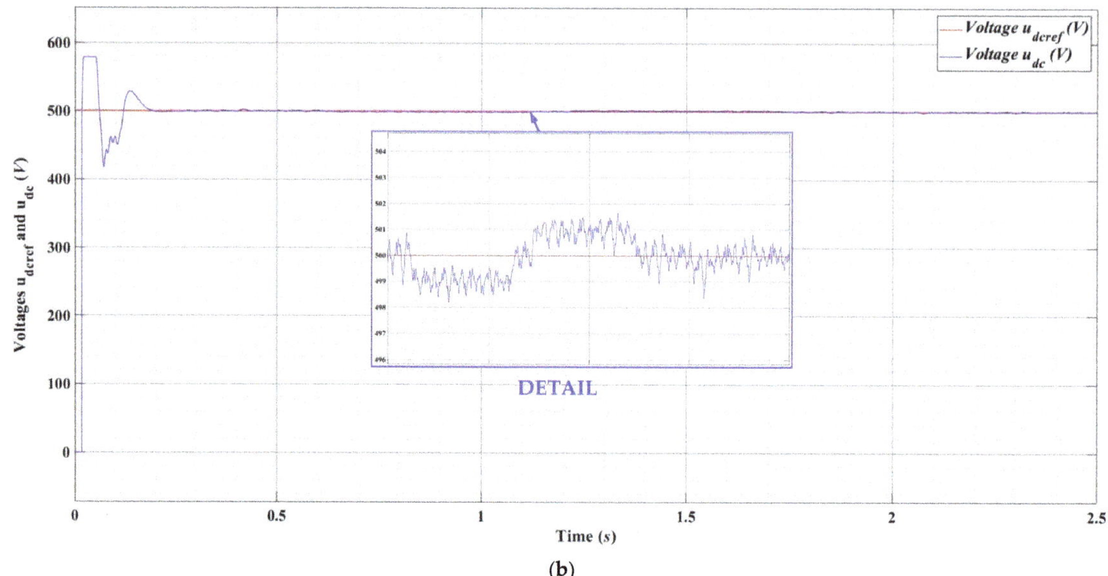

(**b**)

Figure 9. Time evolution of the u_{dc} for the irradiance and temperature signals type 1, at 13 kvar load: (**a**) FO-SMC/FO-SYN controllers; (**b**) classical PI controllers.

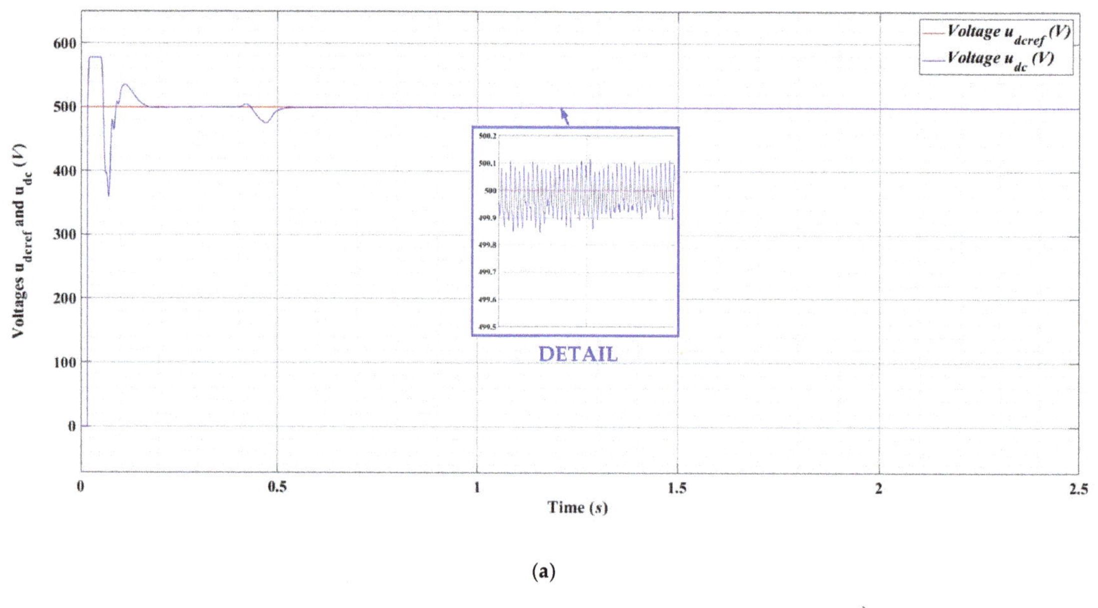

(**a**)

Figure 10. *Cont.*

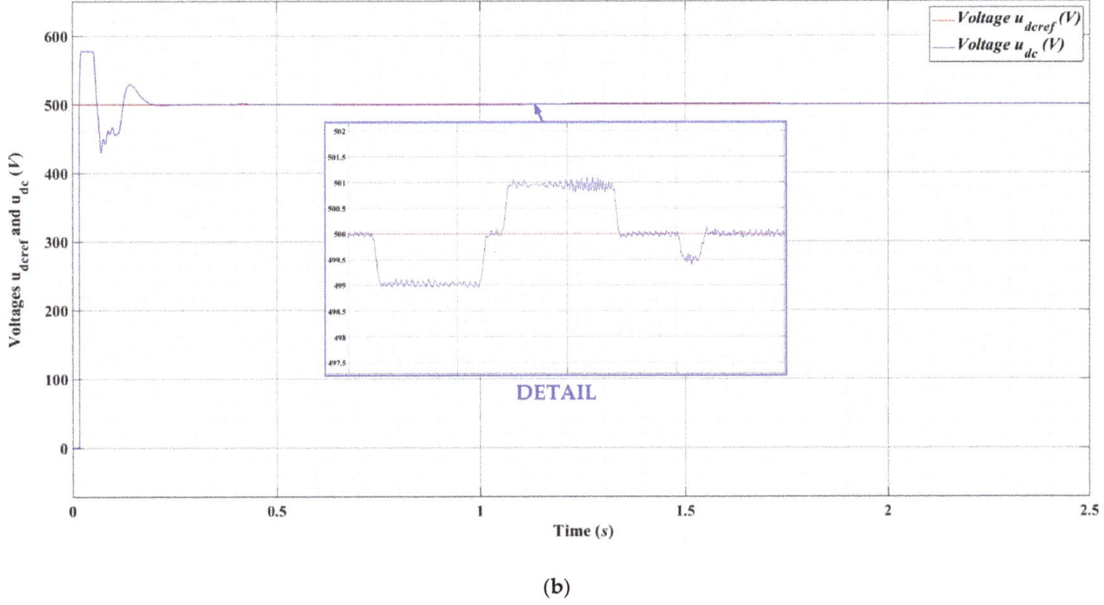

(b)

Figure 10. Time evolution of the u_{dc} for the irradiance and temperature signals type 1, at 7 kvar load: (**a**) FO-SMC/FO-SYN controllers; (**b**) classical PI controllers.

Figure 8a shows the response of the FO-SMC type control system for the control of the DC voltage u_{dc} combined with the FO-synergetic type control system for the control of currents i_d and i_q (FO-SMC/FO-SYN controllers), if the DC voltage reference u_{dcref} = 500 V, and Figure 8b shows the response of the control system where the controllers used for both the control of the DC voltage u_{dc} and for the control of currents i_d and i_q are of the PI type. After the validation of the control system start-up (after 50 ms), the MPPT algorithm start-up occurs (after 100 ms), and the end of the transitory regime (after 250 ms) is noted. In steady state, a steady-state error of 0.1 V, i.e., 0.02%, is noted for the FO-SMC/FO-SYN controllers, while the steady-state error for the PI controller is of 1 V, i.e., 0.2%. If the load is varied by a 30% increase or decrease, reaching 13 or 7 kvar, respectively, the superiority of the control of the grid-connected PV system based on FO-SMC/FO-SYN controllers is noted in Figures 9 and 10.

Regarding the dynamic regime, Figure 11 shows the response of the control of the grid-connected PV system based on FO-SMC/FO-SYN controllers as compared to PI controllers, where at time t = 1 s, the reference signal u_{dcref} undergoes a step variation at 550 V.

The parameters of the PI controller for u_{dc} control are K_p = 7 and K_i = 800 and the parameters of the PI controller for i_d and i_q currents are K_p = 0.3 and K_i = 20 [15].

The parameters of the FO-SMC/FO-SYN controllers (presented in Section 3) are; c_1 = 100; c_2 = 1; c_3 = 1; k = 180; ε = 110; C_2 = 6000e-06; T_1 = 0.01; T_2 = 0.01; K_p = 100; K_d = 0.2; L_3 = 2.5000e-04; R_3 = 0.0019; ω = 2·π·60.

It is noted that the performances in the dynamic regime, as well as those in the stationary regime, are superior in the case of using FO-SMC/FO-SYN controllers, and in Figure 11, an override of 0.2% (1 V) and a response time of 20 ms are noted, compared to an override of 1% (5 V) and a response time of 50 ms in the case of PI controllers.

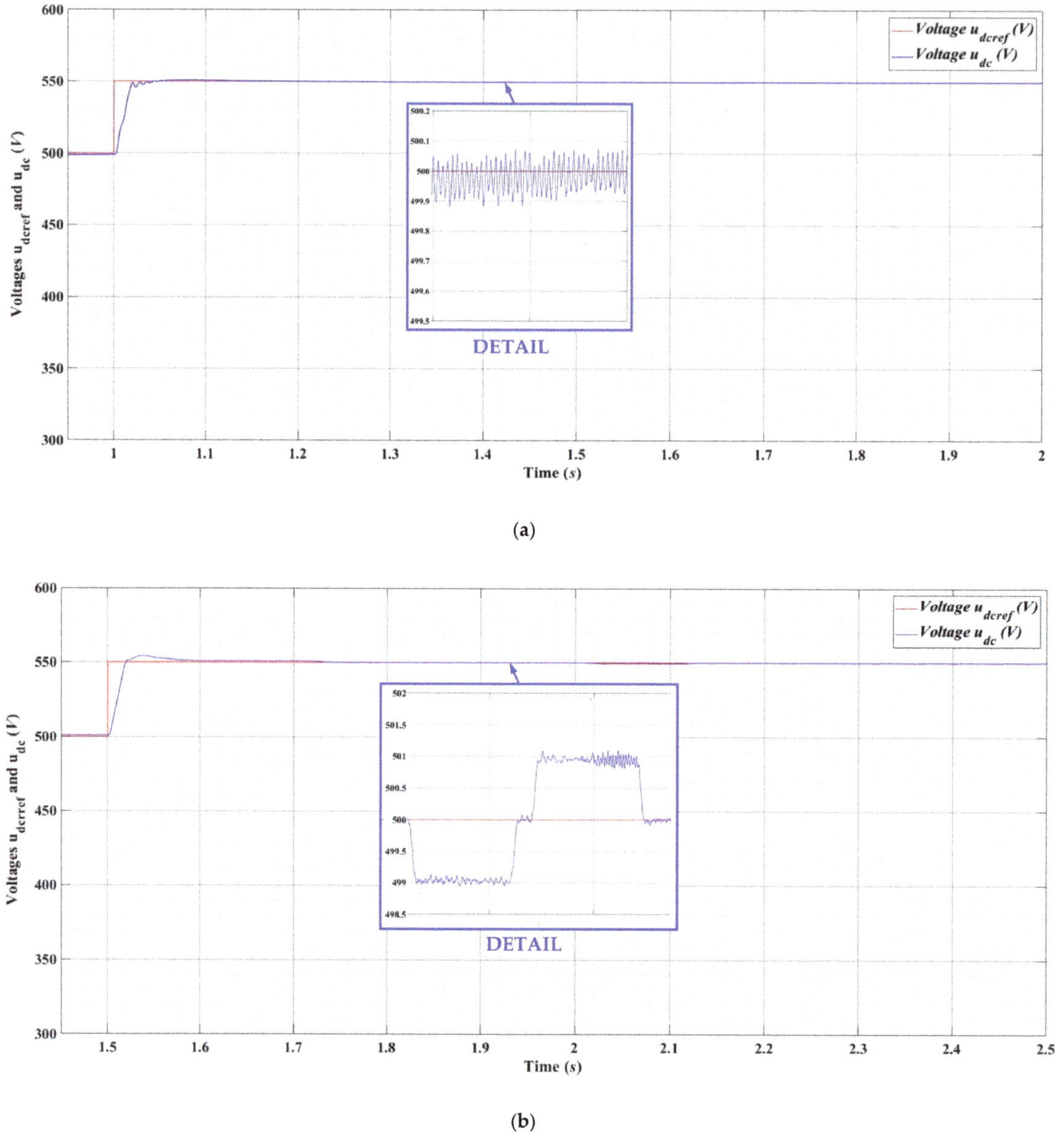

Figure 11. Time evolution of the u_{dc} for the irradiance and temperature signals type 1, at 10 kvar load for a step variation of u_{dcref} from 500 to 550 V: (**a**) FO-SMC/FO-SYN controllers; (**b**) classical PI controllers.

Figures 12–16 show a series of waveform graphs regarding the time evolution of the main inputs of interest in the control of the grid-connected PV system. Thus, Figure 12 shows the time evolution of i_d and i_q currents, where it is noted that i_d follows the i_{dref} reference provided by the FO-SMC controller, while i_q follows the set reference $i_{qref} = 0$. Figure 13 shows the evolution of the average power and voltage of the PV array, P_{mean} and U_{mean}. Further, with regard to the DC-DC converter, Figure 13 presents the time evolution of the duty cycle signal provided by the MPPT algorithm, and with regard to the DC-AC converter, it presents the time evolution of the modulation index, a signal which is supplied by the control system of the VSC controller, which supplies control pulses. Figure 14 shows the evolution over time of the output voltage between two phases of the DC-AC converter. Figure 15 shows the time evolution of the voltage and current on a phase of the transformer for the connection to the main grid.

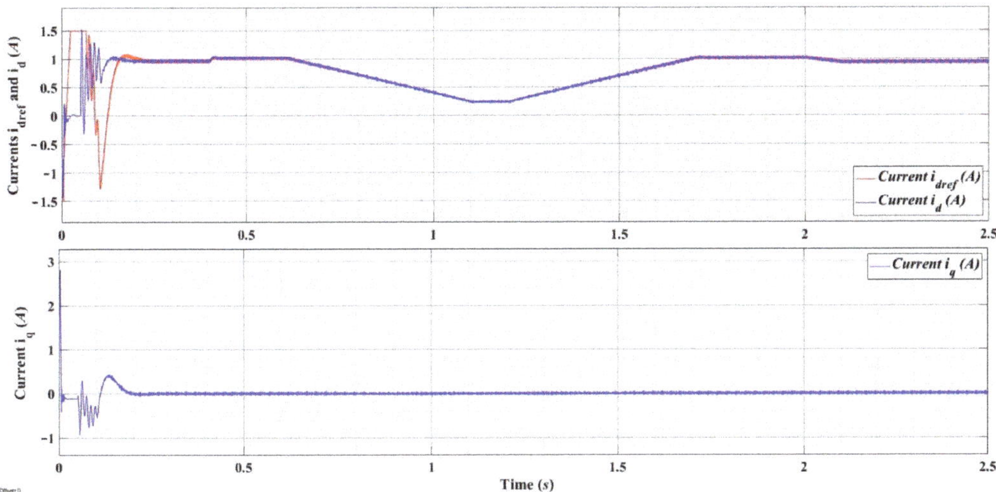

Figure 12. Time evolution of the i_d and i_q currents.

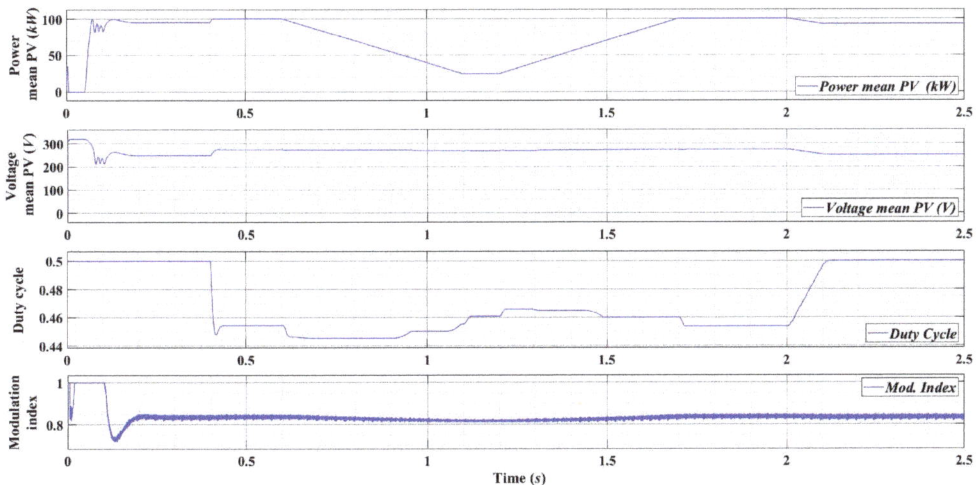

Figure 13. Time evolutions of the power P_{mean} and voltage U_{mean} of the PV, duty cycle of the DC-DC converter, and modulation index of the DC-AC converter.

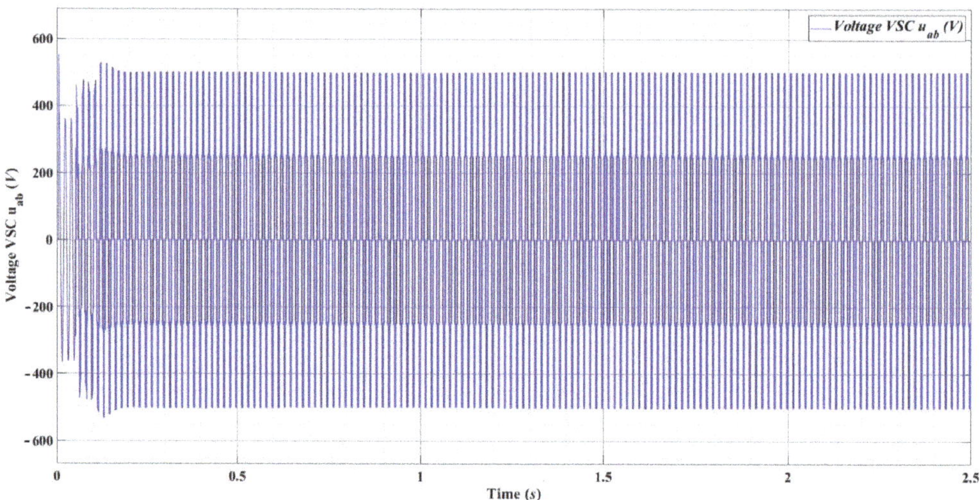

Figure 14. Time evolution of the voltage u_{ab} of the voltage source converter (VSC) controller.

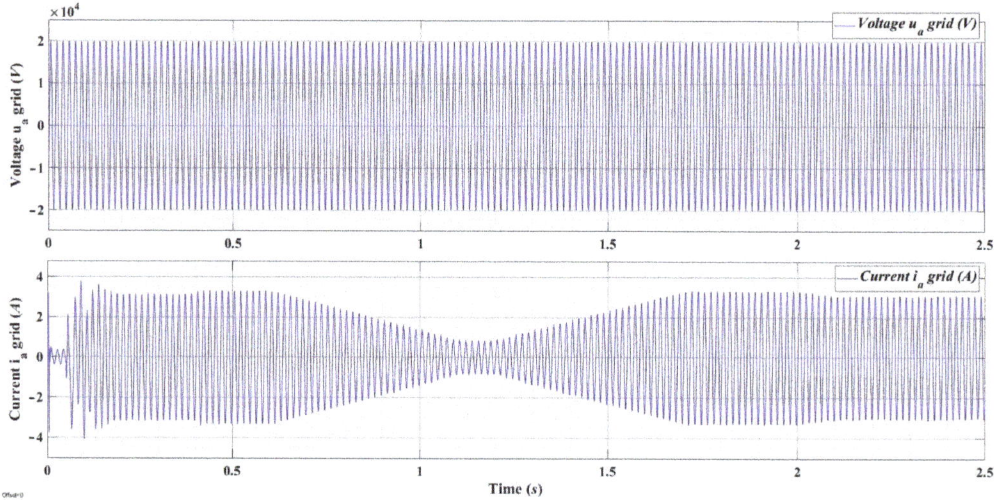

Figure 15. Time evolution of the voltage u_a and current i_a of the main grid.

Moreover, Figure 16 shows the time evolution of the active power which flows between the analyzed system and the main grid.

The control of the grid-connected PV system implemented in [15] is also discussed in [26], in which the control system of u_{dc} voltage is of the SMC type, the control systems of currents i_d and i_q are of the classic PI type, and the time evolution of the irradiance and temperature signals is presented in Figure 17. For the FO-SMC/FO-SYN controllers proposed in this article, Figures 18–20 show the time evolution of the voltage u_{dc} in the DC intermediate circuit, if the reference voltage u_{dcref} is of 500 V. Superior performances are also noted in this case, both for the nominal load of 10 kvar and for its variations to 13 and 7 kvar. It is noted that the steady-state error is maintained at 0.1 V (0.02%).

Due to the fact that the basic model in Matlab/Simulink is complex and has all the aspects regarding the transformation chain from the PV array to the main grid connection and considering that it is used for comparison in other papers [15,26,27,31], this model can be considered as a benchmark for the control of the grid-connected PV system. Thus, the superior performance obtained by using the FO-SMC/FO-SYN controllers proposed in this article can be considered as a validation of the proposed control system.

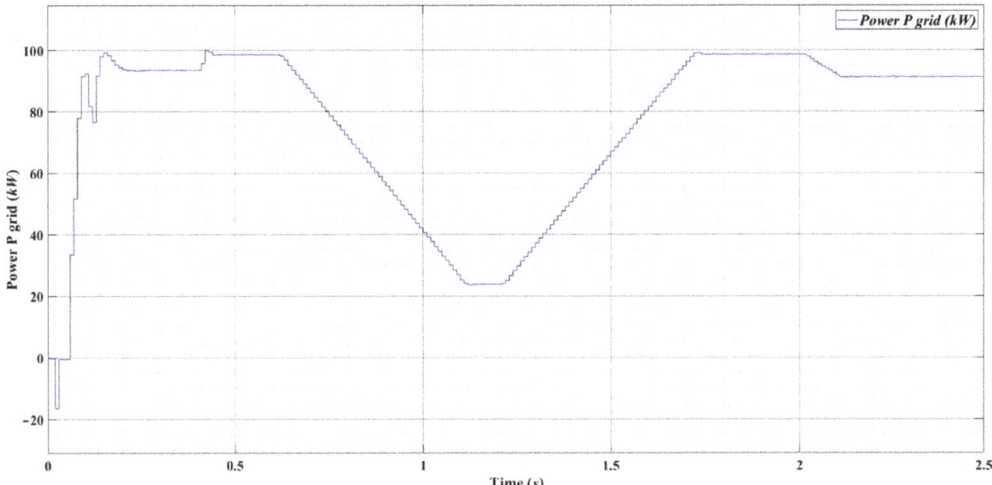

Figure 16. Time evolution of the power flow P between PV system and main grid.

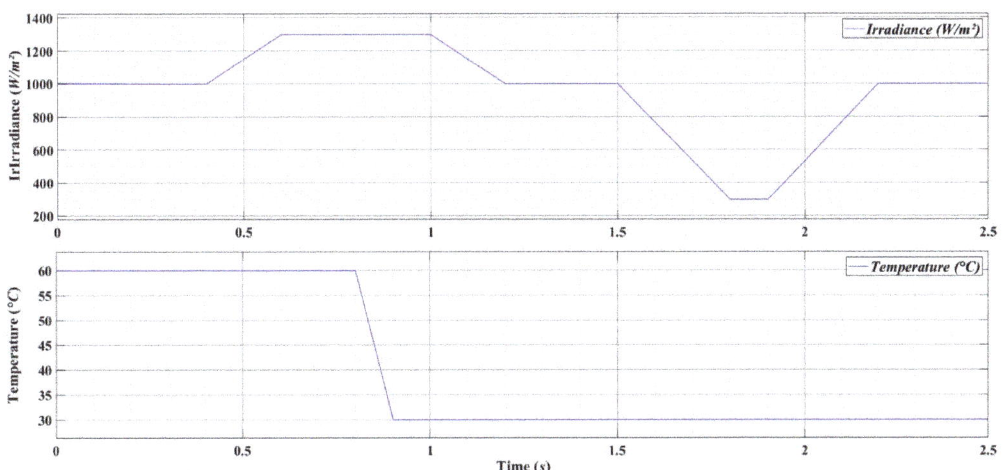

Figure 17. Time evolution of the irradiance and temperature signals type 2.

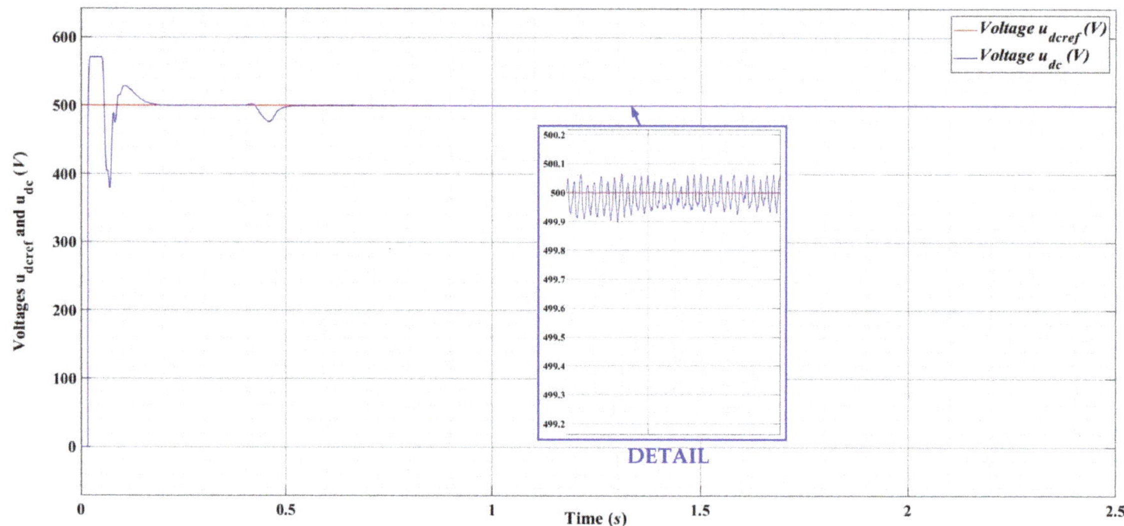

Figure 18. Time evolution of the u_{dc} for the irradiance and temperature signals type 2, at 10 kvar load with FO-SMC/FO-SYN controllers.

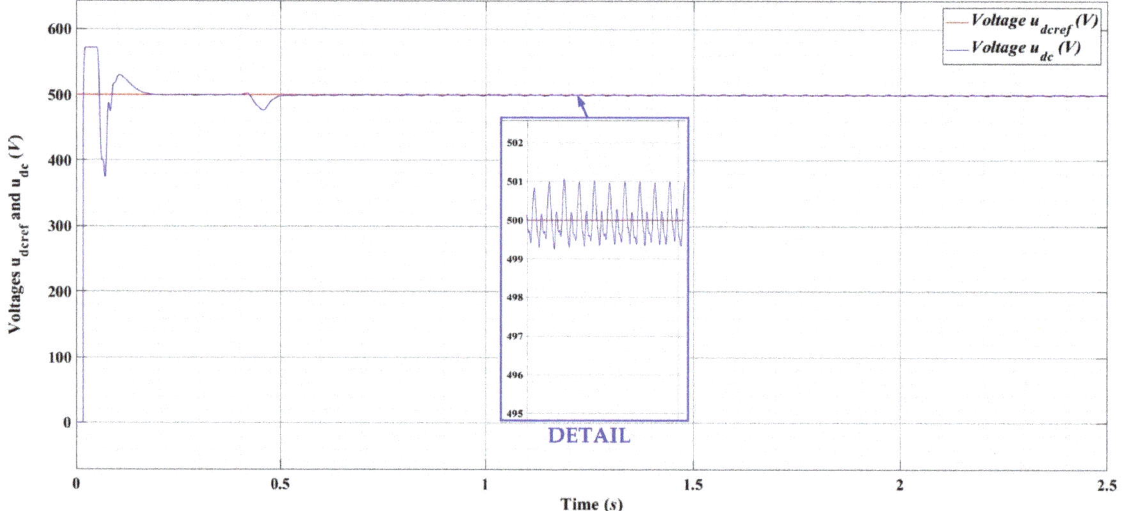

Figure 19. Time evolution of the u_{dc} for the irradiance and temperature signals type 2, at 13 kvar load with FO-SMC/FO-SYN controllers.

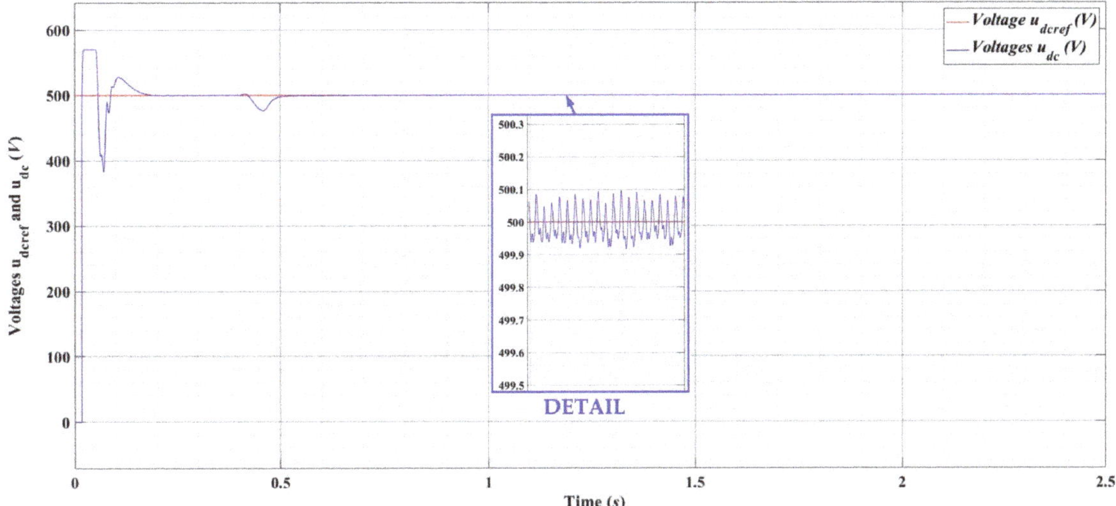

Figure 20. Time evolution of the u_{dc} for the irradiance and temperature signals type 2, at 7 kvar load with FO-SMC/FO-SYN controllers.

5. Conclusions

This article presents the control of a grid-connected PV system using FO-SMC and FO-synergetic controllers. The mathematical model of a PV system connected to the main grid is presented together with the chain of intermediate elements: the DC-DC boost converter, the DC intermediate circuit, the DC-AC converter, the filtering block, and the transformer for the connection to the main grid, together with their control systems. The robustness and superior performance of an SMC-type controller for the control of u_{dc} voltage in the DC intermediate circuit are combined with the advantages provided by the flexibility of using synergetic control for the control of currents i_d and i_q. In addition, these control techniques are suitable for the control of nonlinear systems, and it is not necessary to linearize the controlled system around a static operating point; thus, the control system achieved is robust to parametric variations and provides the required static and dynamic performance. Further, by approaching the synthesis of these controllers using the fractional calculus for integration and differentiation operators, this article proposes a control system based on FO-SMC/FO-SYN controllers. The validation of the synthesis of the proposed control system is achieved by comparing it with a benchmark for the control of a grid-connected PV system implemented in Matlab/Simulink.

Author Contributions: Conceptualization, M.N.; data curation, M.N. and C.-I.N.; formal analysis, M.N. and C.-I.N.; funding acquisition, M.N.; investigation, M.N. and C.-I.N.; methodology, M.N. and C.-I.N.; project administration, M.N.; resources, M.N. and C.-I.N.; software, M.N. and C.-I.N.; supervision, M.N.; validation, M.N. and C.-I.N.; visualization, M.N. and C.-I.N.; writing—original draft, M.N. and C.-I.N.; writing—review and editing, M.N. and C.-I.N. All authors have read and agreed to the published version of the manuscript.

Funding: The paper was developed with funds from the Ministry of Education and Scientific Research—Romania as part of the NUCLEU Program: PN 19 38 01 03.

Institutional Review Board Statement: Not applicable.

Informed Consent Statement: Not applicable.

Data Availability Statement: Data sharing not applicable.

Conflicts of Interest: The authors declare no conflict of interest.

References

1. Tricarico, T.; Gontijo, G.; Neves, M.; Soares, M.; Aredes, M.; Guerrero, J.M. Control Design, Stability Analysis and Experimental Validation of New Application of an Interleaved Converter Operating as a Power Interface in Hybrid Microgrids. *Energies* **2019**, *12*, 437. [CrossRef]
2. Petersen, L.; Iov, F.; Tarnowski, G.C. A Model-Based Design Approach for Stability Assessment, Control Tuning and Verification in Off-Grid Hybrid Power Plants. *Energies* **2020**, *13*, 49. [CrossRef]
3. Veerashekar, K.; Askan, H.; Luther, M. Qualitative and Quantitative Transient Stability Assessment of Stand-Alone Hybrid Microgrids in a Cluster Environment. *Energies* **2020**, *13*, 1286. [CrossRef]
4. Zhao, F.; Yuan, J.; Wang, N.; Zhang, Z.; Wen, H. Secure Load Frequency Control of Smart Grids under Deception Attack: A Piecewise Delay Approach. *Energies* **2019**, *12*, 2266. [CrossRef]
5. Montoya, O.D.; Gil-González, W.; Rivas-Trujillo, E. Optimal Location-Reallocation of Battery Energy Storage Systems in DC Microgrids. *Energies* **2020**, *13*, 2289. [CrossRef]
6. Alshehri, J.; Khalid, M.; Alzahrani, A. An Intelligent Battery Energy Storage-Based Controller for Power Quality Improvement in Microgrids. *Energies* **2019**, *12*, 2112. [CrossRef]
7. Estévez-Bén, A.A.; Alvarez-Diazcomas, A.; Rodríguez-Reséndiz, J. Transformerless Multilevel Voltage-Source Inverter Topology Comparative Study for PV Systems. *Energies* **2020**, *13*, 3261. [CrossRef]
8. Yan, X.; Cui, Y.; Cui, S. Control Method of Parallel Inverters with Self-Synchronizing Characteristics in Distributed Microgrid. *Energies* **2019**, *12*, 3871. [CrossRef]
9. Coppola, M.; Guerriero, P.; Dannier, A.; Daliento, S.; Lauria, D.; Del Pizzo, A. Control of a Fault-Tolerant Photovoltaic Energy Converter in Island Operation. *Energies* **2020**, *13*, 3201. [CrossRef]
10. Khan, K.; Kamal, A.; Basit, A.; Ahmad, T.; Ali, H.; Ali, A. Economic Load Dispatch of a Grid-Tied DC Microgrid Using the Interior Search Algorithm. *Energies* **2019**, *12*, 634. [CrossRef]
11. Cook, M.D.; Trinklein, E.H.; Parker, G.G.; Robinett, R.D., III; Weaver, W.W. Optimal and Decentralized Control Strategies for Inverter-Based AC Microgrids. *Energies* **2019**, *12*, 3529. [CrossRef]
12. Oviedo Cepeda, J.C.; Osma-Pinto, G.; Roche, R.; Duarte, C.; Solano, J.; Hissel, D. Design of a Methodology to Evaluate the Impact of Demand-Side Management in the Planning of Isolated/Islanded Microgrids. *Energies* **2020**, *13*, 3459. [CrossRef]
13. Stadler, M.; Pecenak, Z.; Mathiesen, P.; Fahy, K.; Kleissl, J. Performance Comparison between Two Established Microgrid Planning MILP Methodologies Tested On 13 Microgrid Projects. *Energies* **2020**, *13*, 4460. [CrossRef]
14. Artale, G.; Caravello, G.; Cataliotti, A.; Cosentino, V.; Di Cara, D.; Guaiana, S.; Nguyen Quang, N.; Palmeri, M.; Panzavecchia, N.; Tinè, G. A Virtual Tool for Load Flow Analysis in a Micro-Grid. *Energies* **2020**, *13*, 3173. [CrossRef]
15. MathWorks—Detailed Model of a 100-kW Grid-Connected PV Array. Available online: https://nl.mathworks.com/help/physmod/sps/ug/detailed-model-of-a-100-kw-grid-connected-pv-array.html;jsessionid=29903e2e045151ffb3e27a4920e1 (accessed on 4 November 2020).
16. Hong, W.; Tao, G. An Adaptive Control Scheme for Three-phase Grid-Connected Inverters in Photovoltaic Power Generation Systems. In Proceedings of the Annual American Control Conference (ACC), Milwaukee, WI, USA, 27–29 June 2018; pp. 899–904.
17. Naderi, M.; Khayat, Y.; Bevrani, H. Robust Multivariable Microgrid Control Synthesis and Analysis. *Energy Procedia* **2016**, *100*, 375–387. [CrossRef]
18. Hua, H.; Qin, Y.; Xu, H.; Hao, C.; Cao, J. Robust Control Method for DC Microgrids and Energy Routers to Improve Voltage Stability in Energy Internet. *Energies* **2019**, *12*, 1622. [CrossRef]
19. Villalón, A.; Rivera, M.; Salgueiro, Y.; Muñoz, J.; Dragičević, T.; Blaabjerg, F. Predictive Control for Microgrid Applications: A Review Study. *Energies* **2020**, *13*, 2454. [CrossRef]
20. Zeb, K.; Islam, S.U.; Din, W.U.; Khan, I.; Ishfaq, M.; Busarello, T.D.C.; Ahmad, I.; Kim, H.J. Design of Fuzzy-PI and Fuzzy-Sliding Mode Controllers for Single-Phase Two-Stages Grid-Connected Transformerless Photovoltaic Inverter. *Electronics* **2019**, *8*, 520. [CrossRef]
21. Kamal, T.; Karabacak, M.; Perić, V.S.; Hassan, S.Z.; Fernández-Ramírez, L.M. Novel Improved Adaptive Neuro-Fuzzy Control of Inverter and Supervisory Energy Management System of a Microgrid. *Energies* **2020**, *13*, 4721. [CrossRef]
22. Song, L.; Huang, L.; Long, B.; Li, F. A Genetic-Algorithm-Based DC Current Minimization Scheme for Transformless Grid-Connected Photovoltaic Inverters. *Energies* **2020**, *13*, 746. [CrossRef]
23. Yoshida, Y.; Farzaneh, H. Optimal Design of a Stand-Alone Residential Hybrid Microgrid System for Enhancing Renewable Energy Deployment in Japan. *Energies* **2020**, *13*, 1737. [CrossRef]
24. Younesi, A.; Shayeghi, H.; Siano, P. Assessing the Use of Reinforcement Learning for Integrated Voltage/Frequency Control in AC Microgrids. *Energies* **2020**, *13*, 1250. [CrossRef]
25. Serra, F.M.; Fernández, L.M.; Montoya, O.D.; Gil-González, W.; Hernández, J.C. Nonlinear Voltage Control for Three-Phase DC-AC Converters in Hybrid Systems: An Application of the PI-PBC Method. *Electronics* **2020**, *9*, 847. [CrossRef]
26. Wu, B.; Zhou, X.; Ma, Y. Bus Voltage Control of DC Distribution Network Based on Sliding Mode Active Disturbance Rejection Control Strategy. *Energies* **2020**, *13*, 1358. [CrossRef]

27. Qian, J.; Li, K.; Wu, H.; Yang, J.; Li, X. Synergetic Control of Grid-Connected Photovoltaic Systems. *Int. J. Photoenergy* **2017**, *2107*, 1–11. [CrossRef]
28. Tepljakov, A. Fractional-Order Calculus Based Identification and Control of Linear Dynamic Systems. Master's Thesis, Department of Computer Control, Tallinn University of Technology, Tallinn, Estonia, 2011.
29. Tepljakov, A.; Petlenkov, E.; Belikov, J. FOMCON: Fractional-order modeling and control toolbox for MATLAB. In Proceedings of the 18th International Conference Mixed Design of Integrated Circuits and Systems—MIXDES, Gliwice, Poland, 16–18 June 2011; pp. 684–689.
30. Mehiri, A.; Bettayeb, M.; Hamid, A. Fractional Nonlinear Synergetic Control for Three Phase Inverter Tied to PV System. In Proceedings of the 8th International Conference on Modeling Simulation and Applied Optimization (ICMSAO), Manama, Bahrain, 15–17 April 2019; pp. 1–5.
31. de Brito, M.A.G.; Sampaio, L.P.; Luigi, G.; e Melo, G.A.; Canesin, C.A. Comparative analysis of MPPT techniques for PV applications. In Proceedings of the International Conference on Clean Electrical Power (ICCEP), Ischia, Italy, 14–16 June 2011; pp. 99–104.
32. Nicola, M.; Nicola, C.-I. Sensorless Fractional Order Control of PMSM Based on Synergetic and Sliding Mode Controllers. *Electronics* **2020**, *9*, 1494. [CrossRef]

Article

Variable-Gain Super-Twisting Sliding Mode Damping Control of Series-Compensated DFIG-Based Wind Power System for SSCI Mitigation

Ronglin Ma, Yaozhen Han * and Weigang Pan *

School of Information Science and Electrical Engineering, Shandong Jiaotong University, Jinan 250357, China; maronglin@sdjtu.edu.cn
* Correspondence: hanyz@sdjtu.edu.cn (Y.H.); 205036@sdjtu.edu.cn (W.P.); Tel.: +86-0531-8068-7920 (Y.H. & W.P.)

Abstract: Subsynchronous oscillation, caused by the interaction between the rotor side converter (RSC) control of the doubly fed induction generator (DFIG) and series-compensated transmission line, is an alleged subsynchronous control interaction (SSCI). SSCI can cause DFIGs to go offline and crowbar circuit breakdown, and then deteriorate power system stability. This paper proposes a novel adaptive super-twisting sliding mode SSCI mitigation method for series-compensated DFIG-based wind power systems. Rotor currents were constrained to track the reference values which are determined by maximum power point tracking (MPPT) and reactive power demand. Super-twisting control laws were designed to generate RSC control signals. True adaptive and non-overestimated control gains were conceived with the aid of barrier function, without need of upper bound of uncertainty derivatives. Stability proof of the studied closed-loop power system was demonstrated in detail with the help of the Lyapunov method. Time-domain simulation for 100 MW aggregated DFIG wind farm was executed on MATLAB/Simulink platform. Some comparative simulation results with conventional PI control, partial feedback linearization control, and first-order sliding mode were also obtained, which verify the validity, robustness, and superiority of the proposed control strategy.

Keywords: subsynchronous control interaction; super-twisting sliding mode; variable-gain; doubly fed induction generator

Citation: Ma, R.; Han, Y.; Pan, W. Variable-Gain Super-Twisting Sliding Mode Damping Control of Series-Compensated DFIG-Based Wind Power System for SSCI Mitigation. *Energies* **2021**, *14*, 382. https://doi.org/10.3390/en14020382

Received: 3 December 2020
Accepted: 8 January 2021
Published: 12 January 2021

Publisher's Note: MDPI stays neutral with regard to jurisdictional claims in published maps and institutional affiliations.

Copyright: © 2021 by the authors. Licensee MDPI, Basel, Switzerland. This article is an open access article distributed under the terms and conditions of the Creative Commons Attribution (CC BY) license (https://creativecommons.org/licenses/by/4.0/).

1. Introduction

In order to cope with energy shortage and environmental pollution, countries all over the world are intensively promoting renewable energy development [1]. Wind energy is considered as one of the most promising types of renewable energy. In wind power generation systems, doubly fed induction generators (DFIG) play a leading role due to their distinct advantages [2]. However, DFIG-based wind farms are always far away from the load center and need long-distance transmission, which can weaken power capacity and stability margin [3]. Series-compensated capacitors are generally applied in the DFIG-based wind farms transmission line to enhance the capacity and stability [4].

Series-compensated capacitors method can induce subsynchronous control interaction (SSCI), due to the interaction between DFIG's converter control and series-compensated transmission line [5,6]. In the SSCI, the frequency and attenuation rate are mainly codetermined by parameters of wind turbines and power transmission systems, irrelevant to natural modal frequency of shafting [7]. With no mechanical part involved, SSCI has a small damping effect, and also its divergence speed is faster than that of conventional subsynchronous resonance [8]. Thus, SSCI can cause more severe damage. From public reports, related accidents have been observed in America and China [5,9], causing equipment damage and loss of power generation.

In recent years, many efforts have been made on SSCI issues, e.g., frequency scan, eigenvalue analysis, complex torque coefficient method, and time domain simulation [10]. Based on these SSCI analysis methods, scholars tried to study SSCI damping strategies [11–17]. Papers [5,18–21] carried out many pioneering studies on modeling and supplementary damping control for subsynchronous resonance analysis. Paper [22] discussed multi-input multi-output supplementary damping control for both the rotor-side converter (RSC) and grid-side converter (GSC). In order to reduce the influence of PI control parameters on SSCI, paper [23] presented an optimization algorithm of PI parameters based on the t-distributed stochastic neighbor embedding for enhancing the damping. Paper [24] proposed a SSCI damping control for both the two control channels in the inner current loop of RSC with particle swarm optimized control coefficients. In short, the above studies [18–21,23,24] are all based on conventional double closed-loop PI control, and the design processes for SSCI damping controllers are relatively simple. Yet, these linear control methods can be inoperative when system operating points are changed, since a series-compensated DFIG-based wind power system is a complex and highly nonlinear system, with strong coupling features in both the aerodynamic and electrical parts [25,26]. These nonlinear factors can be dealt with by feedback linearization control. Paper [27] adopted a partial feedback linearization method to design damping controllers for GSC. Considering that RSC control is actually the dominant factor for SSCI mitigation, paper [28] continued the study of [27] to design SSCI damping controllers for RSC, achieving a good damping effect. Paper [29] proposed nonlinear controllers for both GSC and RSC based on the state feedback linearization method and verified its superior performance compared with conventional PI control.

Although feedback linearization control is an effective method for solving nonlinear problems in SSCI mitigation, it is rather sensitive to uncertainties in series-compensated DFIG-based wind power systems. These uncertainties exist in generator parameters, transmission line parameters, series compensation level, wind speed, and multiple series capacitor compensated lines, which can deteriorate subsynchronous oscillation of the system [30]. Hence, robustness is the desired characteristic for the series-compensated DFIG-based wind power control system. There are several attempts in SSCI robust control, such as H_∞ and active disturbance rejection methods [30–32].

The widely adopted sliding mode control [33,34], which possesses invariance property for system disturbances and parameter perturbation, is another good choice for robust control of SSCI. Papers [35,36] discussed SSCI mitigation strategies by combining feedback linearization control with sliding mode method. Paper [9] proposed first-order sliding mode controllers to track the reference rotor currents for damping SSCI. Control chattering of rotor voltage, which can damage electronic components and increase SSCI, is a big obstacle for these conventional sliding mode methods. Furthermore, the upper bounds of system uncertainties derivatives, which are actually hard to calculate beforehand, have to be known in advance for all the above robust control methods.

Consequently, this paper proposes a novel variable-gain super-twisting damping control strategy for SSCI mitigation. It can greatly reduce sliding mode chattering and does not need the unknown upper bounds of uncertainty derivatives. The SSCI mechanism was firstly analyzed with the aid of the presented series-compensated DFIG-based wind farm model. Rotor current dynamics constraint was identified as the dominant factor for SSCI mitigation. Super-twisting control laws were then constructed to track the prescribed rotor currents under dq direction. Adaptive control laws were subsequently conceived via barrier function. Then, the control gains can be self-adjusted following the upper bounds of uncertainty derivatives. SSCI mitigation was achieved without conservative RSC control signals. The performance of the newly designed variable-gain super-twisting sliding mode (VGSTSM) damping control scheme was evaluated under different wind speed, series compensation level, and short circuit fault. Comparative studies with conventional PI method, feedback linearization control, and first-order sliding mode control were also completed to verify the superiority.

This paper is organized as follows. Modeling for series-compensated DFIG-based wind farm is stated in Section 2. Section 3 details SSCI mechanism, design procedure of the proposed control strategy, and stability proof of the closed-loop power system. The demonstration of effectiveness and superiority of the proposed VGSTSM damping control strategy for SSCI mitigation is shown in Section 4. Some conclusions are finally drawn in Section 5.

2. Series-Compensated DFIG-Based Wind Power System Modeling

The model of a series-compensated DFIG-based wind farm is shown in Figure 1 [18,36]. It mainly includes wind turbine, shafting, induction generator, RSC, DC bus link, GSC, and series-compensated transmission line. DFIG represents the 100 MW equivalent lumped model of 50 generators (2 MW for each unit). R_s and L_s are stator resistance and inductance, respectively. R_{RSC} and L_{RSC} are RSC link resistance and inductance, respectively. R_{GSC} and L_{GSC} are GSC link resistance and inductance, respectively. R_L and L_L are equivalent resistance and inductance of series-compensated transmission line, respectively. C_{dc} is DC bus capacitor, and C_{SC} is the series-compensated capacitor. e is the grid voltage. System equations are all analyzed under the synchronous rotating reference frame.

Power fluctuation and subsynchronous rotor current will be induced when subsynchronous disturbance current occurs in the series-compensated transmission line. The affected RSC generates corresponding output voltage, and then injects subsynchronous current into the rotor, and finally induces the superposition of the stator side and the original disturbance. It will increase the original disturbance and form a divergent subsynchronous oscillation if the amplitude of the superimposed current is larger than that of the original disturbance current.

Electrical dynamic of series-compensated DFIG-based wind farms can be deduced via Kirchhoff's laws under a synchronous rotating reference frame.

$$
\begin{cases}
\dfrac{di_{rd}}{dt} = \omega_1 i_{rq} - \dfrac{R'_r}{L'_r} i_{rd} - \dfrac{u_{rd}}{L'_r} + \dfrac{u_{dc}}{L'_r} S_d \\[2mm]
\dfrac{di_{rq}}{dt} = -\omega_1 i_{rd} - \dfrac{R'_r}{L'_r} i_{rq} - \dfrac{u_{rq}}{L'_r} + \dfrac{u_{dc}}{L'_r} S_q \\[2mm]
\dfrac{du_{dc}}{dt} = \dfrac{1}{C_{dc}} i_{dc} - \dfrac{1}{C_{dc}} i_{rd} S_d - \dfrac{1}{C_{dc}} i_{rq} S_q \\[2mm]
\dfrac{di_{sd}}{dt} = \omega_1 i_{sq} + \dfrac{R'_s}{L'_s} i_{sd} + \dfrac{u_{sd}}{L'_s} \\[2mm]
\dfrac{di_{sq}}{dt} = -\omega_1 i_{sd} + \dfrac{R'_s}{L'_s} i_{sq} + \dfrac{u_{sq}}{L'_s} \\[2mm]
\dfrac{di_{gd}}{dt} = -\omega_1 i_{gq} - \dfrac{R_{GSC}}{L_{GSC}} i_{gd} - \dfrac{u_{gd}}{L_{GSC}} \\[2mm]
\dfrac{di_{gq}}{dt} = \omega_1 i_{gd} - \dfrac{R_{GSC}}{L_{GSC}} i_{gq} - \dfrac{u_{gq}}{L_{GSC}} \\[2mm]
\dfrac{di_{Ld}}{dt} = \omega_1 i_{Lq} + \dfrac{R_L}{L_L} i_{Ld} + \dfrac{1}{L_L}(u_{ld} - u_{scd} - E_d) \\[2mm]
\dfrac{di_{Lq}}{dt} = -\omega_1 i_{Ld} + \dfrac{R_L}{L_L} i_{Lq} + \dfrac{1}{L_L}\left(u_{lq} - u_{scq} - E_q\right) \\[2mm]
\dfrac{du_{scd}}{dt} = \omega_1 u_{scq} + \dfrac{1}{C_{SC}} i_{Ld} \\[2mm]
\dfrac{du_{scq}}{dt} = -\omega_1 u_{scd} + \dfrac{1}{C_{SC}} i_{Lq}
\end{cases}
\tag{1}
$$

where u_g, i_g, u_s, i_s, u_r, i_r, u_{dc}, and i_{dc} are the voltages and currents of GRC, stator, rotor, and DC bus capacitor, respectively. u_{SC} is defined as series-compensated voltage. $R'_r = R_r + R_{RSC}$, $R'_s = R_s + R_L$, $L'_s = L_s + L_L - 1/\omega_1^2 C_{SC}$, and $L'_r = L_r + L_{RSC} - L_m^2/L'_s$.

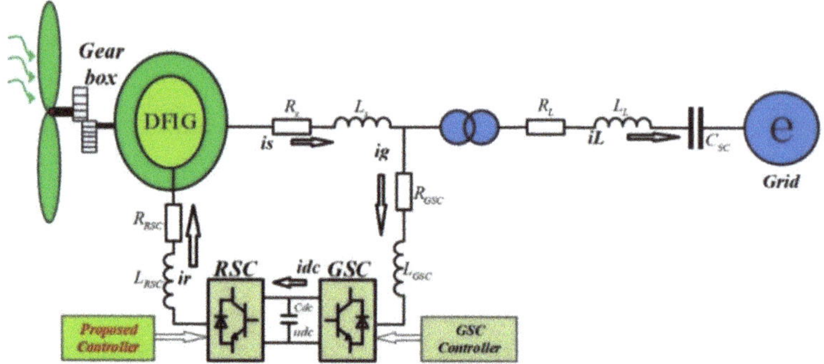

Figure 1. Structure of series-compensated doubly fed induction generator (DFIG)-based wind farm connected to grid.

After the applied stator field-orient, the stator active and reactive power for the DFIG model described in the *dq* reference frame can be represented as:

$$\begin{cases} P_s = -\frac{3L_m}{2L'_s}U_s i_{rq} \\ Q_s = \frac{3U_s}{2L'_s\omega_1}(U_s - \omega_1 L_m i_{rd}) \end{cases} \tag{2}$$

According to Betz theory, mechanical power captured by wind turbine is denoted as:

$$P_T = \tfrac{1}{2}C_p S_w \rho_T v_T^3 \tag{3}$$

where C_p is power coefficient, S_w is the blade sweep area, ρ_T is air density, and v_T is wind speed. As is shown in Equation (3), the mechanical power, P_T, is determined by power coefficient, C_p, under the fixed wind speed. C_p is related to tip speed ratio, λ_T, and blade pitch angle, β_T, with the typical functional relation [27,36]:

$$C_p = 0.5176\left(\frac{116}{\lambda_i} - 0.4\beta_T - 5\right)^{\frac{-21}{\lambda_i}} + 0.0068\lambda_T \tag{4}$$

Two related equations are $\frac{1}{\lambda_i} = \frac{1}{\lambda_T + 0.08\beta_T} - \frac{0.035}{\beta_T^3 + 1}$ and $\lambda_T = \frac{\omega_T R_T}{v_T}$, where ω_T is mechanical angular speed and R_T is the rotor radius of the wind turbine. With the change of λ_T and fixed β_T, power coefficient, C_p, has a maximum value, C_{pmax}, and the corresponding λ_T is optimum tip speed ratio, λ_{Topt}. In other words, for a specific wind speed, the wind turbine can only run under specific mechanical angular speed, ω_T, to achieve maximum power point tracking (MPPT). Thus, generator rotor speed must be regulated timely with the change of wind speed to capture maximum power.

The mechanical drive system of the wind turbine transmits the captured kinetic energy to the generator via the gear box, high speed shaft, and low speed shaft. It is rather complicated, and the mechanical shaft dynamic can be modeled as one mass, two mass, and three mass, according to different modeling methods [18]. The two mass model is sufficient and widely praised in SSCI studies, and its dynamic is represented as:

$$\begin{cases} \frac{d\omega_T}{dt} = \frac{1}{2H_T}(T_T - K_s\theta_s) \\ \frac{d\omega_r}{dt} = \frac{1}{2H_G}(K_s\theta_s - T_e) \\ \frac{d\theta_s}{dt} = 2\pi f_1\left(\omega_T - \frac{\omega_r}{N_g}\right) \end{cases} \tag{5}$$

where H_T and H_G are inertia time constants of wind turbine and generator, respectively; K_s is stiffness coefficient of shafting; θ_s is relative angular displacement of the two mass block; T_T and T_e denote mechanical torque and electromagnetic torque of the wind turbine and generator, respectively; ω_r is rotor angular speed of the generator; N_g represents gear ratio; and f_1 is power frequency.

3. SSCI Analysis and Control Design

3.1. SSCI Mechanism

For conventional double closed-loop PI control of RSC under stator-flux oriented method, the increments of rotor voltage and current under dq reference frame are [22,28]:

$$\begin{cases} \Delta u_{rd} = R_r \Delta i_{rd} - k_1(\omega_s - \omega_r)\Delta i_{rq} + k_1 p \Delta i_{rd} \\ \Delta u_{rq} = R_r \Delta i_{rq} - k_1(\omega_s - \omega_r)\Delta i_{rd} + k_1 p \Delta i_{rq} \\ \Delta i_{rd} = \frac{1}{k_2}\Delta i_{sd} \\ \Delta i_{rq} = \frac{1}{k_2}\Delta i_{sq} \end{cases} \tag{6}$$

where $k_1 = L_r - L_m^2/L_s$ and $k_2 = -L_m/L_s$, and p is the differential operator.

The terminal voltage of DFIG is supposed to be a three-phase symmetric fundamental sinusoidal wave, and phase voltage is expressed as:

$$u_{sa} = \sqrt{2}U_s \sin(\omega_s t + \varphi_{u0}) \tag{7}$$

where φ_{u0} is initial phase of fundamental voltage.

When current disturbance (with resonance angular frequency, ω_n) appears in the fixed series-compensated transmission line, a phase current of DFIG can be expressed as:

$$i_{sa} = \sqrt{2}I_s \sin(\omega_s t + \varphi_{i0}) + \sqrt{2}I_n \sin(\omega_n t + \varphi_{in}) = i_{sa0} + i_{sa\_sub} \tag{8}$$

where I_s and φ_{i0} are effective value and initial phase of fundamental current, i_{sa0}, respectively. I_n, ω_n, and φ_{in} are effective value, angular frequency, and initial phase of subsynchronous current, i_{sa_sub}, respectively.

Under dq reference frame, subsynchronous voltage and current can be expressed as:

$$\begin{cases} u_{sd} = 0 \\ u_{sq} = -\sqrt{3}U_s \end{cases} \tag{9}$$

$$\begin{cases} i_{sd} = -\sqrt{3}I_s \sin(\varphi_{u0} - \varphi_{i0}) - \sqrt{3}I_n \sin[(\omega_s - \omega_n)t + \varphi_i] = i_{sd0} + i_{sd\_sub} \\ i_{sq} = -\sqrt{3}I_s \cos(\varphi_{u0} - \varphi_{i0}) - \sqrt{3}I_n \cos[(\omega_s - \omega_n)t + \varphi_i] = i_{sq0} + i_{sq\_sub} \end{cases} \tag{10}$$

where $\varphi_i = \varphi_{u0} - \varphi_{in}$, i_{sd0}, and i_{sq0} are direct current components of stator current under dq frame, and i_{sd_sub} and i_{sq_sub} are subsynchronous components with frequency $\omega_s - \omega_n$.

It is supposed that fundamental power can be accurately tracked, and variation of instantaneous active and reactive power only contains subsynchronous components.

$$\begin{cases} \Delta p_s = 3U_s I_n \cos[(\omega_s - \omega_n)t + \varphi_i] = -\sqrt{3}U_s i_{sq\_sub} \\ \Delta q_s = 3U_s I_n \sin[(\omega_s - \omega_n)t + \varphi_i] = -\sqrt{3}U_s i_{sd\_sub} \end{cases} \tag{11}$$

As is shown in Formula (11), subsynchronous current with angular frequency, ω_n, can induce power fluctuation with angular frequency, $\omega_s - \omega_n$. Then, Δp_s and Δq_s can enter into the inner current control loop and turn into reference values of the rotor current. Meanwhile, a rotating magnetic field is formed via cutting rotor winding by subsynchronous current of the stator side, then three phase subsynchronous current with angular frequency, $\omega_r - \omega_n$, is induced in rotor winding, which can cause rotor voltage disturbances. These disturbances react upon rotor winding and impose subsynchronous current with angular frequency, $\omega_s - \omega_n$, which eventually cause a new subsynchronous current. Once this new subsynchronous current is added to original current disturbance, $\sqrt{2}I_n \sin(\omega_n t + \varphi_{in})$, the current disturbance will be gradually increased. The DFIG controller and series-compensated transmission line interacts and stimulates each other, which causes diverging oscillation of active and reactive power.

3.2. Control Design

As analyzed above, SSCI can be well suppressed once the rotor current dynamic is constrained by following the prescribed values. The reference values of rotor current can be deduced from (2), where active power, P_s^*, is acquired by maximum power point tracking (MPPT) and reactive power, Q_s^*, is calculated according to grid demand.

$$\begin{cases} i_{rq}^* = -\frac{2L'_s P_s^*}{3L_m U_s} \\ i_{rd}^* = \frac{U_s}{\omega_1 L_m} - \frac{2L'_s Q_s^*}{3L_m U_s} \end{cases} \tag{12}$$

SSCI is mainly caused by the interaction between the RSC control and series-compensated transmission line. Thus, RSC control signals are chosen as control variables. According to (3) and (4), the equation of series-compensated wind power systems can be represented as:

$$\begin{cases} \dot{x} = f(x) + g(x)u \\ y = h(x) \end{cases} \tag{13}$$

$$f(x) = \begin{bmatrix} \omega_1 i_{rq} - \frac{R'_r}{L'_r} i_{rd} \\ -\omega_1 i_{rd} - \frac{R'_r}{L'_r} i_{rq} \\ \frac{1}{C} i_{dc} \\ \omega_1 i_{sq} + \frac{R'_s}{L'_s} i_{sd} + \frac{u_{sd}}{L'_s} \\ -\omega_1 i_{sd} + \frac{R'_s}{L'_s} i_{sq} + \frac{u_{sq}}{L'_s} \\ -\omega_1 i_{gq} - \frac{R_{GSC}}{L_{GSC}} i_{gd} - \frac{u_{gd}}{L_{GSC}} \\ \omega_1 i_{gd} - \frac{R_{GSC}}{L_{GSC}} i_{gq} - \frac{u_{gq}}{L_{GSC}} \\ \omega_1 i_{Lq} + \frac{R_L}{L_L} i_{Ld} + \frac{1}{L_L}(u_{ld} - u_{scd} - E_d) \\ -\omega_1 i_{Ld} + \frac{R_L}{L_L} i_{Lq} + \frac{1}{L_L}\left(u_{lq} - u_{scq} - E_q\right) \\ \omega_1 u_{scq} + \frac{1}{C_{SC}} i_{Ld} \\ -\omega_1 u_{scd} + \frac{1}{C_{SC}} i_{Lq} \\ \frac{1}{2H_T}(T_T - K_s \theta_s) \\ \frac{1}{2H_G}(K_s \theta_s - T_e) \\ 2\pi f_1\left(\omega_T - \frac{\omega_r}{N_g}\right) \end{bmatrix} \tag{14}$$

$$g(x) = \begin{bmatrix} \frac{u_{dc}}{2L'_r} & 0 \\ 0 & \frac{u_{dc}}{2L'_r} \\ -\frac{1}{C_{dc}} i_{rd} & -\frac{1}{C_{dc}} i_{rq} \\ 0 & 0 \\ 0 & 0 \\ 0 & 0 \\ 0 & 0 \\ 0 & 0 \\ 0 & 0 \\ 0 & 0 \\ 0 & 0 \\ 0 & 0 \\ 0 & 0 \\ 0 & 0 \end{bmatrix} \tag{15}$$

where state vector is $x = [i_{rd}\ i_{rq}\ u_{dc}\ i_{sd}\ i_{sq}\ i_{gd}\ i_{gq}\ i_{Ld}\ i_{Lq}\ u_{scd}\ u_{scq}\ \omega_T\ \omega_r\ \theta_s]^T$, control variables are $u = [S_{rd}\ S_{rq}]^T$, and output equations are $y = [i_{rd} - i_{rd}^*\ i_{rq} - i_{rq}^*]^T$.

To choose sliding mode function:

$$\sigma_{rd} = (i_{rd} - i_{rd}^*) + c_1 \int (i_{rd} - i_{rd}^*)dt \tag{16}$$

$$\sigma_{rq} = \left(i_{rq} - i_{rq}^*\right) + c_2 \int \left(i_{rq} - i_{rq}^*\right)dt \tag{17}$$

where positive constants, c_1 and c_2, are weight coefficients of integral sliding mode items. This can help to remove steady state error. To calculate first-order derivatives of σ_{rd} and σ_{rq}:

$$\dot{\sigma}_{rd} = \omega_1 i_{rq} - \frac{R'_r}{L'_r}i_{rd} - \dot{i}_{rd}^* + c_1(i_{rd} - i_{rd}^*) + \frac{u_{dc}}{2L'_r}S_{rd} \tag{18}$$

$$\dot{\sigma}_{rq} = -\omega_1 i_{rd} - \frac{R'_r}{L'_r}i_{rq} - \dot{i}_{rq}^* + c_2\left(i_{rq} - i_{rq}^*\right) + \frac{u_{dc}}{2L'_r}S_{rq} \tag{19}$$

Observed from (18) and (19), the relative degrees with respect to σ_{rd} and σ_{rq} are both 1. They are less than the system order, which is 14. System dynamics can be divided into external dynamics and internal dynamics, according to zero dynamics stability theory. External dynamics are normally demanded to be stable and have good dynamic quality, while internal dynamics can only satisfy asymptotic stability. This paper will not go into details about asymptotic stability of internal dynamics for series-compensated DFIG wind power systems, which has been stated in papers [28,29]. Next, the design procedure for VGSTSM control law will be presented in detail. Here, i_{rq} control design is taken as an example because the design procedure is similar for i_{rd}.

Considering parameter perturbation, measuring error, and external disturbance, the lumped uncertainty is represented by Δd_q. Then, formula (19) was rewritten as:

$$\dot{\sigma}_{rq} = -\omega_1 i_{rd} - \frac{R'_r}{L'_r}i_{rq} - \dot{i}_{rq}^* + c_2\left(i_{rq} - i_{rq}^*\right) + \frac{u_{dc}}{2L'_r}S_{rq} + \Delta d_q \tag{20}$$

Taking state feedback control into account, this gave:

$$S_{rq} = \frac{2L'_r}{u_{dc}}\left(\omega_1 i_{rd} + \frac{R'_r}{L'_r}i_{rq} + \dot{i}_{rq}^* - c_2\left(i_{rq} - i_{rq}^*\right) + v_{rq}\right) \tag{21}$$

Then:

$$\dot{\sigma}_{rq} = v_{rq} + \Delta d_q \tag{22}$$

The next step was to design auxiliary control law, v_{rq}, for (22). RSC control chattering can be rather serious if conventional first-order sliding mode method is adopted. Thus, super-twisting algorithm with continuous control effect and small chattering was employed to construct v_{rq}:

$$\begin{cases} v_{rq} = -\alpha_q \gamma_q |\sigma_{rq}|^{1/2} sign(\sigma_{rq}) + v_{rq2} \\ \dot{v}_{rq2} = -\beta_q \gamma_q^2 sign(\sigma_{rq}) \end{cases} \tag{23}$$

The upper bound D_{qup} of $\Delta \dot{d}_q$ was demanded to be known in this control law. If control parameters, α_q and β_q, were chosen as 1.5 and 1.1, and γ_q was set as D_{qup}, then finite time stability and second-order sliding mode with respect to σ_{rq} can be established [37]. However, this upper bound D_{qup} is hard to acquire in series-compensated DFIG-based wind power systems. In case the value for D_{qup} is conservative, RSC will produce excessive control effect, increase unnecessary chattering, and damage electromechanical devices. Therefore, the super-twisting control gains should be constructed as adaptive ones. Control gains can increase or decrease according to upper bound of uncertainty derivatives. This adaptive strategy does not only satisfy control requirement, but also restrains chattering, and the superiority of super-twisting algorithm is fully developed.

Considering the characteristic of barrier function [37], adaptive gain super-twisting sliding mode control law is designed as:

$$\begin{cases} v_{rq} = -1.5\sqrt{\gamma_q}\,|\sigma_{rq}|^{1/2}sign(\sigma_{rq}) + v_{rq2} \\ \dot{v}_{rq2} = -1.1\gamma_q sign(\sigma_{rq}) \end{cases} \tag{24}$$

Adaptive control gain is constructed as:

$$\begin{cases} \dot{\gamma}_q = \gamma_{q0}, & if\ 0 < t \le t_{rs} \\ \gamma_q = \dfrac{b_q \varepsilon_q}{\varepsilon_q - |\sigma_{rq}|}, & if\ t_{rs} < t \end{cases} \tag{25}$$

where t_{rs} is the time that $|\sigma_{rq}|$ reaches $\varepsilon_q/2$. γ_{q0} and b_q, ε_q are positive constants. Then, for any $\varepsilon_q > 0$, $t_{rs} > 0$ for any initial status, $\sigma_{rq}(0)$. When $t \ge t_{rs}$, then $|\sigma_{rq}| < \varepsilon_q$ is satisfied. It was indicated that i_{rq} can converge to the error range of its reference value in finite time and achieve actual tracking for i_{rq}^*.

The proof for the above conclusion is followed below. Firstly, let us prove that $|\sigma_{rq}|$ can reach $\varepsilon_q/2$ in finite time, t_{rs}. It is supposed that $\left|\sigma_{rq}(0)\right| > \frac{\varepsilon_q}{2}$ is satisfied, then adaptive control gain is determined by $\dot{\gamma}_q = \gamma_{q0}$, according to (25).

Consider the following variable transformation:

$$\begin{cases} z_{q1} = \dfrac{\sigma_q}{\gamma_q^2} \\ z_{q2} = \dfrac{\dot{\sigma}_q}{\gamma_q^2} \end{cases} \tag{26}$$

The derivatives for z_{q1} and z_{q2} can be denoted as:

$$\begin{cases} \dot{z}_{q1} = -\alpha_q |z_{q1}|_{1/2} sign(z_{q1}) + z_{q2} - \dfrac{2\dot{\gamma}_q}{\gamma_q} z_{q1} \\ \dot{z}_{q2} = -\beta_q sign(z_{q1}) - \dfrac{\Delta \dot{d}_q}{\gamma_q^2} - \dfrac{2\dot{\gamma}_q}{\gamma_q} z_{q2} \end{cases} \tag{27}$$

Choose Lyapunov function:

$$V_{q1} = \chi_q^T(t) P_q \chi_q(t) \tag{28}$$

where P_q is constant symmetric positive definite matrix, $\chi_q^T(t) = \begin{bmatrix} |z_{q1}|_{1/2} sign(z_{q1}) & z_{q2} \end{bmatrix}$. Then, time derivative of $\chi_q(t)$ can be deduced as:

$$\dot{\chi}_q(t) = \dfrac{1}{2|z_{q1}|^{1/2}} K_q \chi_q + \dfrac{\dot{\gamma}_q}{\gamma_q} \Lambda_q \chi_q - \dfrac{M_q}{\gamma_q^2} \tag{29}$$

where $K_q = \begin{bmatrix} -\alpha_q/2 & 1/2 \\ -\beta_q & 0 \end{bmatrix}$, $\Lambda_q = \begin{bmatrix} -1/2 & 0 \\ 0 & -1 \end{bmatrix}$, and $M_q = \begin{bmatrix} 0 \\ \Delta \dot{d}_q \end{bmatrix}$.

Time derivative of V_{q1} is:

$$\dot{V}_{q1} = -\dfrac{1}{2|z_{q1}|^{1/2}} \chi_q^T Q_q \chi_q - \dfrac{\dot{\gamma}_q}{\gamma_q} \chi_q^T R_q \chi_q - \dfrac{2\Delta \dot{d}_q}{\gamma_q^2} P_q \chi_q \tag{30}$$

where $K_q^T P_q + P_q K_p = -Q_p$ and $\Lambda_q^T P_q + P_q \Lambda_q = -R_q$. Then, thanks to the studies in [12], symmetric positive definite matrix, P_q, exits, and then Q_p are R_q are positive definite. Then:

$$\dot{V}_{q1} \leq -k_{q1} V_{q1}^{\frac{1}{2}} + 2k_{q3} \frac{D_{qup}}{\gamma_q^2} V_{q1}^{\frac{1}{2}} - \frac{\dot{\gamma}_q}{\gamma_q} k_{q2} V_{q1} \tag{31}$$

where $k_{q1} = \frac{\lambda_{\min}(Q_q)}{2\sqrt{p_{11}}\lambda_{\max}(P_q)}$, $k_{q2} = \frac{\lambda_{\min}(R_q)}{\lambda_{\max}(P_q)}$, and $k_{q3} = \frac{\lambda_{\max}(R_q)}{\lambda_{\min}(P_q)^{\frac{1}{2}}}$. $\lambda_{\min}$ and $\lambda_{\max}$ are minimum eigenvalue and maximum eigenvalue of the relative matrix, respectively. p_{11} is the first element of matrix P_q.

The first item of the right side in (31) is negative, while the second item is positive. Here, the second item will decrease following increasement of adaptive control gain. Adaptive control gain becomes big enough to conquer uncertainties. Thus, the second item becomes very small. The third item will be negative and further reduced when $\dot{\gamma}_q$ is negative.

As discussed, the first item will be bigger than the second one, and the third item will become smaller. Then, the right side of (31) will be negative and $\dot{V}_{q1} \leq -a_q V_{q1}^{1/2}$ satisfied, which means finite time stability is achieved. V_{q1} will continue to decrease and then $|\sigma_{rq}|$ can reach $\varepsilon_q/2$.

It was proved above that $|\sigma_{rq}|$ can reach $\varepsilon_q/2$ when the time is $t = t_{rs}$. The second step is to prove that $|\sigma_{rq}| \leq \varepsilon_q$ can be satisfied after $t \geq t_{rs}$.

Choose Lyapunov function:

$$V_{q2} = \frac{1}{2}\sigma_{rq}^2 \tag{32}$$

Then:

$$\dot{V}_{q2} = \sigma_{rq}\dot{\sigma}_{rq} = \sigma_{rq}\left(-\alpha_q \gamma_q |\sigma_{rq}|^{1/2} \text{sign}(\sigma_{rq}) + \sigma_{rq2}\right) \tag{33}$$

where $\sigma_{rq2} = v_{rq2} + \Delta d_q$. Then:

$$\dot{V}_{q2} \leq |\sigma_{rq}|\left(-\alpha_q \gamma_q |\sigma_{rq}|^{1/2} \text{sign}(\sigma_{rq}) + |\sigma_{rq2}|\right) \tag{34}$$

According to the barrier function, $\gamma_q = \frac{b_q \varepsilon_q}{\varepsilon_q - |\sigma_{rq}|}$, of (25):

$$\begin{aligned}
\dot{V}_{q2} &\leq \frac{|\sigma_{rq}|}{\varepsilon_q - |\sigma_{rq}|}\left(\alpha_q \varepsilon_q b_q |\sigma_{rq}|^{1/2} - |\sigma_{rq2}|\varepsilon_q + |\sigma_{rq}||\sigma_{rq2}|\right) \\
&= -\frac{|\sigma_{rq}||\sigma_{rq2}|}{\varepsilon_q - |\sigma_{rq}|}\left(\frac{\alpha_q \varepsilon_q b_q |\sigma_{rq}|^{1/2}}{|\sigma_{rq2}|} - \varepsilon_q + |\sigma_{rq}|\right)
\end{aligned} \tag{35}$$

Take note of the right side of (35), define:

$$F_q = \frac{\alpha_q \varepsilon_q b_q |\sigma_{rq}|^{1/2}}{|\sigma_{rq2}|} - \varepsilon_q + |\sigma_{rq}| \tag{36}$$

$F_q = 0$ is a quadratic equation, and the two roots are:

$$|e_{11}|^{1/2} = \frac{1}{2}\left(\frac{-\alpha_q \varepsilon_q b_q}{|\sigma_{rq2}|} + \left(\left(\frac{\alpha_q \varepsilon_q b_q}{|\sigma_{rq2}|}\right)^2 + 4\varepsilon_q\right)^{1/2}\right) \tag{37}$$

$$|e_{12}|^{1/2} = \frac{1}{2}\left(\frac{-\alpha_q \varepsilon_q b_q}{|\sigma_{rq2}|} - \left(\left(\frac{\alpha_q \varepsilon_q b_q}{|\sigma_{rq2}|}\right)^2 + 4\varepsilon_q\right)^{1/2}\right) \tag{38}$$

It can be easily observed that the second root is negative, and then only the second root needed to be paid more attention. According to (37):

$$e_{11} = \pm \frac{1}{4} \left(\frac{-\alpha_q \varepsilon_q b_q}{|\sigma_{rq2}|} + \left(\left(\frac{\alpha_q \varepsilon_q b_q}{|\sigma_{rq2}|} \right)^2 + 4\varepsilon_q \right)^{1/2} \right)^2 \tag{39}$$

According to the well-known inequation, $a^2 + b^2 \leq (a+b)^2$:

$$\left(\frac{\alpha_q \varepsilon_q b_q}{|\sigma_{rq2}|} \right)^2 + \left(2\varepsilon_q^{1/2} \right)^2 \leq \left(\frac{\alpha_q \varepsilon_q b_q}{|\sigma_{rq2}|} + 2\varepsilon_q^{1/2} \right)^2 \tag{40}$$

With the aid of (40), the upper bound of $|e_{11}|$ can be written as:

$$|e_{11}| \leq \left(\frac{\left(\frac{-\alpha_q \varepsilon_q b_q}{|\sigma_{rq2}|} + \left(\left(\frac{\alpha_q \varepsilon_q b_q}{|\sigma_{rq2}|} \right)^2 + 2\varepsilon_q^{1/2} \right)^2 \right)^{\frac{1}{2}}}{2} \right)^2 = \left(\frac{\frac{-\alpha_q \varepsilon_q b_q}{|\sigma_{rq2}|} + \left(\frac{\alpha_q \varepsilon_q b_q}{|\sigma_{rq2}|} \right) + 2\varepsilon_q^{1/2}}{2} \right)^2 = \left(\frac{2\varepsilon_q^{1/2}}{2} \right)^2 = \varepsilon_q \tag{41}$$

Finally, the inequation from (41) can be deduced as:

$$|e_{11}| \leq \left(\frac{2\varepsilon_q^{1/2}}{2} \right)^2 \leq \varepsilon_q \tag{42}$$

If $|\sigma_q(t)| \geq |e_{11}|$, then F_q is positive definite. Consequently, $\dot{V}_{q2} < 0$ is satisfied for $|e_{11}| \leq |\sigma_q(t)| < \varepsilon_q$. Hence, $\sigma_q(t)$ will always satisfy $|\sigma_q(t)| < |e_{11}|$ the rest of the time, and $|e_{11}|$ is smaller than ε_q for any derivative of Δd_q.

Therefore, real sliding mode, with respect to $\sigma_q(t)$, is established in finite time. The q-axis rotor current, i_{rq}, allows us to track for the prescribed i_{rq}^* with unknown upper bound of uncertainty derivative.

Adaptive gain control law for σ_{rd} can be designed in a similar way. State feedback control is:

$$S_{rd} = \frac{2L'_r}{u_{dc}} \left(-\omega_1 i_{rq} + \frac{R'_r}{L'_r} i_{rd} + \dot{i}_{rd}^* - c_1 (i_{rd} - i_{rd}^*) + v_{rd} \right) \tag{43}$$

Sliding mode control law and adaptive control gain are:

$$\begin{cases} v_{rd} = -1.5\sqrt{\gamma_d} |\sigma_{rd}|^{1/2} sign(\sigma_{rd}) + v_{rd2} \\ \dot{v}_{rd2} = -1.1\gamma_d sign(\sigma_{rd}) \end{cases} \tag{44}$$

$$\begin{cases} \dot{\gamma}_d = \gamma_{d0}, & if \ 0 < t \leq t_{rs1} \\ \gamma_d = \frac{b_d \varepsilon_d}{\varepsilon_d - |\sigma_{rd}|}, & if \ t_{rs1} < t \end{cases} \tag{45}$$

i_{rd} can converge to the demanded neighborhood in finite time. As mentioned above, the internal dynamics of the system are asymptotically stable, and the external dynamics are finite time stable. Thus, the stability of the whole control system is guaranteed.

4. Time-Domain Simulation

Time-domain simulation is one of the best measures for dynamic stability analysis of a power system. Nonlinear mathematical models can be employed in time-domain simulations, which is very suitable for the nonlinear and complex characteristics of the DFIG power system. The 100 MW aggregated model was adopted to verify effectiveness. Superiority of the proposed control strategy was also compared with PI [19], feedback linearization [27], and conventional sliding mode methods [9] under MATLAB/Simulink. Simulation parameters for series-compensated DFIG-based wind power systems is referred to Table 1. The schematic diagram of the proposed VGSTSM damping control scheme is depicted as Figure 2. Firstly, the sliding mode functions were calculated, then the auxiliary control quantities were obtained according to the adaptive law and super-twisting sliding mode control laws, and finally the RSC control signals were obtained through the feedback control. Control parameters were chosen as $\varepsilon_q = 0.001$, $\gamma_q = 2.2$, $b_q = 2.0$, $\varepsilon_d = 0.001$, $\gamma_d = 2.5$, and $b_d = 2.3$.

Table 1. Series-compensated DFIG-based wind power system parameters.

Quantity	Value
Nominal power	100 Mw
Rater voltage	690 V
R_s	0.0084 pu
L_s	0.167 pu
H_T	2.5 pu
H_G	0.5 pu
K_s	0.15 pu
R_{RSC}	0.0083 pu
L_{RSC}	0.1323 pu
R_{GSC}	0.0015 pu
L_{GSC}	0.151 pu
DC-lin capacitance	10 mF
Nominal DC-link voltage	1150 V
R_L	0.02 pu
L_L	0.0016 pu
C_{SC} (at 45% compensation)	42.61 uF

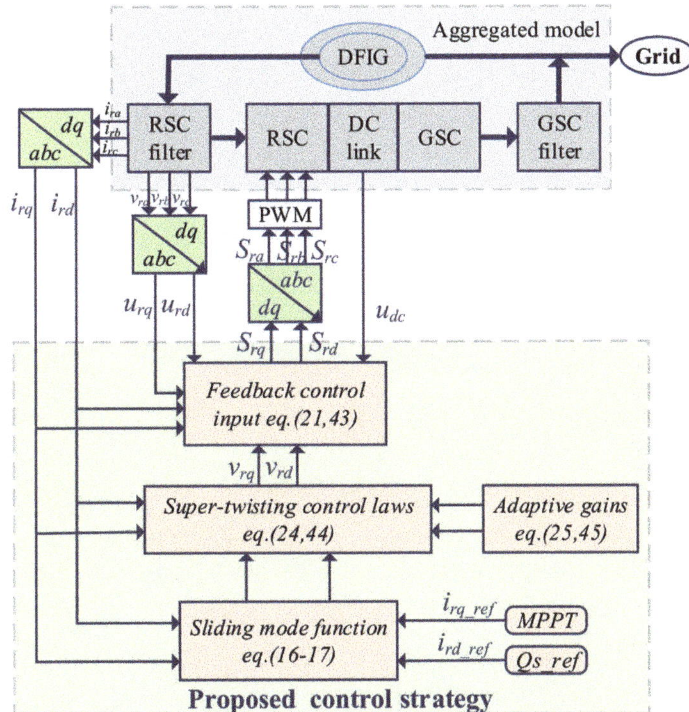

Figure 2. Schematic diagram of the proposed damping control scheme.

The wind speed was set as 9 m/s. Series-compensated capacitor was injected into the DFIG-based wind power transmission line at 2.5 s, forming 40% series-compensated level. Transient responses of active power, reactive power, electromagnetic torque, rotor angular speed, DC bus voltage, and transmission line current are shown in Figures 3 and 4. As observed, all of these variables start oscillation when this switch capacitor is put into the system, they can then be rapidly stabilized under the proposed control strategy. When the series-compensated level is increased to 85%, the variables can still converge to steady state, as shown in Figures 5 and 6, though the oscillation time somewhat increased. Figures 3–6 indicate that the proposed strategy was effective for SSCI mitigation under different series-compensated level.

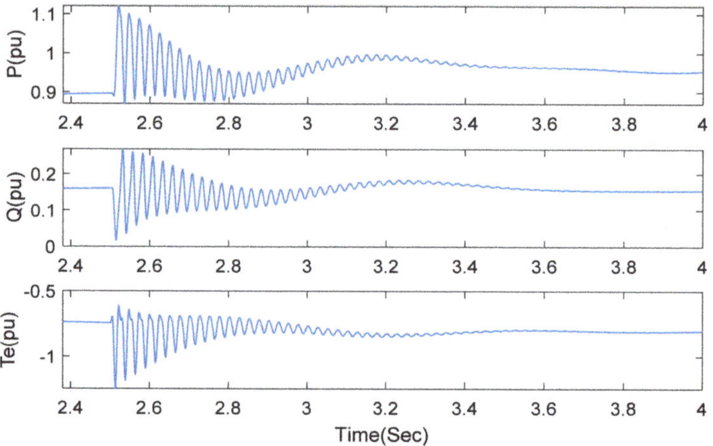

Figure 3. Transient responses of active power, reactive power, and electromagnetic torque after 40% series compensation is switched.

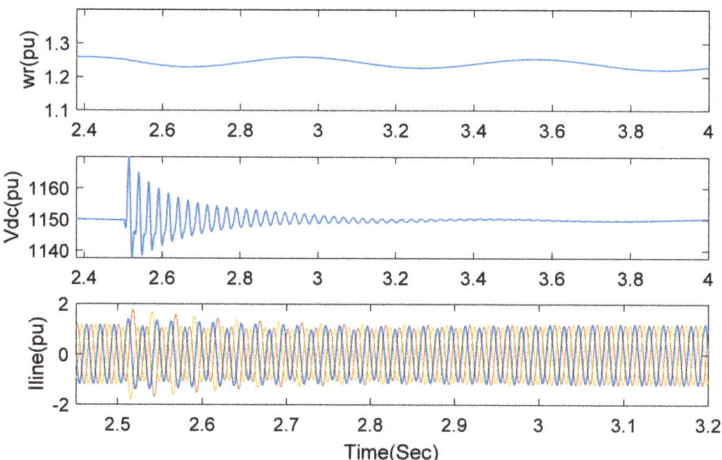

Figure 4. Transient responses of rotor angular speed, DC bus voltage, and transmission line current after 40% series compensation is switched.

Wind speed increased to 11 m/s under 85% compensation to evaluate controller performances for different wind speeds. The responses of these variables are demonstrated in Figure 7. By comparing Figure 7 with Figures 5 and 6, it was observed that SSCI mitigation was better when wind speed was higher.

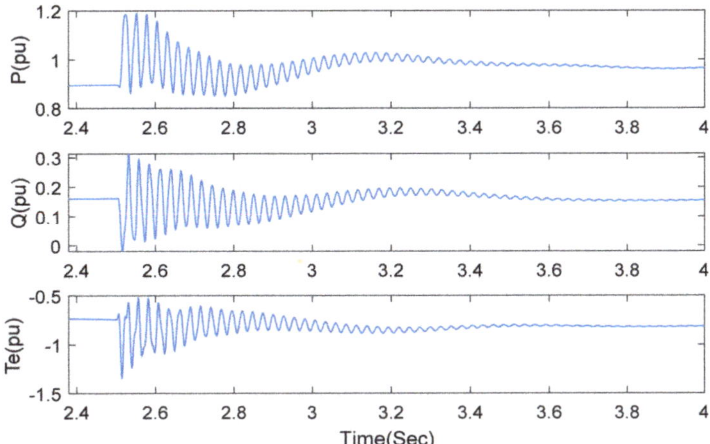

Figure 5. Transient responses of active power, reactive power, and electromagnetic torque after 85% series compensation is switched.

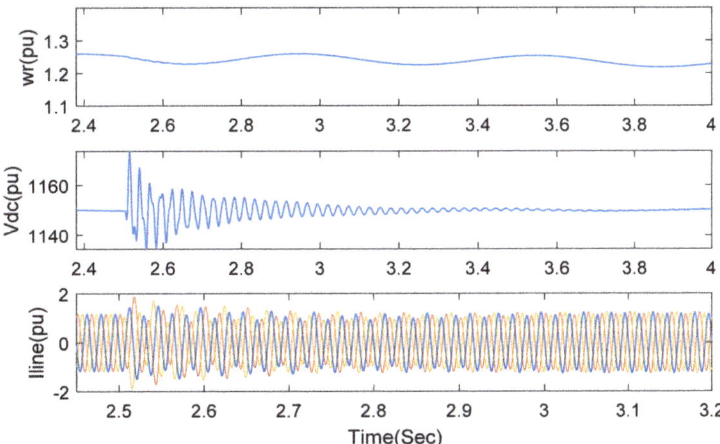

Figure 6. Transient responses of rotor angular speed, DC bus voltage, and transmission line current after 85% series compensation is switched.

After the system entered steady state, three-phase short circuit fault occurred at the high voltage side of transformer at t = 5 s to verify capacity for fault ride-through of the proposed control method. The duration of the fault was 20 ms. As shown in Figure 8, dynamic responses of active power, reactive power, and electromagnetic torque can all return to normal after a short transient fluctuation. This indicates that SSCI can be quickly suppressed under three-phase short circuit fault and the capacity for fault ride-through was enhanced under the proposed control method.

The performance for SSCI mitigation was compared to that of other control means based on PI control, partial feedback linearization, and first-order sliding mode. Figure 9 is the control structure of the classical double closed-loop PI scheme. The symbol * means reference value. Control parameters for PI controllers are $K_p = 0.1$, $K_Q = 0.83$, $K_{iq} = 1.2$, $K_{id} = 5$, $T_p = 0.05$, $T_{iq} = 0.005$, $T_Q = 0.025$, and $T_{id} = 0.0025$. Figure 10 shows active and reactive power responses under PI (Proportional Integral) controller [19] and the proposed method, when wind speed is 7 m/s and capacitance compensation is 60%. Growing oscillations are observed under PI control, while the effect for SSCI mitigation is good under the proposed method.

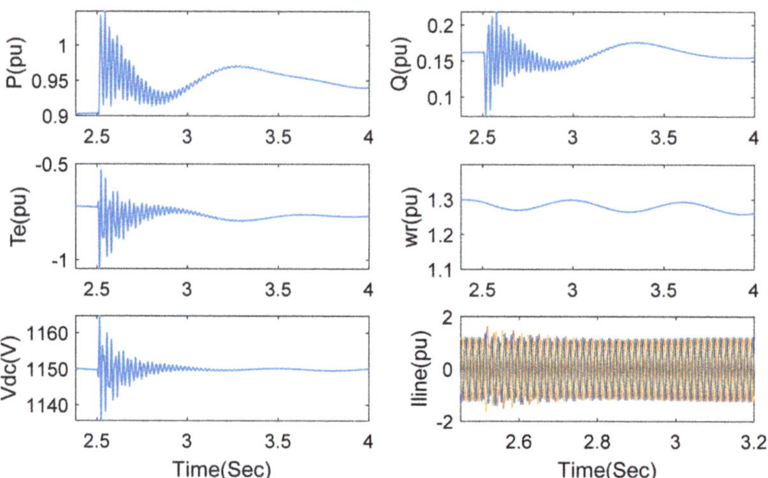

Figure 7. Transient responses under wind speed of 11 m/s and series-compensated level of 85%.

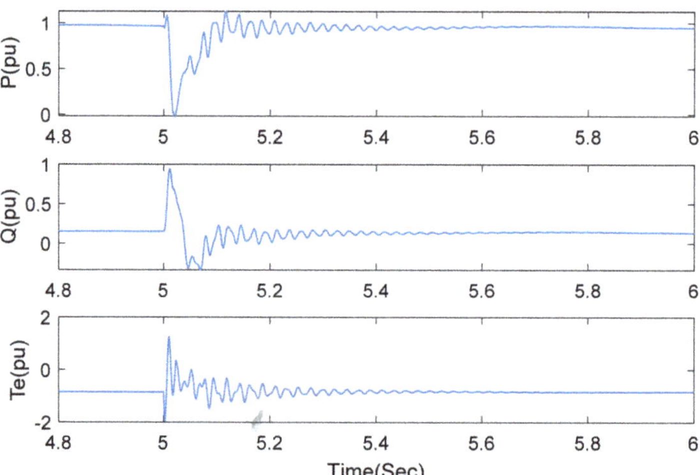

Figure 8. Dynamic responses of active power, reactive power, and electromagnetic torque under three-phase short circuit fault.

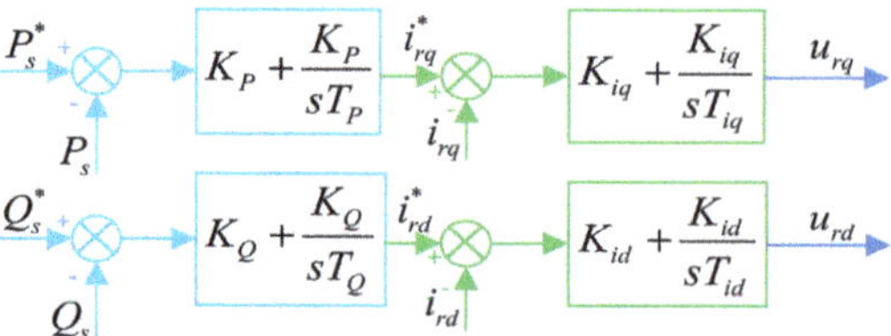

Figure 9. Control structure of PI scheme.

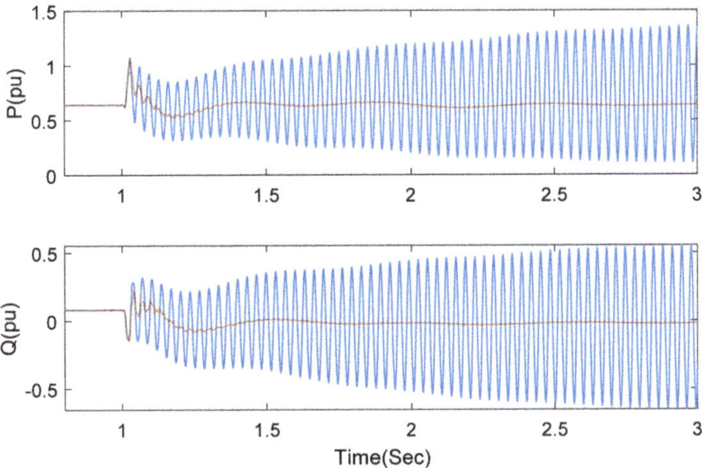

Figure 10. Active and reactive responses under PI (blue) and the proposed method (red), with wind speed of 7 m/s and series-compensated level of 60%.

To compare the control performance under the proposed method and partial feedback linearization method [27], the implementation block diagram of the damping controller based on the partial feedback linearization method is shown in Figure 11. The relative control laws are represented as:

$$\begin{cases} S_q = \dfrac{L_{rf}}{u_{dc}}\left(v_1 + \omega_1 i_{rd} + \dfrac{R_{rf}}{L_{rf}} i_{rq} + \dfrac{v_{rq}}{L_{rf}}\right) \\ S_d = \dfrac{L_{rf}}{u_{dc}}\left(v_2 - \omega_1 i_{rq} + \dfrac{R_{rf}}{L_{rf}} i_{rd} + \dfrac{v_{rd}}{L_{rf}}\right) \end{cases} \tag{46}$$

$$\begin{cases} v_1 = k_{1p}\left(i_{rq\_ref} - i_{rq}\right) + k_{1i}\int_0^t \left(i_{rq\_ref} - i_{rq}\right)dt \\ v_2 = k_{2p}\left(i_{rd\_ref} - i_{rd}\right) + k_{2i}\int_0^t \left(i_{rd\_ref} - i_{rd}\right)dt \end{cases} \tag{47}$$

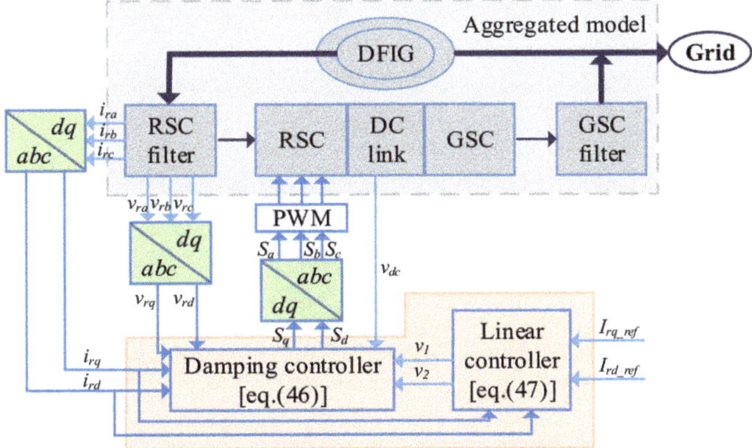

Figure 11. Implementation block diagram of the damping controller using partial feedback linearization method.

Figure 12 shows active power responses at 6 m/s wind speed and 60% compensation, and Figure 13 show the responses when parameter perturbation is considered. The variation ranges of L_m, L_s, L_{RSC}, and R_{RSC} are $\pm 50\%$ of the nominal values with combination of sine and cosine functions. The curves barely changed under the proposed method while it seems to be greatly affected under the partial feedback linearization method. This verified robustness to parameter perturbation of the proposed method.

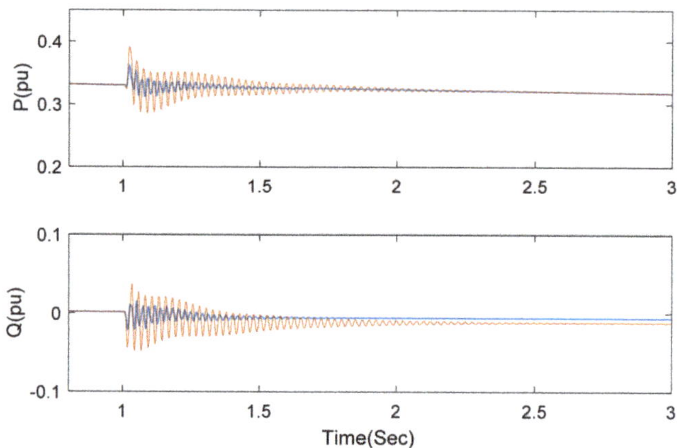

Figure 12. Active responses under the proposed method (blue) and partial feedback linearity (red) with wind speed of 6 m/s and compensation of 60%.

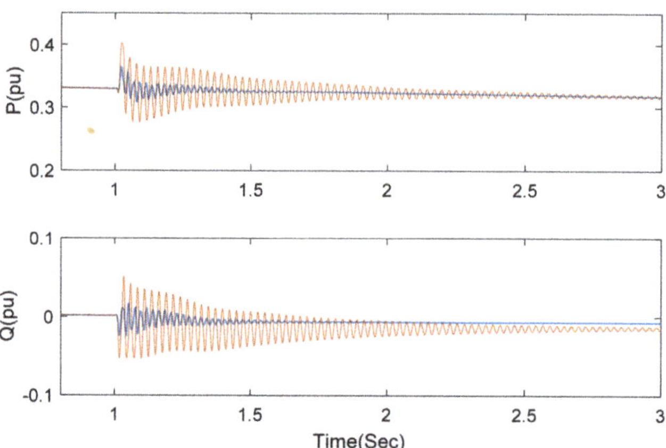

Figure 13. Active responses under the proposed method (blue) and partial feedback linearity (red) at 6 m/s wind speed and 60% compensation with parameter perturbation.

Conventional first-order sliding mode control (SMC) damping scheme [9] is shown as Figure 14. The control laws are:

$$
\begin{aligned}
S_{rq} = {} & \frac{2}{L_s u_{dc}}\left(-i_{rd}\omega_1(L_m^2 - L_{rr}L_s) - L_m(i_{sq}R_s - u_{sq} + L_s i_{sd}\omega_r) + L_s i_{rq}R_{rr} - L_{rr}L_s i_{rd}\omega_r \right. \\
& \left. -\dot{i}_{rq}^{*}(L_m^2 - L_{rr}L_s) - \rho_{i_{rq}}sign(\sigma_{i_{rq}})(L_{rr}L_s - L_m^2)
\end{aligned}
\tag{48}
$$

$$
\begin{aligned}
S_{rd} = {} & \frac{2}{L_s u_{dc}}\left(-i_{rq}\omega_1(L_m^2 - L_{rr}L_s) + L_m(-i_{sd}R_s + u_{sd} + L_s i_{sq}\omega_r) + L_s i_{rd}R_{rr} + L_{rr}L_s i_{rq}\omega_r \right. \\
& \left. -\dot{i}_{rd}^{*}(L_m^2 - L_{rr}L_s) - \rho_{i_{rd}}sign(\sigma_{i_{rd}})(L_{rr}L_s - L_m^2)
\end{aligned}
\tag{49}
$$

where $\rho_{i_{rq}} = 8.5 \times 10^5$ and $\rho_{i_{rd}} = 2.3 \times 10^6$.

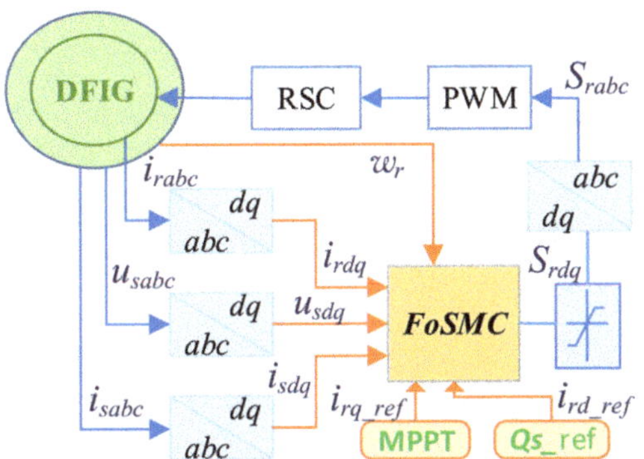

Figure 14. Conventional first-order SMC scheme.

The transient responses of active power, electromagnetic torque, and DC link voltage under both the proposed method and first-order sliding mode method [9] are shown as Figure 15 when wind speed is 10 m/s and compensation is 60%. The oscillation processes both quickly disappeared after about 0.5 s, and SSCI mitigation was achieved. However, under the proposed method, the control chattering of RSC control input was greatly restrained, and the upper bounds of uncertainty derivatives were not needed. Figure 16 shows the regulating process of adaptive gains of super-twisting control.

For evaluating the control performance, two indices have been defined as follows:

$$RMS_{e_i} = \sqrt{\frac{1}{n_s}\sum_{k=1}^{n_s} e_i^2}, \ RMS_{u_i} = \sqrt{\frac{1}{n_s}\sum_{k=1}^{n_s} u_i^2} \tag{50}$$

where n_s, e_i, and u_i are the number of samples, root mean square (RMS) of tracking errors, and control quantities, respectively.

When the studied system works with series-compensated level of 60% and wind speed of 10 m/s, the RMS for tracking errors and control quantities are shown in Table 2, which exhibit superiority of the proposed method.

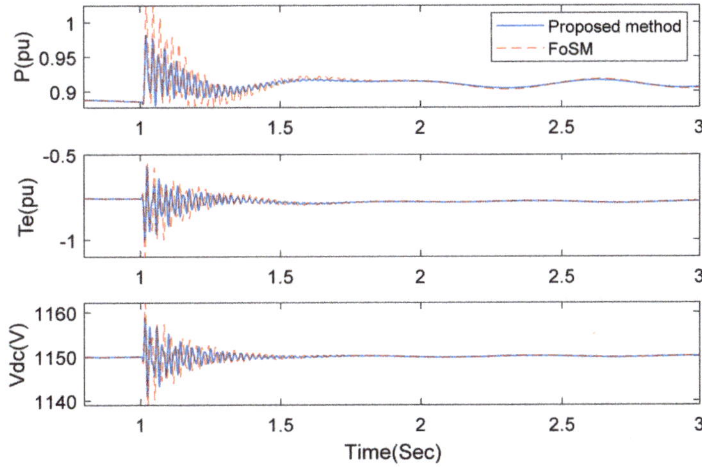

Figure 15. The transient responses of active power, electromagnetic torque, and DC link voltage under both the proposed method and first-order sliding mode method.

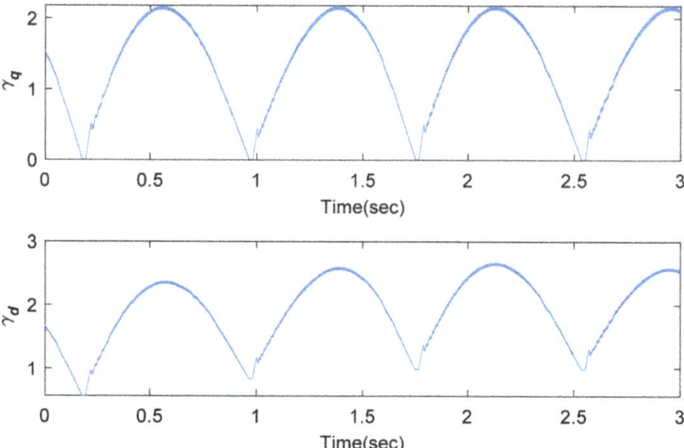

Figure 16. Adaptive gains.

Table 2. Root mean square (RMS) for tracking errors and control quantities.

	RMSu$_{rq}$	**RMSu$_{rd}$**	**RMSe$_{irq}$**	**RMSe$_{ird}$**
PI	0.9061	0.8906	0.0591	0.0698
SMC	0.8037	0.7091	0.0506	0.0593
VGSTSM	0.6071	0.4921	0.0365	0.0361

5. Conclusions

SSCI can severely influence stability of series-compensated DFIG-based wind power systems and damage electrical devices. This study proposes a novel VGSTSM damping control strategy for SSCI mitigation. After analyzing the series-compensated model and SSCI mechanism, it is assumed that SSCI were mainly caused by the interaction between current dynamics in the RSC side and the transmission line. Rotor current dynamics are desired to track the prescribed values from MPPT and grid demand. State feedback was firstly carried out in the control law design, and then super-twisting control algorithm was designed. Adaptive control gain was conceived via barrier function. Stability for the series-compensated system was proved, based on the Lyapunov function. The proposed method can achieve chattering suppression of RSC control signals. More importantly, control gains can indeed be adjusted according to uncertainty variation. It did not require the upper bound of uncertainty derivative in advance. Contrastive control simulations were verified to show superiority. Future works will focus on experimental realization.

Author Contributions: Conceptualization, R.M. and Y.H.; Funding acquisition, Y.H. and W.P.; Methodology, R.M.; Supervision, R.M.; Writing—original draft, R.M.; Writing—review & editing, Y.H. and W.P. All authors have read and agreed to the published version of the manuscript.

Funding: This research was funded by National Natural Science Foundation of China under Grant 61803230,61773015; A Project of Shandong Province Higher Educational Science and Technology Program under Grant J18KA330; University Outstanding Youth Innovation Team Development Plan of Shandong Province under Grant 2019KJN023; and Key R & D project of Shandong Province under Grant 2019GSF109076.

Institutional Review Board Statement: Not applicable.

Informed Consent Statemen: Not applicable.

Data Availability Statement: Data sharing not applicable.

Conflicts of Interest: The authors declare no conflict of interest.

References

1. Mahalakshmi, R.; Thampatty, K.C.S. Design and implementation of modified RSC controller for the extenuation of sub-synchronous resonance oscillations in series compensated DFIG-based WECS. *Int. Trans. Electr. Energy Syst.* **2020**, *30*, e12396. [CrossRef]
2. Han, Y.; Ma, R. Adaptive-Gain second-order sliding mode direct power control for wind-turbine-driven dfig under balanced and unbalanced grid voltage. *Energies* **2019**, *12*, 3886. [CrossRef]
3. Suppioni, V.P.; Grilo, A.P.; Teixeira, J.C. Coordinated control for the series grid side converter-based DFIG at subsynchronous operation. *Electr. Power Syst. Res.* **2019**, *173*, 18–28. [CrossRef]
4. Huang, P.H.; El Moursi, M.S.; Xiao, W.; Kirtley, J.L. Subsynchronous resonance mitigation for series-compensated DFIG-based wind farm by using two-degree-of-freedom control strategy. *IEEE Trans. Power Syst.* **2014**, *30*, 1442–1454. [CrossRef]
5. Fan, L.; Zhu, C.; Miao, Z.; Hu, M. Modal analysis of a DFIG-based wind farm interfaced with a series compensated network. *IEEE Trans. Energy Convers.* **2011**, *26*, 1010–1020. [CrossRef]
6. Jiao, Y.; Li, F.; Dai, H.; Nian, H. Analysis and mitigation of sub-synchronous resonance for doubly fed induction generator under VSG control. *Energies* **2020**, *13*, 1582. [CrossRef]
7. Revel, G.; Alonso, D.M. Subsynchronous interactions in power networks with multiple DFIG-based wind farms. *Electr. Power Syst. Res.* **2018**, *165*, 179–190. [CrossRef]
8. Yang, J.-W.; Sun, X.-F.; Chen, F.-H.; Lao, K.; He, Z.-Y. Subsynchronous resonance suppression strategy for doubly fed induction generators based on phase-shift average of rotor current. *Int. Trans. Electr. Energy Syst.* **2019**, *29*, e2831. [CrossRef]
9. Karunanayake, C.; Ravishankar, J.; Dong, Z.Y. Nonlinear SSR damping controller for DFIG Based wind generators interfaced to series compensated transmission systems. *IEEE Trans. Power Syst.* **2020**, *35*, 1156–1165. [CrossRef]
10. Ghaffarzadeh, H.; Mehrizi-Sani, A. Mitigation of subsynchronous resonance induced by a type III wind system. *IEEE Trans. Sustain. Energy* **2020**, *11*, 1717–1727. [CrossRef]
11. Shair, J.; Xie, X.; Yan, G. Mitigating subsynchronous control interaction in wind power systems, Existing techniques and open challenges. *Renew. Sustain. Energy Rev.* **2019**, *108*, 330–346. [CrossRef]
12. Velpula, S.; Thirumalaivasan, R.; Janaki, M. A Review on subsynchronous resonance and its mitigation techniques in DFIG based wind farms. *Int. J. Renew. Energy Res.* **2018**, *8*, 2275–2288.
13. Wang, L.; Sun, X.; You, Y. DFIG wind farm modeling for subsynchronous control interaction Analysis. *IEEJ Trans. Electr. Electron. Eng.* **2018**, *13*, 253–261. [CrossRef]
14. Yousuf, V.; Ahmad, A. Unit template-based control design for alleviation and analysis of SSR in power system using STATCOM. *Electr. Power Compon. Syst.* **2019**, *47*, 1805–1813. [CrossRef]
15. Tung, D.D.; Dai, V.L.; Quyen, L.C. Subsynchronous resonance and facts-novel control strategy for its mitigation. *J. Eng.* **2019**, 1–14. [CrossRef]
16. Bhushan, R.; Chatterjee, K. Effects of parameter variation in DFIG-based grid connected system with a FACTS device for small-signal stability analysis. *IET Gener. Transm. Distrib.* **2017**, *11*, 2762–2777. [CrossRef]
17. Boopathi, V.P.; Devi, R.P.K.; Ramanujam, R.; Dasan, S.G.B. Investigation of subsynchronous oscillations in grid connected Type-2 wind farm and its mitigation using STATCOM. *J. Electr. Eng.* **2017**, *17*, 10.
18. Fan, L.; Kavasseri, R.; Miao, Z.L.; Zhu, C. Modeling of DFIG-based wind farms for SSR analysis. *IEEE Trans. Power Deliv.* **2010**, *25*, 2073–2082. [CrossRef]
19. Fan, L.; Miao, Z. Mitigating SSR using DFIG-based wind generation. *IEEE Trans. Sustain. Energy* **2012**, *3*, 349–358. [CrossRef]
20. Li, Y.; Fan, L.; Miao, Z. Replicating real-world wind farm SSR events. *IEEE Trans. Power Deliv.* **2020**, *35*, 339–348. [CrossRef]
21. Ghafouri, M.; Karaagac, U.; Karimi, H.; Jensen, S.; Mahseredjian, J.; Faried, S.O. An LQR controller for damping of subsynchronous interaction in DFIG-based wind farms. *IEEE Trans. Power Syst.* **2017**, *32*, 4934–4942. [CrossRef]
22. Leon, A.E.; Solsona, J.A. Sub-Synchronous interaction damping control for DFIG wind turbines. *IEEE Trans. Power Syst.* **2014**, *30*, 419–428. [CrossRef]
23. Chen, A.; Xie, D.; Zhang, D.; Gu, C.; Wang, K. PI parameter tuning of converters for sub-synchronous interactions existing in grid-connected DFIG wind turbines. *IEEE Trans. Power Electron.* **2019**, *34*, 6345–6355. [CrossRef]
24. Yao, J.; Wang, X.; Li, J.; Liu, R.; Zhang, H. Sub-Synchronous resonance damping control for series-compensated DFIG-based wind farm with improved particle swarm optimization algorithm. *IEEE Trans. Energy Convers.* **2019**, *34*, 849–859. [CrossRef]
25. Rahimi, M. Dynamic performance assessment of DFIG-based wind turbines: A review. *Renew. Sustain. Energy Rev.* **2014**, *37*, 852–866. [CrossRef]
26. Djoudi, A.; Bacha, S.; Iman-Eini, H.; Rekioua, T. Sliding mode control of DFIG powers in the case of unknown flux and rotor currents with reduced switching frequency. *Int. J. Electr. Power Energy Syst.* **2017**, *96*, 347–356. [CrossRef]
27. Chowdhury, M.A.; Mahmud, M.A.; Shen, W.; Pota, H.R. Nonlinear controller design for series-compensated DFIG-based wind farms to mitigate subsynchronous control interaction. *IEEE Trans. Energy Convers.* **2017**, *32*, 707–719. [CrossRef]
28. Chowdhury, M.A.; Shafiullah, G.M. SSR mitigation of series-compensated DFIG wind farms by a nonlinear damping controller using partial feedback linearization. *IEEE Trans. Power Syst.* **2018**, *33*, 2528–2538. [CrossRef]
29. Penghan, L.; Jie, W.; Fei, W.; Li, H. Nonlinear controller based on state feedback linearization for series-compensated DFIG-based wind power plants to mitigate sub-synchronous control interaction. *Int. Trans. Electr. Energy Syst.* **2019**, *29*, e2682. [CrossRef]
30. Wang, Y.; Wu, Q.; Yang, R.; Tao, G.; Liu, Z. H∞ current damping control of DFIG based wind farm for sub-synchronous control interaction mitigation. *Int. Trans. Electr. Energy Syst.* **2018**, *98*, 509–519. [CrossRef]

31. Ghafouri, M.; Karaagac, U.; Karimi, H.; Mahseredjian, J. Robust subsynchronous interaction damping controller for DFIG-based wind farms. *J. Mod. Power Syst. Clean Energy* **2019**, *7*, 1663–1674. [CrossRef]
32. Xu, Y.; Zhao, S. Mitigation of subsynchronous resonance in series-compensated DFIG wind farm using active disturbance rejection control. *IEEE Access* **2019**, *7*, 68812–68822. [CrossRef]
33. Liu, X.; Han, Y. Decentralized multi-machine power system excitation control using continuous higher-order sliding mode technique. *Int. J. Electr. Power Energy Syst.* **2016**, *82*, 76–86. [CrossRef]
34. Susperregui, A.; Herrero, J.M.; Martinez, M.I.; Tapia-Otaegui, G.; Blasco, X. Multi-Objective optimisation-based tuning of two second-order sliding-mode controller variants for dfigs connected to non-ideal grid voltage. *Energies* **2019**, *12*, 3782. [CrossRef]
35. Li, P.; Xiong, L.; Wu, F.; Ma, M.; Wang, J. Sliding mode controller based on feedback linearization for damping of sub-synchronous control interaction in DFIG-based wind power plants. *Int. J. Electr. Power Energy Syst.* **2019**, *107*, 239–250. [CrossRef]
36. Li, P.; Xiong, L.; Wang, Z.; Ma, M.; Wang, J. Fractional-Order sliding mode control for damping of subsynchronous control interaction in DFIG-based wind farms. *Wind Energy* **2020**, *23*, 749–762. [CrossRef]
37. Obeid, H.; Fridman, L.M.; Laghrouche, S.; Harmouche, M. Barrier function-based adaptive sliding mode control. *Automatica* **2018**, 540–544. [CrossRef]

Article

Real-Time Digital Twin of a Wound Rotor Induction Machine Based on Finite Element Method

Sami Bouzid *, Philippe Viarouge and Jérôme Cros

Department of Electrical and Computer Engineering, Laval University, Quebec, QC G1V 0A6, Canada;
philippe.viarouge@gel.ulaval.ca (P.V.); jerome.cros@gel.ulaval.ca (J.C.)
* Correspondence: sami.bouzid.1@outlook.com; Tel.: +1-8197639153

Received: 30 September 2020; Accepted: 12 October 2020 ; Published: 16 October 2020

Abstract: Monitoring and early fault prediction of large electrical machines is important to maintain a sustainable and safe power system. With the ever-increasing computational power of modern processors, real-time simulation based monitoring of electrical machines is becoming a topic of interest. This work describes the development of a real-time digital twin (RTDT) of a wound rotor induction machine (WRIM) using a precomputed finite element model fed with online measurements. It computes accurate outputs in real-time of electromagnetic quantities otherwise difficult to measure such as local magnetic flux, current in bars and torque. In addition, it considers space harmonics, magnetic imbalance and fault conditions. The development process of the RTDT is described thoroughly and outputs are compared in real-time to measurements taken from the actual machine in rotation. Results show that they are accurate with harmonic content respected.

Keywords: digital twin; doubly-fed induction generator, electrical machines; finite elements method; monitoring; real-time; wound rotor induction machine

---

1. Introduction

Real-time simulation based monitoring of physical systems through digital twins is getting more attention due to the ever-increasing computational power of modern processors [1,2]. The term "digital twin" holds many definitions, each of them being more or less different, but it is commonly known as a set of accurate models capable of representing a physical system throughout all of its life cycle, from design phase to operation [3,4]. Real-time monitoring with a digital twin fed with sensor data coming from the actual system has many benefits, such as fault detection, output optimization and downtime planning. The challenge lies in developing accurate multi-physical models able to interact between each others.

For electrical machines, the core model is the electromagnetic model, which computes electrical quantities such as currents, voltages and electromagnetic torque. As of now, dq models are used extensively for real-time simulation due to their relative simplicity and small computation time, mainly for hardware-in-the-loop testing purpose [5] and grid simulation [6]. A computationally efficient model is needed because real-time constraints impose that the simulation time step be larger than the computation time it takes in real-world clock [7]. However, dq models are based on numerous assumptions; phenomenons like space harmonics or magnetic imbalances are neglected. Furthermore, they only provide information about the quantities seen at the machine terminals.

That said, the only way to obtain a high order model capable of representing accurately any machine structure and computing local quantities as magnetic densities and current densities is by using the well-known finite element method (FEM). It computes locally the Maxwell's equation solution for any given machine geometry considering even local magnetic saturation. The downside of this technique is the enormous computation time it takes for magnetodynamic resolution which make it challenging for real-time implementation.

Several detailed models of electrical machines have been developed as an alternative to FEM in an effort to make compromises between accuracy and computational requirement for implementation on real-time simulators [8]. Among these, magnetic equivalent circuits (MEC) models and lumped circuit models using winding functions are the most prominent.

MEC [9] is a model based on a network of flux tubes capable of representing the unique geometry and winding distribution of a machine. In [10], an induction motor was implemented for real-time execution on a high clock speed FPGA with a time step of 500 μs. A time step of 150 μs is attained in [11] for a switched reluctance motor. While adaquate accuracy can be achieved at a fraction of the time required by FEM, the amount of calculations required in one time step remains high.

On another hand, lumped circuit models using winding function calculations [12] were also implemented for real-time. In [13], a synchronous machine model was implemented at 20 μs time step using a CPU-based real-time digital simulator. These models considers the geometry of the machine to a certain extent only, because simplifications are made to make analytical calculations manageable and efficient.

However, it was demonstrated in [14] that, by using a magnetically coupled circuit approach using precomputed inductance functions with a FEM software, it is possible to reach relatively small computation times while keeping the accuracy of FEM. This model is called CFE-CC—Combination of Finite Element with Coupled Circuits [15]. It has three major strong points:

1. It considers space harmonics and magnetic imbalances since any machine geometry can be modeled in a FEM software;
2. Any number of electrical circuits can be added to the models, including phase windings of course, but also bars, dampers and even search coils. This means the model can output actual currents flowing in bars and dampers and also local magnetic fluxes by using search coils.
3. It can compute distribution of power losses inside the machine [16], allowing calculations of local temperature elevation if used with a thermal model [17].

For these reasons, the CFE-CC model is very close to its physical counterpart. All it takes for building it is the FEM model, which is available from the design phase.

This work presents the methodology to implement for real-time execution the CFE-CC model fed with measurements taken from the real machine, thus allowing effective online monitoring of internal electromagnetic quantities. The proposed electromagnetic model can then be used alongside thermal and mechanical models to complement this real-time digital twin (RTDT). The machine used as case-study to realize the proposed RTDT is a wound rotor induction machine (WRIM), also known as doubly-fed induction machine. This particular type of machine was chosen because all of its electrical circuits are accessible for measurements, thus making it convenient for signal waveform validation. It is important to note however that the described method herein is applicable to every type of electrical machine. After a detailed explanation on the model's construction and implementation on digital real-time simulator (DRTS), results are validated by comparing in real-time the RTDT's outputs to the physical machine's measurements.

2. CFE-CC Model Construction of a Wound Rotor Induction Machine

2.1. Model's Equations

The model used for the purpose of this work does not include magnetic saturation. Iron losses as well as capacitive effects are also neglected.

A wound rotor induction machine, as well as many other types of electrical machines, comprises n electrical circuits where current can flow, such as the phases and bars. The total magnetic flux ϕ at a given circuit is the sum of the magnetic fluxes received from all other circuits and itself. It is worth noting that every quantity referenced in this manuscript is instantaneous and no referential transformation is used.

$$\phi_i = \sum_{j=1}^{n} \phi_{ij} \tag{1}$$

The magnetic flux received at a given circuit from another one depends on the current i flowing in the sender and the magnetic coupling L between the two. The coupling is a function of the rotor angular position θ.

$$\phi_{ij} = L_{ij}(\theta)i_j \tag{2}$$

All aforementioned equations are included into a single matrix equation:

$$[\phi] = [L(\theta)][i] \tag{3}$$

Flux variation seen by any circuit creates a voltage v given by:

$$[v] = [R][i] + \frac{d[\phi]}{dt} \tag{4}$$

By combining Equations (3) and (4), the dynamic system of Equation (5) is obtained.

$$\frac{d[\phi]}{dt} = -[R][L(\theta)]^{-1}[\phi] + [v] \tag{5}$$

As for electromagnetic torque T_{em}, it can be computed using:

$$T_{em} = 0.5[i]^T \frac{d[L(\theta)]}{d\theta}[i] \tag{6}$$

For search coils or any electrical circuit always left open, Equation (7) is used instead to compute back-emf v_w for better numerical stability of ordinary differential equations solvers. Matrix $[L_w]$ defines the coupling between the search coils and the other circuits carrying a current.

$$[v_w] = \frac{d[L_w][i]}{dt} \tag{7}$$

To sum up, identification of the model requires to know the inductance matrix function $[L(\theta)]$ which can then be inverted and differentiated. What makes the CFE-CC model versatile is the identification process of $[L(\theta)]$ with an FEM software. It uses a lookup table to store the inverse and derivative inductance matrix for many discrete positions of the rotor.

As in [14], it is possible to consider magnetic saturation with the precomputed inductance functions by using a correction coefficient afterward on computed magnetic flux linkages. Performances and computing time differ depending on type of machine and chosen quantities to compute the coefficient.

2.2. Identification of Electrical Circuits

Table 1 gathers specifications of the WRIM used in this work.

Table 1. Specifications of the WRIM.

Parameter	Value	Unit
Power rating	1864	W
Rated voltage	208	V
Number of phases	3	
Number of poles	4	
Number of stator slots	48	
Number of rotor slots	36	
Frequency	60	Hz
Speed	1690	RPM
Power factor	0.81	

As depicted in Figure 1b, there are three windings on the stator (a_s, b_s, c_s) and three windings on the rotor (a_r, b_r, c_r). A small search coil is also added around one stator tooth (w_s) in order to measure the voltage across it during operation. These seven circuits make up the CFE-CC model.

The resistance matrix $[R]$ of the WRIM is diagonal as shown below. Each terminal is accessible so resistance values can easily be measured. The search coil is not included because it is left as an open circuit, thus Equation (7) is used instead of (5) for this particular case.

$$[R] = \begin{bmatrix} R_{as} & 0 & 0 & 0 & 0 & 0 \\ 0 & R_{bs} & 0 & 0 & 0 & 0 \\ 0 & 0 & R_{cs} & 0 & 0 & 0 \\ 0 & 0 & 0 & R_{ar} & 0 & 0 \\ 0 & 0 & 0 & 0 & R_{br} & 0 \\ 0 & 0 & 0 & 0 & 0 & R_{cr} \end{bmatrix} \tag{8}$$

As for $[L]$, for any given rotor position, it has the form shown below. Magnetic couplings of the search coil are also included in a separate matrix $[L_w]$.

$$[L] = \begin{bmatrix} L_{asas} & L_{asbs} & L_{ascs} & L_{asar} & L_{asbr} & L_{ascr} \\ L_{bsas} & L_{bsbs} & L_{bscs} & L_{bsar} & L_{bsbr} & L_{bscr} \\ L_{csas} & L_{csbs} & L_{cscs} & L_{csar} & L_{csbr} & L_{cscr} \\ L_{aras} & L_{arbs} & L_{arcs} & L_{arar} & L_{arbr} & L_{arcr} \\ L_{bras} & L_{brbs} & L_{brcs} & L_{brar} & L_{brbr} & L_{brcr} \\ L_{cras} & L_{crbs} & L_{crcs} & L_{crar} & L_{crbr} & L_{crcr} \end{bmatrix} \tag{9}$$

$$[L_w] = \begin{bmatrix} L_{wsas} & L_{wsbs} & L_{wscs} & L_{wsar} & L_{wsbr} & L_{wscr} \end{bmatrix} \tag{10}$$

Figure 1. (**a**) Photo of the wound rotor induction machine (WRIM) showing its stator and rotor; (**b**) circuits of the WRIM used to build the model.

2.3. Computation of the Inductance Matrix Using FEM

Computation of $[L(\theta)]$ is performed with an FEM software and accurate plans of the machine geometry and windings. A 3D FEM software is an ideal choice as it consider geometries that are impossible to model in a 2D environment. However, a 2D FEM software is normally easier to use and less computationally expensive. It should be noted that computation time required to solve the finite element domain is not related to the CFE-CC model's computation time at each time step, because results of the FEM are used as precomputed lookup tables.

For the present work, the authors opted for a 2D software to quickly compute $[L(\theta)]$. The geometry of the WRIM makes it convenient to use a 2D space, but a few issues cannot be accounted in 2D alone such as rotor skewing and coil ends. Their respective effects are added to the model by altering the computed inductance matrix. Details of this procedure are presented at Appendix A. Figure 2 shows the FEM model of the WRIM in FLUX2D, the FEM software used for this work.

The domain can be reduced to only half of the complete machine because the machine has two pairs of poles and presents a symmetry. All magnetic materials are assumed to be linear, so saturation is not considered. Meshing is performed using the software's automatic mesh tool. Mesh size was selected by comparing the solution obtained using a fine mesh to more and more coarse ones, until the error starts increasing noticeably. Once the FEM model is ready, the procedure for identifying magnetic couplings is launched. Inductance matrix $[L]$ is computed for many rotor positions to build the lookup table.

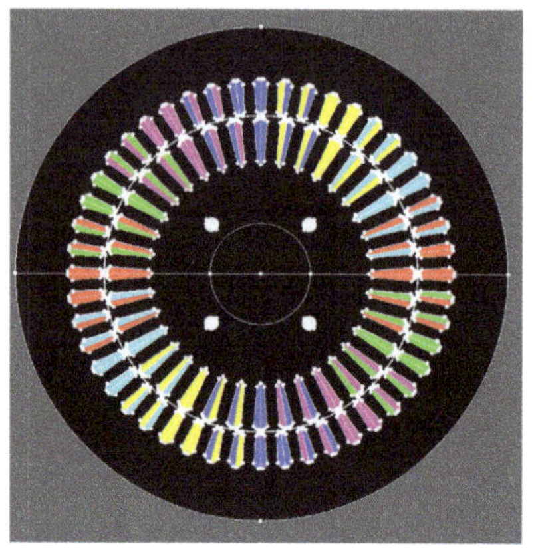

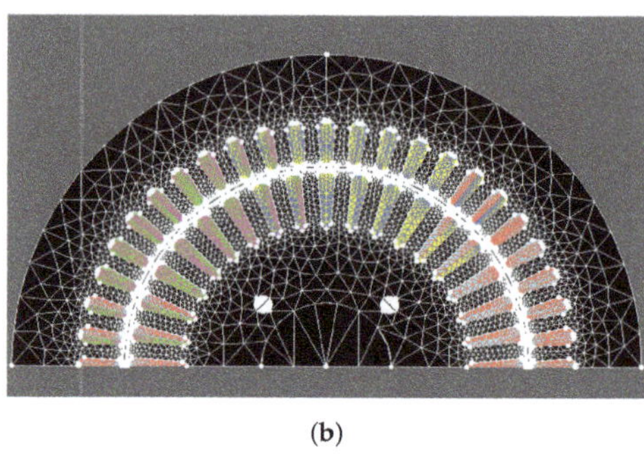

(b)

(a)

Figure 2. (a) WRIM rendered in FLUX2D (b) Meshed domain.

Inductance matrix $[L(\theta)]$ is obtained by first computing the inductance matrix of the machine without windings, i.e., the inductance matrix of the slots $[L_{slot}(\theta)]$. That way, one may generate any winding configuration without restarting the FEM process simply by applying the following matrix transformation:

$$[L(\theta)] = [N]^T [L_{slot}(\theta)][N] \tag{11}$$

$[N]$ is the winding configuration matrix. It contains the number of turns N of each electrical circuit in each slot. Equation (12) shows how to build the matrix, where n_c is the total number of circuits and n_s is the total number of slot.

$$N = \begin{bmatrix} N_{1,1} & N_{1,2} & \cdots & \cdots & N_{1,n_c} \\ N_{2,1} & N_{2,2} & & & \\ \vdots & & \ddots & & \\ \vdots & & & \ddots & \\ N_{n_s,1} & & & & N_{n_s,n_c} \end{bmatrix} \tag{12}$$

In total, 1440 positions were computed evenly for a 180° rotation, i.e., 60 positions per stator slot pitch or an angular step of 0.125°. It is plenty to have a good resolution. Impact of inductance discretization is described thoroughly in [18]. At this point, it is important to take into consideration the physical memory to store the lookup table. Since the WRIM has six circuits without the search coil, the total number of elements in the lookup table for the inverse inductance matrix is 51,840. Another lookup table of the same size is necessary for the derivative if torque computation is included.

3. Hardware Setup

3.1. Real-Time Simulator And RT-Lab

The processing unit of the digital twin is an OP4510 entry-level DRTS of OPAL-RT. Real-time computations are performed by a dedicated multi-core CPU capable of time step of 2 μs. For applications requiring smaller time step, one may rather opt for a higher-end DRTS or FPGA based platforms instead. Table 2 sums up the main technical specifications of the DRTS used in this work.

Table 2. Technical specifications of the OP4510 real-time digital simulator.

CPU	Intel Xeon E3 v5 CPU (4 core, 8 MB cache, 2.1 or 3.5 GHz)
Dynamic memory	16 GB RAM
Storage	128 GB SSD
Digital outputs	32 channels, 5 V to 30 V adjustable by a user-supplied voltage
Digital inputs	32 channels, 4 V to 50 V
Analog inputs	16 channels, 16 bits, ±20 V true differential
Analog outputs	16 channels, 16 bits, ±16 V

The OP4510 needs to be connected via an ethernet cable to a personal computer (PC) with RT-LAB installed. RT-LAB is a software environment for real-time simulation to interface with OPAL-RT's DRTS. Using a PC and RT-LAB, user can:

1. Modify execution state of the program (run, stop, pause);
2. Change parameters in real-time of the running model;
3. Receive data from the running model.

3.2. Voltage and Current Measurement

Measured voltages are the inputs of the model. Currents are also measured, but they are used for validation purposes to compare with the RTDT's outputs. A printed circuit board (PCB) was designed for voltage and current acquisition from the machine. The signals are then sampled using the DRTS's analog input channels.

3.3. Angular Position Sensor

The angular position of the rotor is needed to retrieve the inductance matrix from the lookup table. It is measured using an absolute shaft encoder. The measured position is encoded in a 12 bits signal sent through a parallel communication interface. Each bit is acquired by a digital input channel of the DRTS and binary to decimal conversion is performed in software.

3.4. Setup Overview

Figure 3 shows an overview of the hardware setup. The CFE-CC model is executed in real-time along with the actual machine in operation. All estimated waveforms can be displayed on an oscilloscope using the DRTS's analog output channels.

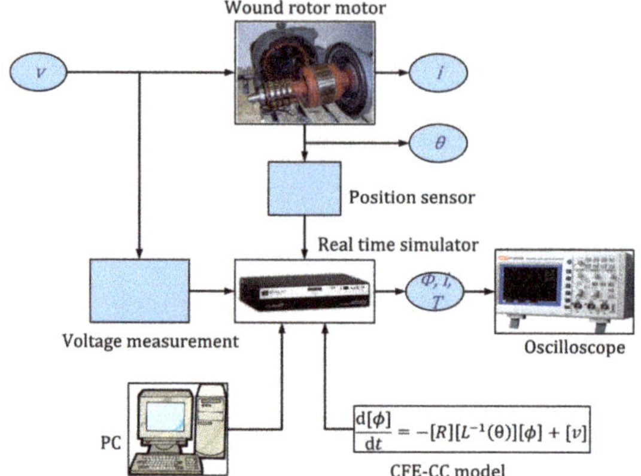

Figure 3. Hardware setup overview.

4. Implementation on Real-Time Simulator

4.1. Programming Method

Programming of the RTDT was realized entirely within the MATLAB/Simulink environment using RT-LAB. It generated C code from a Simulink block diagram and sent it to the DRTS for compilation and execution.

4.2. Position Tracking Scheme

The acquired position signal cannot be fed directly to the CFE-CC model after decimal conversion because it is a noisy discrete signal. The model would see speed impulsions that would impact its estimations. In order to filter the position signal, the control loop shown at Figure 4 was programmed in the DRTS.

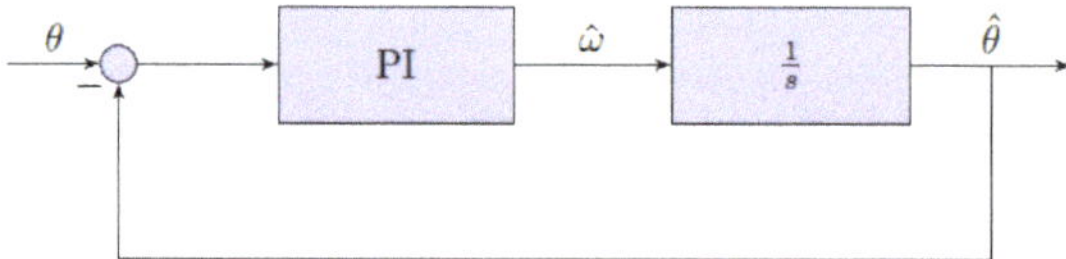

Figure 4. Control loop to filter out position sensor's noise.

The resulting closed loop transfer function is:

$$H(s) = \frac{\hat{\theta}}{\theta} = \frac{K_p s + K_i}{s^2 + K_p s + K_i} \tag{13}$$

where K_p and K_i are the PI controller's proportional and integral gains. The measured position was transformed beforehand from a sawtooth waveform into a ramp. The difference between measured and estimated position was sent to a PI controller and an integrator. Hence, the error in steady state was theoretically reduced to zero. The controller's gains should be chosen high enough to keep track of the speed variations of the machine.

4.3. CFE-CC Model Implementation for Real-Time Execution

Resolution of the global system was achieved using the State-Space-Nodal (SSN) solver of OPAL-RT available in the ARTEMIS package [19]. It allowed us to solve power systems equations with a hybrid method between pure state-space and nodal approach [20]. It was built upon the SimPowerSystems (SPS) package, thus the Simulink diagram was designed the same way by using SPS blocks (resistances, sources, inductances, etc.).

A custom block made for SSN resolution was implemented for the CFE-CC model. Figure 5 shows the WRIM bloc containing the CFE-CC model's equations as well as external components connected to it. The stator's terminals were connected to controlled voltage sources, which were the measured stator voltages of the physical machine. The rotor's terminals were connected to external resistors adjustable in real-time from the PC. By placing the machine in a separate nodal group than the external components, it had a decoupling effect with SSN resolution, thus greater stability was achieved even with large external resistances [21].

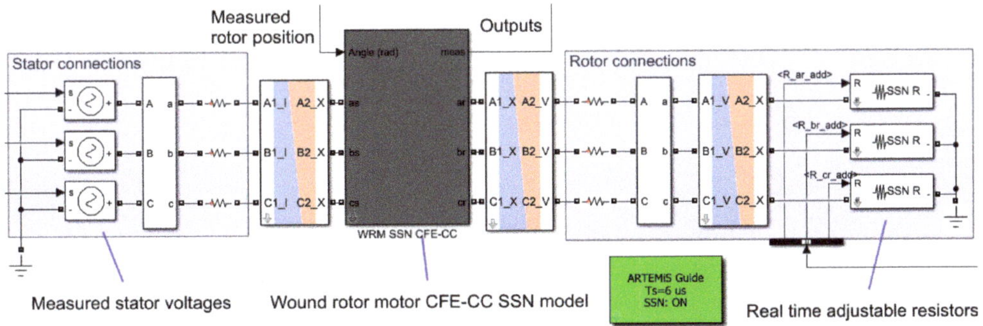

Figure 5. Real-time digital twin (RTDT) implementation overview in Simulink using ARTEMIS package.

In order to implement such a model for SSN resolution, one must first select which circuits are available to the user as external connections, i.e., the Simulink block's terminals of the machine. The WRIM had six circuits, all available for external connection, thus vectors $[v_{in}]$ and $[i_{out}]$ were each six units large. They were respectively the input voltages and output currents at the Simulink block's terminals. The search coil w_s was not treated as a circuit because no current could flow through.

$$[\phi] = \begin{bmatrix} \phi_{as} & \phi_{bs} & \phi_{cs} & \phi_{ar} & \phi_{br} & \phi_{cr} \end{bmatrix}^T \tag{14}$$

$$[v_{in}] = \begin{bmatrix} v_{as} & v_{bs} & v_{cs} & v_{ar} & v_{br} & v_{cr} \end{bmatrix}^T \tag{15}$$

$$[i_{out}] = \begin{bmatrix} i_{as} & i_{bs} & i_{cs} & i_{ar} & i_{br} & i_{cr} \end{bmatrix}^T \tag{16}$$

Dynamical system of Equation (5) is rearranged as standard state-space form. To facilitate reading of the following equations, the θ dependency is omitted.

$$[\dot{\phi}] = [A][\phi] + [B][v_{in}] \tag{17}$$

$$[i_{out}] = [C][\phi] \tag{18}$$

With

$$[A] = -[R][L]^{-1} \tag{19}$$

$$[B] = [I] \tag{20}$$

$$[C] = [L]^{-1} \tag{21}$$

Matrices $[B]$ and $[C]$ depend on the electrical terminals of the model. For the WRIM, $[B]$ is equal to the identity matrix $[I]$ since every circuit is also a terminal. For the same reason, matrix $[C]$ is directly equal to $[L]^{-1}$.

SSN solver discretizes Equation (17) for a fixed time step T_s using trapezoidal rule [21]:

$$[\phi]_{k+1} = [A_d][\phi]_k + [B_d][v]_k + [B_d][v]_{k+1} \tag{22}$$

$$[A_d] = ([I] + [A]T_s/2)/([I] - [A]T_s/2) \tag{23}$$

$$[B_d] = ([B]T_s/2)/([I] - [A]T_s/2) \tag{24}$$

SSN solver needs the following two quantities from the CFE-CC model to operate:

$$[Y_{CDM}] = [C][B_d] \tag{25}$$

$$[i_{hist}] = [C]([A_d][\phi_k] + [B_d][v_k]) \tag{26}$$

Matrices $[A_d(\theta)]$, $[B_d(\theta)]$ and $[L(\theta)]^{-1}$ were stored in a lookup table. At each time step, the exact matrices were computed as the result of an interpolation between two adjacent matrices of the lookup table by using the angular position measurement. The same type of storage and interpolation was used to retrieve the derivative inductance matrix for torque computation using Equation (6).

Now regarding the search coil, since no current, i.e., no state, was associated with it, it was not included in the dynamical system. Instead, it was computed using Equation (7).

4.4. Time Step Selection

The execution time step of the program was chosen to be as small as possible without any risk of overrun. An overrun occurred if the DRTS did not finish all computations within the specified time step. In that case, a time step was skipped resulting in erroneous estimation. RT-LAB gives information regarding computational requirement of the program. Based on it, a time step of 6 µs was chosen. It was important to have a small time step particularly when a circuit of the CFE-CC model was connected to a large external resistance. A large external resistance created a stiff state-space system, thus requiring a smaller time step to keep the solver stable. The SSN solver increased stability compared to the traditional approach but accuracy decreases when the time step was too large.

5. Validation

Validation of the RTDT was performed by comparing its real-time outputs to measurements available on the physical WRIM during operation. Waveforms shown below are real-time raw data acquired by an oscilloscope. For comparison purpose, a dq model of the WRIM was also implemented on another core of the DRTS for parallel execution alongside the CFE-CC model.

For the first setup, shown in Figure 6a, the WRIM's stator windings were connected to a three-phase 60 Hz autotransformer. It was used as an induction motor, so rotor terminals were short-circuited. The WRIM was mechanically coupled to a synchronous generator loaded by resistors. Figures 7 and 8 show waveform comparisons between measurements, CFE-CC model and dq model. Only winding *a* is shown because b and c were only 120° phase-shifted. Both models yielded good accuracy, but it was possible to draw more conclusions from the frequency decomposition. Tables 3–5 show the notables frequencies of the spectra. As expected, the CFE-CC model approximated well higher frequencies which were related to the geometry of the machine, whereas the dq model only computed harmonics related to input voltage.

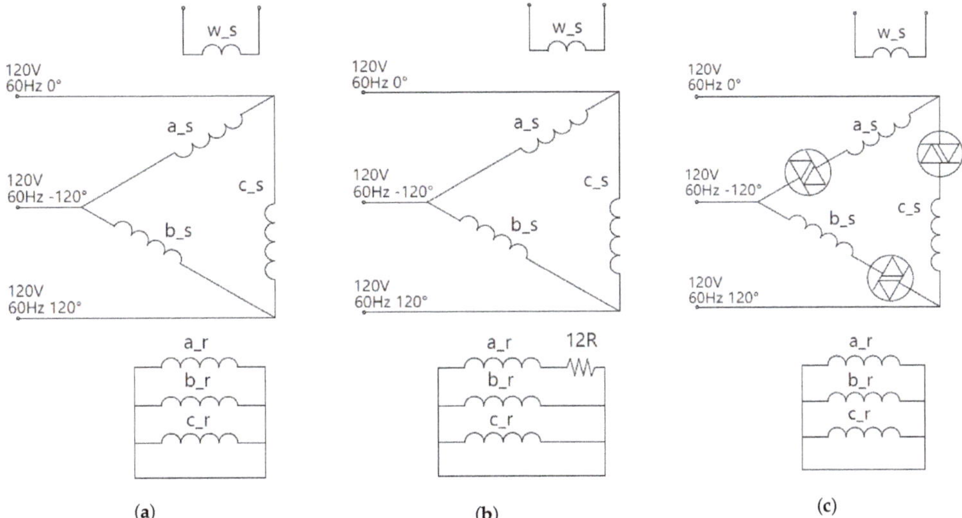

Figure 6. WRIM setups used for validation. (**a**) Balanced; (**b**) With rotor unbalance; (**c**) With TRIACs.

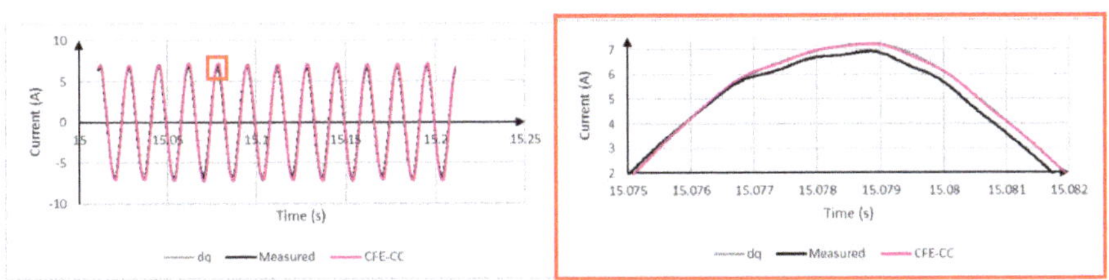

Figure 7. Stator current in winding a_s during load operation, enlarged on right side.

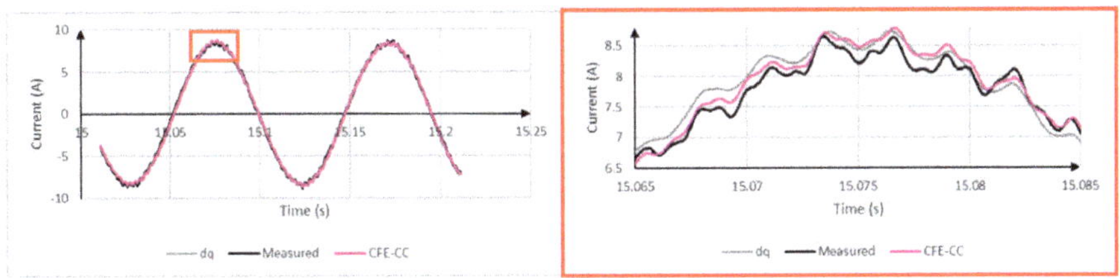

Figure 8. Rotor current in winding a_r during load operation, enlarged on right side.

Table 3. Main frequency components of stator input voltage.

Frequency (Hz)	Measurements (V)
60	156.5
180	1.0
300	3.9
420	1.8

Table 4. Main frequency components in stator current.

Frequency (Hz)	Measurements (A)	CFE-CC (A)	dq (A)
60	5.780	5.865	5.916
180	0.050	0.068	0.102
300	0.119	0.102	0.102
420	0.051	0.051	0.051
588	0.010	0.010	n/a
913	0.015	0.013	n/a
1033	0.017	0.015	n/a

Table 5. Main frequency components in rotor current.

Frequency (Hz)	Measurements (A)	CFE-CC (A)	dq (A)
6.4	7.130	7.286	7.384
114	0.183	0.116	0.117
354	0.139	0.125	0.129
367	0.055	0.044	0.045
642	0.014	0.013	n/a
655	0.014	0.015	n/a
967	0.018	0.015	n/a
979	0.023	0.021	n/a

Geometric irregularities of the machine had an impact on the torque as well, as seen from Figure 9. By analysing the electromagnetic torque of a machine, one can draw many conclusions about its condition.

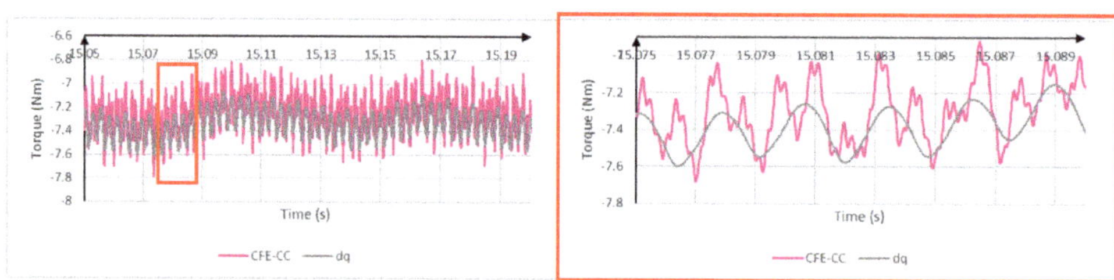

Figure 9. Electromagnetic torque during load operation, enlarged on right side.

Figure 10 shows the clear congruence between real-time output of the search coil voltage and its measured counterpart. Search coil voltage gave a good indication of local magnetic flux linkage. Using Equation (11), any number of search coils could be directly added to the machine's winding configuration without restarting the FEM process. This also proved to be an interesting tool for online fault monitoring.

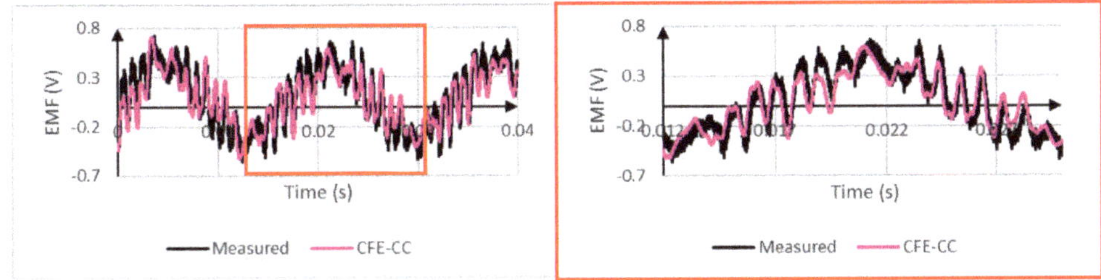

Figure 10. Search coil voltage during load operation, enlarged on right side.

Figure 11 shows the dynamic response of rotor currents when the mechanical load was suddenly disconnected from the shaft. Here, the position tracking control loop played an important role. It needed to be fast enough to minimize the error during speed variations.

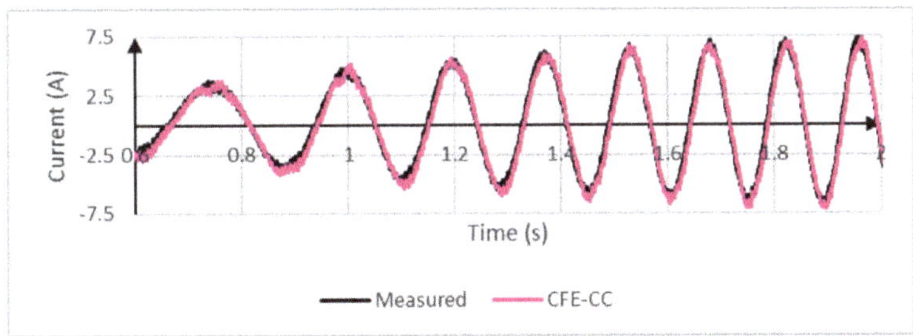

Figure 11. Rotor current in winding a_r starting from load operation and suddenly disconnecting the load.

The following Figures 12–14 show the RTDT's current outputs when a phase imbalance was introduced. The setup is depicted at Figure 6b. A resistor was connected to winding a_r of the physical machine, and the same was done to the RTDT. While the frequency content was accurate, a small phase discrepancy was present at winding b_r and c_r. The cause could be related to the methodology employed to compute $[L]$ in Appendix A, such as the way the coil end inductances were added. Many assumptions were made. Magnetic saturation could also play a role since it was not taken into account in the model.

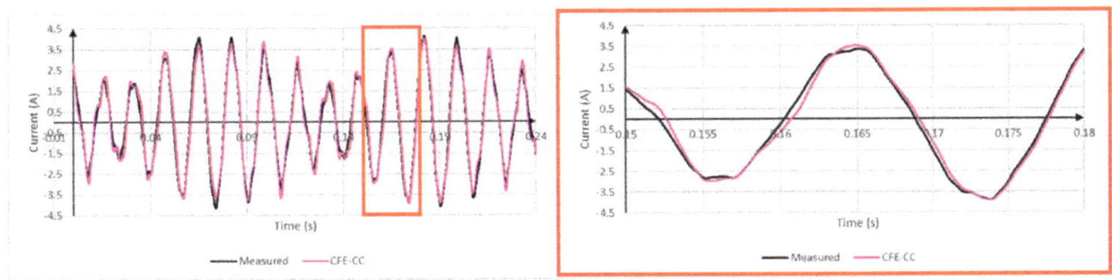

Figure 12. Stator current in winding a_s during load operation with 12 ohm resistor on winding a_r, enlarged on right side.

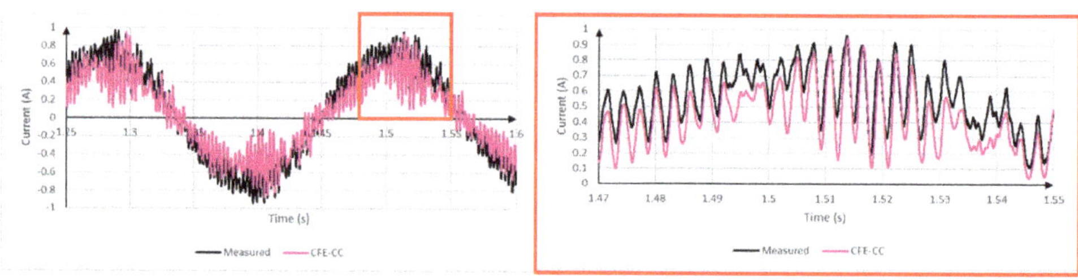

Figure 13. Rotor current in winding a_r during load operation with 12 ohm resistor on winding a_r, enlarged on right side.

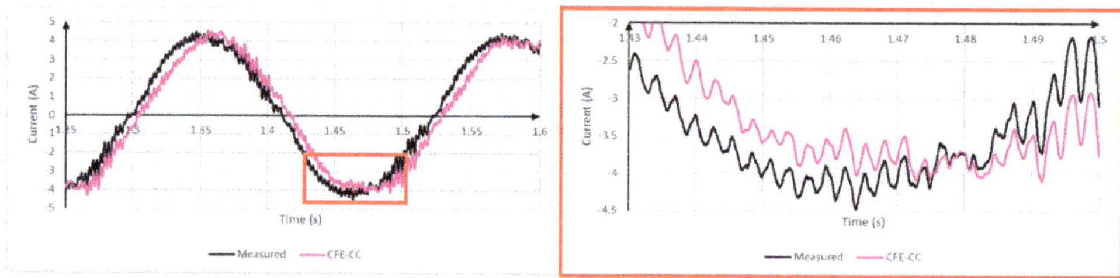

Figure 14. Rotor current in winding b_r during load operation with 12 ohm resistor on winding a_r, enlarged on right side.

Finally, TRIACs were added inside the delta configuration as depicted in Figure 6c. It demonstrated the RTDT's behavior under switching converter's voltage input for an unusual phase configuration. Figures 15 and 16 show currents flowing in windings a_s and a_r. The RTDT yielded accurate estimations.

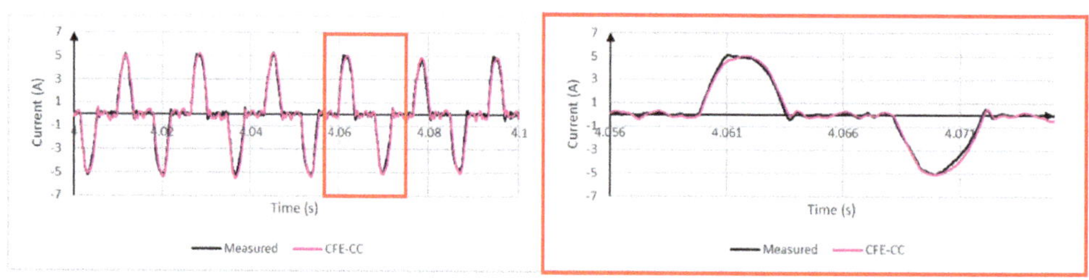

Figure 15. Stator current in winding a_s during load operation and fed through TRIACs, enlarged on right side.

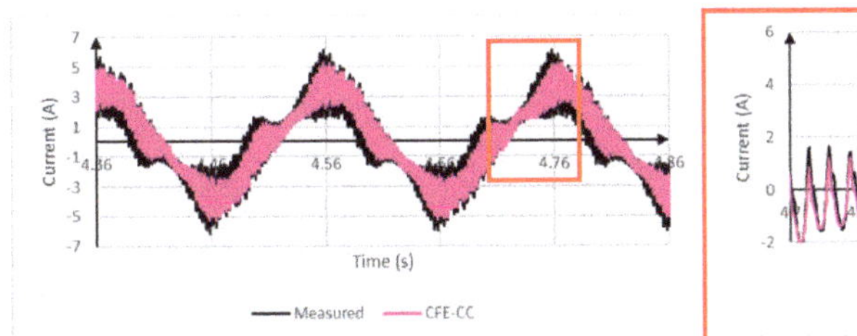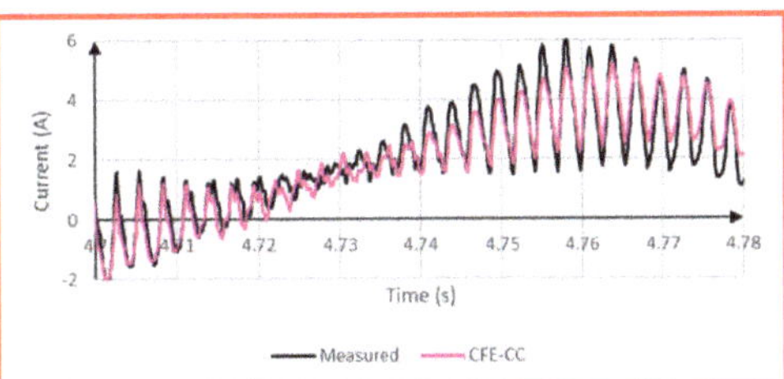

Figure 16. Rotor current in winding a_r during load operation and fed through TRIACs, enlarged on right side.

6. Conclusions

The detailed CFE-CC model of a WRIM was implemented in real-time on an entry-level DRTS of OPAL-RT with a time step of 6 µs. This model considers the machine's particular geometry and winding configuration can be changed easily using a simple matrix transformation. Any number of search coils can be added to the model, as well as faults such as broken bar and eccentricity. Comparisons with real-time measurements have shown that the CFE-CC model is accurate and computationally efficient. This work paves the way to new real-time monitoring techniques for electrical machines using the CFE-CC model alongside others in a digital twin. However, the machine studied in this work has only seven circuits. Including more electrical circuits in the CFE-CC model results in more computations at each time step and more memory requirement for lookup tables. Future works include studying the feasibility of real-time execution of the CFE-CC model for large synchronous machines comprising many electrical circuits.

Author Contributions: Conceptualization, S.B.; methodology, S.B.; software, S.B.; validation, S.B.; formal analysis, S.B., P.V. and J.C.; investigation, S.B.; resources, S.B.; data curation, S.B.; writing—original draft preparation, S.B.; writing—review and editing, S.B., P.V. and J.C.; supervision, P.V.; project administration, P.V. and J.C.; funding acquisition, P.V. and J.C. All authors have read and agreed to the published version of the manuscript.

Funding: This research received no external funding.

Acknowledgments: The authors would like to thank OPAL-RT for their support in this work.

Conflicts of Interest: The authors declare no conflict of interest.

Abbreviations

The following abbreviations are used in this manuscript:

WRIM	Wound rotor induction machine
DRTS	Digital real-time simulator
RTDT	Real-time digital twin
CFE-CC	Combination of finite element and coupled circuits
FEM	Finite element method
SSN	State-Space-Nodal
SPS	SimPowerSystems

Appendix A. Alteration of Inductance Matrix to Consider Rotor Skew and Coil Ends

Appendix A.1. Rotor Skewing

The 2D FEM software used in this work suppose a uniform magnetic field intensity along the third dimension. However, the WRIM and most asynchronous machines have a skewed rotor to prevent cogging and reduce torque oscillations. The skew angle θ_{sk} is generally given by the manufacturer. Fortunately, it is possible to add this effect into the inductance matrix $[L(\theta)]$ previously identified in 2D [22]. By dividing an unskewed machine along its length into M several slices, or sub-machines, the inductance matrix $[L_m(\theta)]$ of each sub-machine is simply the initial one $[L(\theta)]$ divided by M.

$$[L(\theta)] = [L_1(\theta)] + [L_2(\theta)] + \ldots [L_M(\theta)] \tag{A1}$$

$$[L_m(\theta)] = [L(\theta)]/M, \quad m = 1, 2, \ldots M \tag{A2}$$

Now, it is possible to replicate the skewing effect by shifting the rotor gradually along the length. Each of the inductance matrices $[L_1(\theta)] \ldots [L_M(\theta)]$ needs to be shifted by $\theta_{sk}/(M-1)$. Finally, they can be added all together according to Equation (A1) to obtain the skewed $[L(\theta)]$ of the complete machine.

As for the WRIM of this work, θ_{sk} is equal to $7.5°$, which corresponds to a slot pitch. The machine was divided into 61 sub-machines. Figure A1 shows the impact of the added skew effect on rotor currents during load operation, with stator windings fed with perfectly sinusoidal voltages and rotor windings short-circuited. Current ripples are reduced as expected.

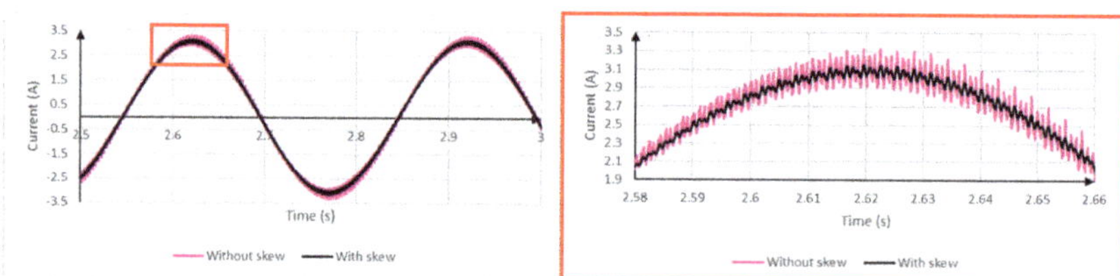

Figure A1. Rotor current with and without added rotor skewing effect, enlarged on right side.

Appendix A.2. Coil Ends

Coil ends can be modeled in a 3D FEM software, whereas it cannot be in 2D. In the latter case such as in the present work, their inductive effect must be added to $[L(\theta)]$ by other means. For simplicity, we ignore the impact on mutual inductances. The technique used is to conduct a standstill frequency response (SSFR) test [14] and add a self-inductance to the windings in order to make the experimental response fit the simulated one. The test was conducted while rotor windings are short-circuited. Figure A2 compares the response of the FEM model to the experiment with and without coil ends. Inductances of 0.01 H and 0.00047 H were added to stator and rotor windings respectively, which is less than 10% of the magnetizing inductance.

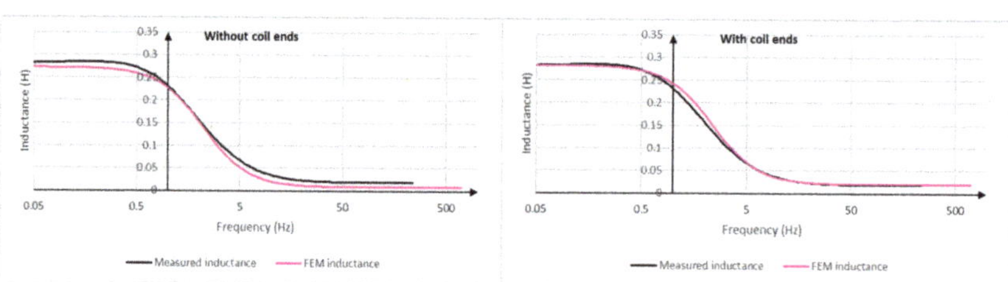

Figure A2. Inductance seen from the stator versus frequency with and without coil ends, rotor terminals short-circuited.

References

1. Ebrahimi, A. Challenges of developing a digital twin model of renewable energy generators. In Proceedings of the 2019 IEEE 28th International Symposium on Industrial Electronics (ISIE), Vancouver, BC, Canada, 12–14 June 2019; pp. 1059–1066.
2. Wang, J.; Ye, L.; Gao, R.X.; Li, C.; Zhang, L. Digital Twin for rotating machinery fault diagnosis in smart manufacturing. *Int. J. Prod. Res.* **2019**, *57*, 3920–3934. [CrossRef]
3. Boschert, S.; Heinrich, C.; Rosen, R. Next generation digital Twin. In Proceedings of the TMCE, Las Palmas de Gran Canaria, Spain, 7–11 May 2018.
4. Fuller, A.; Fan, Z.; Day, C.; Barlow, C. Digital Twin: Enabling Technologies, Challenges and Open Research. *IEEE Access* **2020**, *8*, 108952–108971.CrossRef]
5. Roshandel Tavana, N.; Dinavahi, V. A General Framework for FPGA-Based Real-Time Emulation of Electrical Machines for HIL Applications. *IEEE Trans. Ind. Electron.* **2015**, *62*, 2041–2053. [CrossRef]
6. Kłosowski, Z.; Cieślik, S. Real-time simulation of power conversion in doubly fed induction machine. *Energies* **2020**, *13*, 673. [CrossRef]
7. Sidwall, K.; Forsyth, P. Advancements in Real-Time Simulation for the Validation of Grid Modernization Technologies. *Energies* **2020**, *13*, 4036. [CrossRef]
8. Mojlish, S.; Erdogan, N.; Levine, D.; Davoudi, A. Review of Hardware Platforms for Real-Time Simulation of Electric Machines. *IEEE Trans. Transp. Electrif.* **2017**, *3*, 130–146. [CrossRef]
9. Sudhoff, S.D.; Kuhn, B.T.; Corzine, K.A.; Branecky, B.T. Magnetic Equivalent Circuit Modeling of Induction Motors. *IEEE Trans. Energy Convers.* **2007**, *22*, 259–270. [CrossRef]
10. Fleming, F.E.; Edrington, C.S. Real-Time Emulation of Switched Reluctance Machines via Magnetic Equivalent Circuits. *IEEE Trans. Ind. Electron.* **2016**, *63*, 3366–3376. [CrossRef]
11. Jandaghi, B.; Dinavahi, V. Real-Time HIL Emulation of Faulted Electric Machines Based on Nonlinear MEC Model. *IEEE Trans. Energy Convers.* **2019**, *34*, 1190–1199. [CrossRef]
12. Luo, X.; Liao, Y.; Toliyat, H.; El-Antably, A.; Lipo, T.A. Multiple Coupled Circuit Modeling of Induction Machines; In Proceedings of Conference Record of the 1993 IEEE Industry Applications Conference Twenty-Eighth IAS Annual Meeting, Toronto, ON, Canada, 2–8 October 1993; pp. 203–210. [CrossRef]
13. Dehkordi, A.B.; Neti, P.; Gole, A.; Maguire, T. Development and validation of a comprehensive synchronous machine model for a real-time environment. *IEEE Trans. Energy Convers.* **2009**, *25*, 34–48. doi:10.1109/PES.2009.5275300. [CrossRef]
14. Gbégbé, A.Z.; Rouached, B.; Cros, J.; Bergeron, M.; Viarouge, P. Damper Currents Simulation of Large Hydro-Generator Using the Combination of FEM and Coupled Circuits Models. *IEEE Trans. Energy Convers.* **2017**, *32*, 1273–1283. [CrossRef]
15. Quéval, L.; Ohsaki, H. Study on the implementation of the phase-domain model for rotating electrical machines. In Proceedings of the 2012 15th International Conference on Electrical Machines and Systems (ICEMS), Sapporo, Japan, 21–24 October 2012; pp. 1–6.
16. Rouached, B.; Cros, J.; Clénet, S.; Viarouge, P. *Estimation of Damper Bars Losses in Large Synchronous Alternator using Bar Current Waveforms*; Electrimacs: Toulouse, France, 2017.
17. Demetriades, G.D.; de la Parra, H.Z.; Andersson, E.; Olsson, H. A Real-Time Thermal Model of a Permanent-Magnet Synchronous Motor. *IEEE Trans. Power Electron.* **2010**, *25*, 463–474. [CrossRef]
18. Mathault, J.; Bergeron, M.; Rakotovololona, S.; Cros, J.; Viarouge, P. *Influence of Discrete Inductance Curves on the Simulation of a Round Rotor Generator Using Coupled Circuit Method*; Electrimacs: Valencia, Spain, 2014.
19. Dufour, C.; Mahseredjian, J.; Belanger, J. A Combined State-Space Nodal Method for the Simulation of Power System Transients. *IEEE Trans. Power Deliv.* **2011**, *26*, 928–935. doi:10.1109/TPWRD.2010.2090364. [CrossRef]
20. Wang, L.; Jatskevich, J.; Dinavahi, V.; Dommel, H.W.; Martinez, J.A.; Strunz, K.; Rioual, M.; Chang, G.W.; Iravani, R. Methods of Interfacing Rotating Machine Models in Transient Simulation Programs. *IEEE Trans. Power Deliv.* **2010**, *25*, 891–903. [CrossRef]

21. Dufour, C.; Nasrallah, D.S. State-space-nodal rotating machine models with improved numerical stability. In Proceedings of the IECON 2016—42nd Annual Conference of the IEEE Industrial Electronics Society, Florence, Italy, 23–26 October 2016; pp. 4368–4375. [CrossRef]
22. Mohr, M.; Bíró, O.; Stermecki, A.; Diwoky, F. A Finite Element-Based Circuit Model Approach for Skewed Electrical Machines. *IEEE Trans. Magn.* **2014**, *50*, 837–840. [CrossRef]

Publisher's Note: MDPI stays neutral with regard to jurisdictional claims in published maps and institutional affiliations.

 © 2020 by the authors. Licensee MDPI, Basel, Switzerland. This article is an open access article distributed under the terms and conditions of the Creative Commons Attribution (CC BY) license (http://creativecommons.org/licenses/by/4.0/).

Article

Hybrid Multimodule DC-DC Converters for Ultrafast Electric Vehicle Chargers

Mena ElMenshawy * and Ahmed Massoud

Department of Electrical Engineering, Qatar University, Doha 2713, Qatar; Ahmed.massoud@qu.edu.qa
* Correspondence: MAlmenshawy@qu.edu.qa

Received: 16 August 2020; Accepted: 15 September 2020; Published: 21 September 2020

Abstract: To increase the adoption of electric vehicles (EVs), significant efforts in terms of reducing the charging time are required. Consequently, ultrafast charging (UFC) stations require extensive investigation, particularly considering their higher power level requirements. Accordingly, this paper introduces a hybrid multimodule DC-DC converter-based dual-active bridge (DAB) topology for EV-UFC to achieve high-efficiency and high-power density. The hybrid concept is achieved through employing two different groups of multimodule converters. The first is designed to be in charge of a high fraction of the total required power, operating at a relatively low switching frequency, while the second is designed for a small fraction of the total power, operating at a relatively high switching frequency. To support the power converter controller design, a generalized small-signal model for the hybrid converter is studied. Also, cross feedback output current sharing (CFOCS) control for the hybrid input-series output-parallel (ISOP) converters is examined to ensure uniform power-sharing and ensure the desired fraction of power handled by each multimodule group. The control scheme for a hybrid eight-module ISOP converter of 200 kW is investigated using a reflex charging scheme. The power loss analysis of the hybrid converter is provided and compared to conventional multimodule DC-DC converters. It has been shown that the presented converter can achieve both high efficiency (99.6%) and high power density (10.3 kW/L), compromising between the two other conventional converters. Simulation results are provided using the MatLab/Simulink software to elucidate the presented concept considering parameter mismatches.

Keywords: ultra-fast chargers; input-series input-parallel output-series output-parallel multimodule converter; cross feedback output current sharing; reflex charging

---

1. Introduction

Despite the fact that internal combustion engines (ICEs) have been a mature technology for the past 100 years, it is expected that electric vehicles (EVs) will break the monopoly of conventional vehicles using only ICEs because of their performance and superior fuel economy [1]. Due to the strict regulations on global warming and energy resources constraints, and on reducing fossil fuel prices as well as gas emissions, environmental awareness has led to a high interest in EVs as an alternative solution for further improvement compared to ICEs [2,3]. To increase the adoption of EVs, significant efforts in terms of reducing the charging time are required. Consequently, to allow massive market penetration of EVs, the concept of ultrafast charging (UFC) requires more investigation. In this regard, several research studies targeting fast chargers and UFC for EVs have been provided in the literature [4–8]. UFC technology is a high power charging technology ($\geq$ 400 kW) that can replace or substitute the ICE technology and can charge EVs' battery packs in $\leq$ 10 min [9–11].

Advanced power electronics converters are considered as key enabling technologies for realizing EV UFC, where high-power DC-DC converters are needed. The critical requirements for designing EV battery chargers are high efficiency, low cost, high power density, and galvanic isolation [12].

Furthermore, one of the UFC stations' requirements is to design the DC-DC converters in a modular manner to offer easy maintenance, as well as scalability, redundancy, and fault ride-through capability [13–15]. In modular power converters, each unit handles a small portion of the total input power. Accordingly, the selected power switches are of lower voltage and/or current ratings; therefore, higher switching frequency capability, consequently, reduced weight and size [16–18]. Multimodule DC-DC converters can provide a bidirectional power flow through employing submodules that are based on dual active bridge (DAB), dual half bridge (DHB), or series resonant topologies [19,20].

The DAB configuration, shown in Figure 1, consists of two active bridges that are connected via a medium/high-frequency AC transformer. DAB can be constructed using a single-phase bridge or a three-phase bridge depending on the design criteria. The 2L-DAB, shown in Figure 1, usually operates in a square wave mode. The intermediate transformer leakage inductance limits the maximum power flow and is used as the energy transferring element. This topology is capable of bidirectional power flow that can be achieved by controlling the phase shift between the two bridges and the magnitude of the output voltage per bridge. The switches can be switched at zero voltage switching (ZVS) and/or zero current switching (ZCS). Accordingly, switching losses are reduced, and the power efficiency is increased.

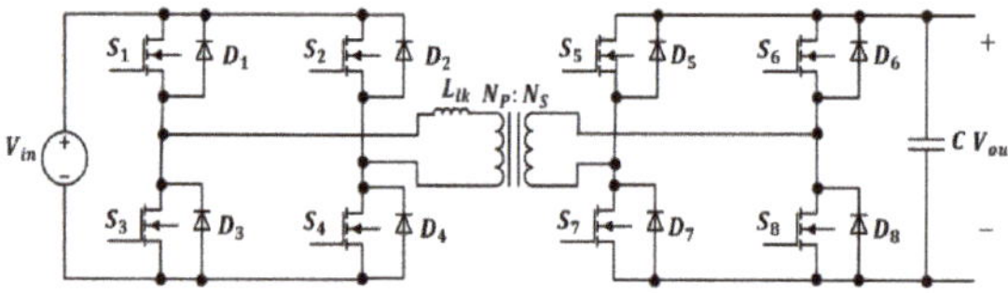

Figure 1. DAB converter circuit diagram.

Figure 2 presents a block diagram for a typical EV UFC that involves an AC-DC stage and a DC-DC stage. This paper will focus on the DC-DC stage employed in EV UFC applications. In the literature, many research studies have been introduced the two stages. In [21], to realize medium-voltage EV UFC stations, a multiport power converter has been proposed. In [22], a bidirectional fast charging system control strategy consisting of two cascaded stages has been proposed, where two DABs are connected in parallel at the battery side. However, in [23], an isolated DAB-based single-stage AC-DC converter has been presented. The charger in [23] contains a single stage that includes the PFC and ensures ZVS over the full load range. In [24,25], a frequency modulated CLLC-R-DAB has been proposed. In this topology, the converter operates over a considerable variation of the input voltage while maintaining soft-switching capability. A smaller switching frequency range is used to modulate the CLLC-R-DAB converter when compared to SR-DAB. In [26], a full-bridge phase-shifted DC-DC converter that combines the characteristics of the double inductor rectifier and the conventional hybrid switching converter is introduced for EV fast chargers. In [27], a medium-voltage high-power isolated DC-DC converter for EVs fast chargers is presented. In [28], the AC-DC and DC-DC stages of an EV charger are studied where the DC-DC stage utilizes interleaved DC-DC converters.

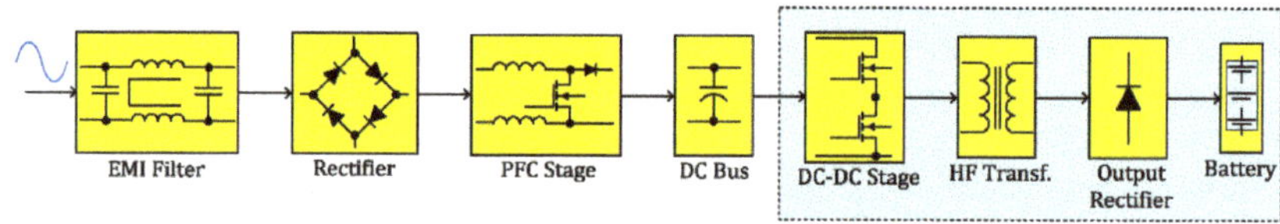

Figure 2. Block diagram for a typical EV UFC.

Multimodule converters are considered a suitable choice for realizing the high power and high voltage requirements of the UFC charger. However, an increased number of modules with low power would increase system complexity, cost and losses, which reduces the cooling requirements and consequently the weight, volume, and cost. However, reduced switching losses can be achieved via soft-switching [29–35]. Nonetheless, introducing a low number of modules with high power would reduce the switching frequency capabilities; therefore, reducing the power density, which increases size and weight.

Accordingly, the main contribution of this work is to introduce a hybrid multimodule DC-DC converter-based DAB topology as the DC-DC stage for EV UFC to achieve high efficiency, high power density, and reduced weight and cost. The hybrid concept is achieved through employing two different groups of multimodule converters. The first group is designed to be in charge of a high fraction of the total required power while operating relatively at a low switching frequency. Nevertheless, the second group is designed for a low fraction of the total power operating relatively at a high switching frequency. The work presented in this paper includes a generalized small-signal model for the presented converter as well as the control strategy required in achieving uniform power-sharing between the employed modules. Besides, a power loss evaluation has been conducted to compare the proposed converter with the other two options.

To verify the presented concept, the number of modules needed to achieve the required ratings is calculated for both; conventional multimodule DC-DC converters and hybrid multimodule DC-DC converters. In addition, the power loss analysis of the hybrid multimodule converter is provided. To support the power converter controller design, a generalized small-signal model for the hybrid multimodule DC-DC converter is studied in detail. Besides, to ensure equal power-sharing among the employed modules, the control scheme for the hybrid multimodule DC-DC converter with the aforementioned specifications is studied. The main contribution of the paper can be summarized as follows:

- Development of a hybrid multimodule DC-DC converter-based DAB topology for EV UFC along with providing generalized small-signal modeling to support the design of the power converter controller. The presented generalized small-signal model of the hybrid multimodule DC-DC converter-based DAB is considered as the primary contribution of this paper.
- Examining the cross feedback output current sharing (CFOCS) for the hybrid Input-series output-parallel (ISOP) multimodule power converters to ensure uniform power-sharing among the employed modules and ensure the desired fraction of power handled by each multimodule group.

This paper is structured as follows: Section 2 presents the hybrid input-series input-parallel output-series output-parallel (ISIP-OSOP) multimodule power converter and the generalized small-signal modeling. Section 3 presents a 200 kW hybrid eight-module ISOP converter. In Section 3, the small-signal model of the presented converter is derived using the analysis provided in Section 2. Section 4 presents the number of modules needed to achieve the required ratings for both; conventional multimodule DC-DC converters and hybrid multimodule DC-DC converters. In addition, the power loss analysis of the conventional and hybrid multimodule converters is provided. Section 5 discusses the control strategy for the proposed hybrid ISOP multimodule DC-DC converters. Section 6 discusses the MatLab/Simulink model and the simulation results. Finally, Section 7 summarizes the key findings of this paper.

2. Generalized Small-Signal Analysis for Dual Series/Parallel Input-Output (ISIP-OSOP) Hybrid Multimodule Converters

In this section, the generalized small-signal modeling for dual series/parallel ISIP-OSOP hybrid multimodule DC-DC converter is introduced.

2.1. Hybrid ISIP-OSOP Generic DC-DC Converter Circuit Configuration

The hybrid ISIP-OSOP generic DC-DC converter configuration, shown in Figure 3, consists of n modules that are connected in series and/or parallel at the input side and in series and/or parallel at the output side. These n modules consist of two different multimodule groups. The primary group consists of L isolated DC-DC converters that are in charge of a high fraction of the total required power operating relatively at a low switching frequency. The secondary group consists of M isolated DC-DC converters that are designed for a small fraction of the total power operating relatively at a high switching frequency. Accordingly, it can be said that the summation of L and M power converters results in a total of n DAB units. To differentiate between the primary and secondary multimodule DC-DC converter in the small signal analysis, the set of equations representing the primary group is black colored while the set of equations representing the secondary group is blue colored. In addition, the red colored symbols reflect the parameters defined for the input side, while the blue colored symbols reflects the parameters defined for the output side, as presented in the following.

By ensuring input current sharing (ICS) and input voltage sharing (IVS) for the primary group, the input current for each module in the primary group is reduced to $\frac{I_{inL}}{\alpha_{L1}}$, and the input voltage for each module in the primary group is reduced to $\frac{V_{inL}}{\beta_{L1}}$. However, by ensuring ICS and IVS for the secondary group, the input current for each module in the secondary group is reduced to $\frac{I_{inM}}{\alpha_{M1}}$, and the input voltage for each module in the secondary group is reduced to $\frac{V_{inM}}{\beta_{M1}}$ group is reduced to $\frac{V_{inM}}{\beta_{M1}}$, in which, I_{inL} and V_{inL} are the input current and the input voltage for the primary group that consists of L number of modules, respectively. I_{inM} and V_{inM} are the input current and the input voltage for the secondary group that consists of M number of modules, respectively. α_{M1} represents the number of modules connected in parallel in the secondary group at the input side. β_{M1} represents the number of modules connected in series the secondary group at the input side.

Similarly, by ensuring output current sharing (OCS) and output voltage sharing (OVS) for the primary group, the output current per module in the primary group is $\frac{I_{oL}}{a_{L1}}$, and the output voltage per module in the primary module is reduced to $\frac{V_{oL}}{b_{L1}}$. However, by ensuring OCS and OVS for the secondary group, the output current per module is $\frac{I_{oM}}{a_{M1}}$, and the output voltage for each module in the secondary group is reduced to $\frac{V_{oM}}{b_{M1}}$. In which, I_{oL} and V_{oL} are the output current and the output voltage for the primary group that consists of L number of modules, respectively. I_{oM} and V_{oM} are the output current and the output voltage for the secondary group that consists of M number of modules, respectively. a_{M1} represents the number of modules connected in parallel in the secondary group at the output side. b_{M1} represents the number of modules connected in series in the secondary group at the output side.

The L isolated modules are responsible for delivering a portion of K_L of the total required power, while the M isolated modules are responsible for delivering a portion of K_M of the total required power, where $K_L + K_M = 1$ pu. The input voltages, input currents, output currents, and output voltages are represented in terms of the total input voltage, total input current, total output voltage, and total output current would result in Table 1.

Table 1. Individual module system parameters representation in terms of the overall system ratings.

Parameters	Representation	Value
		Primary group
Input current	$\dfrac{\alpha_{L2}\, I_{in}}{\alpha_{L1}}$	$\alpha_{L2} = K_L$ if the two groups are connected in parallel at the input side, otherwise $\alpha_{L2} = 1$
Input voltage	$\dfrac{\beta_{L2}\, V_{in}}{\beta_{L1}}$	$\beta_{L2} = K_L$ if the two groups are connected in series at the input side, otherwise $\beta_{L2} = 1$
Output current	$\dfrac{a_{L2}\, I_{o}}{a_{L1}}$	$a_{L2} = K_L$ if the two groups are connected in parallel at the output side, otherwise $a_{L2} = 1$
Output voltage	$\dfrac{b_{L2}\, V_{o}}{b_{L1}}$	$b_{L2} = K_L$ if the two groups are connected in series at the output side, otherwise $b_{L2} = 1$
		Secondary group
Input current	$\dfrac{\alpha_{M2}\, I_{in}}{\alpha_{M1}}$	$\alpha_{M2} = K_M$ if the two groups are connected in parallel at the input side, otherwise $\alpha_{M2} = 1$
Input voltage	$\dfrac{\beta_{M2}\, V_{in}}{\beta_{M1}}$	$\beta_{M2} = K_M$ if the two groups are connected in series at the input side, otherwise $\beta_{M2} = 1$
Output current	$\dfrac{a_{M2}\, I_{oL}}{a_{M1}}$	$a_{M2} = K_M$ if the two groups are connected in parallel at the output side, otherwise $a_{M2} = 1$
Output voltage	$\dfrac{b_{M2}\, V_{o}}{b_{M1}}$	$b_{M2} = K_M$ if the two groups are connected in series at the output side, otherwise $b_{M2} = 1$

2.2. Hybrid ISIP-OSOP DC-DC Converter Small-Signal Modeling

Using the model in [36], and expanding the study of the multimodule DC-DC converters in [37–41], the small-signal model for the hybrid multimodule ISIP-OSOP converter shown in Figure 4 is derived. Since each group is responsible for delivering a particular portion of the overall required power, where this portion is defined according to the overall system power ratings. Accordingly, it is worth mentioning that the equivalent load resistance seen by each group of multimodule converters is different.

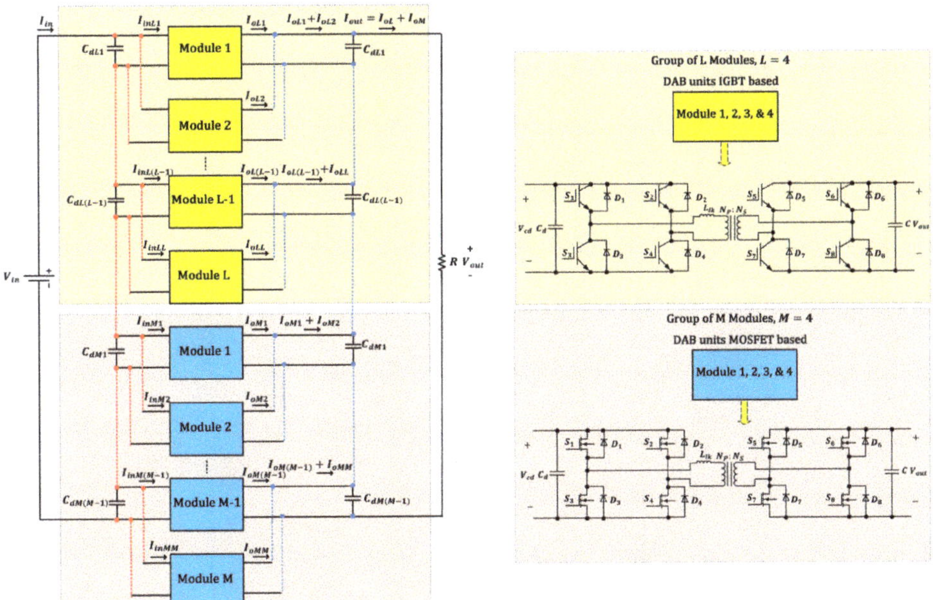

Figure 3. Generalized hybrid multimodule DC-DC converter configuration.

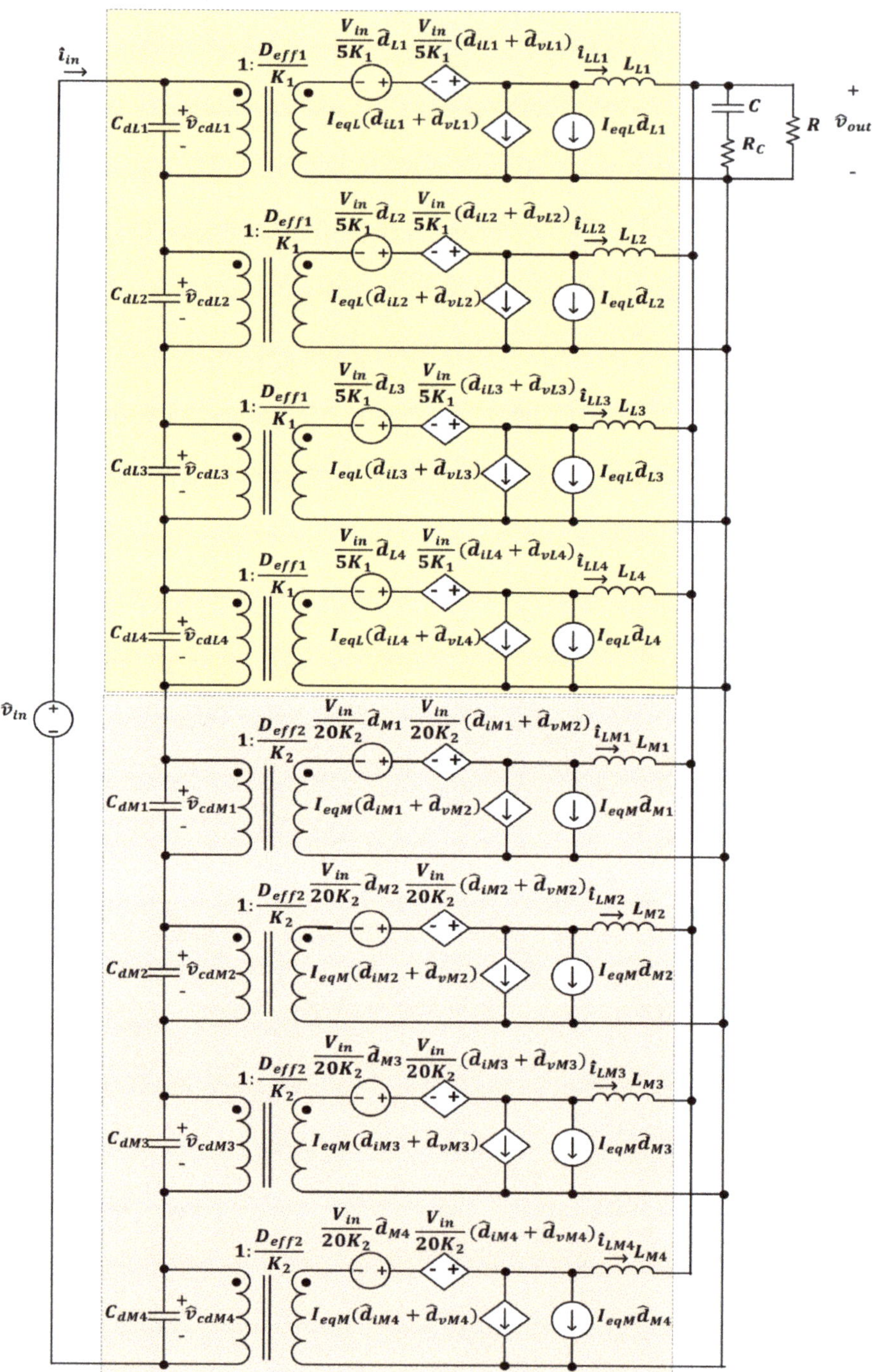

Figure 4. Small signal model for the generalized hybrid ISIP-OSOP multimodule DC-DC converter.

Since the input current and input voltage for each module in the primary group are $\frac{\alpha_{L2}}{\alpha_{L1}}I_{in}$ and $\frac{\beta_{L2}}{\beta_{L1}}V_{in}$, respectively, and the output current and output voltage for each module in the primary group are $\frac{a_{L2}}{a_{L1}}I_o$ and $\frac{b_{L2}}{b_{L1}}V_o$, respectively. Therefore, the load resistance for each module in the primary group is $\frac{a_{L1}b_{L2}}{a_{L2}b_{L1}}R$. Accordingly, $\hat{d}_{ijL}$, $\hat{d}_{vLj}$ which are the effect of changing the filter inductor current and the effect of changing the input voltage on the duty cycle modulation for the primary group and I_{eqL} presented in Figure 4 can be defined as:

$$\hat{d}_{iLj} = -\frac{4\beta_{L1}L_{lkL}f_{sL}}{\beta_{L2}\,K_1V_{in}}\hat{i}_{LLj}, \; j = 1,2,\ldots,L \tag{1}$$

Equation (1) can be written as:

$$\hat{d}_{iLj} = -\frac{\beta_{L1}K_1R_{dL}}{\beta_{L2}\,V_{in}}\hat{i}_{LLj}, \; j = 1,2,\ldots,L \tag{2}$$

where $R_{dL} = \frac{4L_{lkL}f_{sL}}{K_1^2}$.

$$\hat{d}_{vLj} = \frac{4a_{L2}b_{L1}\beta_{L1}L_{lkL}f_{sL}D_{eff1}}{a_{L1}b_{L2}\beta_{L2}\,K_1^2RV_{in}}\hat{v}_{cdLj}, \; j = 1,2,\ldots,L \tag{3}$$

Equation (3) can be written as:

$$\hat{d}_{vLj} = \frac{a_{L2}b_{L1}\beta_{L1}R_{dL}D_{eff1}}{a_{L1}b_{L2}\beta_{L2}\,RV_{in}}\hat{v}_{cdLj}, \; j = 1,2,\ldots,L \tag{4}$$

$$I_{eqL} = \frac{a_{L2}b_{L1}\beta_{L2}V_{in}}{a_{L1}b_{L2}\beta_{L1}\,K_1R} \tag{5}$$

Since the input current and input voltage for each module in the primary group is $\frac{\alpha_{M2}}{\alpha_{M1}}I_{in}$ and $\frac{\beta_{M2}}{\beta_{M1}}V_{in}$, respectively, and the output current and output voltage for each module in the primary group is $\frac{a_{M2}}{a_{M1}}I_o$ and $\frac{b_{M2}}{b_{M1}}V_o$, respectively. Therefore, the load resistance for each module in the primary group is $\frac{a_{M1}b_{M2}}{a_{M2}b_{M1}}R$. Accordingly, $\hat{d}_{ijM}$, $\hat{d}_{vMj}$ which are the effect of changing the filter inductor current and the effect of changing the input voltage on the duty cycle modulation for the primary group and I_{eqM} presented in Figure 4 can be defined as:

$$\hat{d}_{iMj} = -\frac{4\beta_{L1}L_{lkM}f_{sM}}{\beta_{L2}\,K_2V_{in}}\hat{i}_{LMj}, \; j = 1,2,\ldots,M \tag{6}$$

Equation (6) can be written as:

$$\hat{d}_{iMj} = -\frac{\beta_{L1}K_2R_{dM}}{\beta_{L2}\,V_{in}}\hat{i}_{LMj}, \; j = 1,2,\ldots,M \tag{7}$$

where; $R_{dM} = \frac{4L_{lkM}f_{sM}}{K_2^2}$.

$$\hat{d}_{vMj} = \frac{4a_{L2}b_{L1}\beta_{L1}L_{lkM}f_{sM}D_{eff2}}{a_{L1}b_{L2}\beta_{L2}\,K_2^2RV_{in}}\hat{v}_{cdMj}, \; j = 1,2,\ldots,M \tag{8}$$

Equation (8) can be written as:

$$\hat{d}_{vMj} = \frac{a_{L2}b_{L1}\beta_{L1}R_{dM}D_{eff2}}{a_{L1}b_{L2}\beta_{L2}\,RV_{in}}\hat{v}_{cdMj}, \; j = 1,2,\ldots,M \tag{9}$$

$$I_{eqM} = \frac{a_{L2} b_{L1} \beta_{L2} V_{in}}{a_{L1} b_{L2} \beta_{L1} K_2 R} \tag{10}$$

Based on the feature of modularity and to simplify the analysis, it is assumed that all modules in the primary group and all modules in the secondary group have an equal effective duty cycle, transformer turns ratio, capacitor, and inductor values. Accordingly, $K_{L1} = K_{L2} = \cdots = K_{LL} = K_1$, $K_{M1} = K_{M2} = \cdots = K_{MM} = K_2$, $C_{L1} = C_{L2} = \cdots = C_{LL} = C_L$, $C_{M1} = C_{M2} = \cdots = C_{MM} = C_M$, $C_{dL1} = C_{dL2} = \cdots = C_{dLL} = C_{dL}$, $C_{dM1} = C_{dM2} = \cdots = C_{dMM} = C_{dM}$, $L_{L1} = L_{L2} = \cdots = L_{LL} = L_L$ and $L_{M1} = L_{M2} = \cdots = L_{MM} = L_M$. In addition, it is also assumed that all modules in the primary group share the same input voltage and that all modules in the secondary group share the same input voltage. Accordingly, the DC input voltage of each module in the primary group is $\frac{\beta_{L2}}{\beta_{L1}} V_{in}$ and the DC input voltage of each module in the secondary group is $\frac{\beta_{M2}}{\beta_{M1}} V_{in}$. Although each module has a different duty cycle perturbation, it is assumed that all the DAB units have an equal normalized time shift. Besides, the ESR of the output capacitance is considered in this model. However, the ESR can be neglected compared to the load.

The following equations are obtained from Figure 4:

$$\begin{cases}
\dfrac{D_{eff1}}{K_1}\hat{v}_{cdL1} + \dfrac{\beta_{L2}}{\beta_{L1}}\dfrac{V_{in}}{K_1}\left(\hat{d}_{iL1} + \hat{d}_{vL1} + \hat{d}_{L1}\right) = sL_L\hat{i}_{LL1} + \hat{v}_{outL1} \\[2mm]
\dfrac{D_{eff1}}{K_1}\hat{v}_{cdL2} + \dfrac{\beta_{L2}}{\beta_{L1}}\dfrac{V_{in}}{K_1}\left(\hat{d}_{iL2} + \hat{d}_{vL2} + \hat{d}_{L2}\right) = sL_L\hat{i}_{LL2} + \hat{v}_{outL2} \\[2mm]
\quad\vdots \\[2mm]
\dfrac{D_{eff1}}{K_1}\hat{v}_{cdLL} + \dfrac{\beta_{L2}}{\beta_{L1}}\dfrac{V_{in}}{K_1}\left(\hat{d}_{iLL} + \hat{d}_{vLL} + \hat{d}_{LL}\right) = sL_L\hat{i}_{LLL} + \hat{v}_{outLL} \\[2mm]
\dfrac{D_{eff2}}{K_2}\hat{v}_{cdM1} + \dfrac{\beta_{L2}}{\beta_{L1}}\dfrac{V_{in}}{K_2}\left(\hat{d}_{iM1} + \hat{d}_{vM1} + \hat{d}_{M1}\right) = sL_M\hat{i}_{LM1} + \hat{v}_{outM1} \\[2mm]
\dfrac{D_{eff2}}{K_2}\hat{v}_{cdM2} + \dfrac{\beta_{L2}}{\beta_{L1}}\dfrac{V_{in}}{K_2}\left(\hat{d}_{iM2} + \hat{d}_{vM2} + \hat{d}_{M2}\right) = sL_M\hat{i}_{LM2} + \hat{v}_{outM2} \\[2mm]
\quad\vdots \\[2mm]
\dfrac{D_{eff2}}{K_2}\hat{v}_{cdMM} + \dfrac{\beta_{L2}}{\beta_{L1}}\dfrac{V_{in}}{K_2}\left(\hat{d}_{iMM} + \hat{d}_{vMM} + \hat{d}_{MM}\right) = sL_M\hat{i}_{LMM} + \hat{v}_{outMM}
\end{cases} \tag{11}$$

$$\begin{cases}
\hat{i}_{LL11} + \hat{i}_{LL21} + \cdots + \hat{i}_{LL a_{L1} 1} = \dfrac{sC_L}{sR_{cL}C_L + 1}\hat{v}_{outL1} + \dfrac{\hat{v}_{outL}}{R} \\[2mm]
\hat{i}_{LL12} + \hat{i}_{LL22} + \cdots + \hat{i}_{LL a_{L1} 2} = \dfrac{sC_L}{sR_{cL}C_L + 1}\hat{v}_{outL2} + \dfrac{\hat{v}_{outL}}{R} \\[2mm]
\quad\vdots \\[2mm]
\hat{i}_{LL1 b_{L1}} + \hat{i}_{LL2 b_{L1}} + \cdots + \hat{i}_{LL a_{L1} b_{L1}} = \dfrac{sC_L}{sR_{cL}C_L + 1}\hat{v}_{outLL} + \dfrac{\hat{v}_{outL}}{R} \\[2mm]
\hat{i}_{LM11} + \hat{i}_{LM21} + \cdots + \hat{i}_{LM a_{M1} 1} = \dfrac{sC_M}{sR_{cM}C_M + 1}\hat{v}_{outM1} + \dfrac{\hat{v}_{outM}}{R} \\[2mm]
\hat{i}_{LM12} + \hat{i}_{LM22} + \cdots + \hat{i}_{LM a_{M1} 2} = \dfrac{sC_M}{sR_{cM}C_M + 1}\hat{v}_{outM2} + \dfrac{\hat{v}_{outM}}{R} \\[2mm]
\quad\vdots \\[2mm]
\hat{i}_{LM1 b_{L1}} + \hat{i}_{LM2 b_{L1}} + \cdots + \hat{i}_{LM a_{M1} b_{L1}} = \dfrac{sC_M}{sR_{cM}C_M + 1}\hat{v}_{outMM} + \dfrac{\hat{v}_{outM}}{R}
\end{cases} \tag{12}$$

Adding equations representing the primary multimodule group in (12):

$$\sum_{i=1}^{a_{L1}} \sum_{j=1}^{b_{L1}} \hat{i}_{LLij} = \frac{sC_L}{sR_{cL}C_L + 1}\hat{v}_{outL} + \frac{b_{L1}\hat{v}_{outL}}{R} \tag{13}$$

Equation (13) can be written as:

$$\sum_{i=1}^{a_{L1}} \sum_{j=1}^{b_{L1}} \hat{i}_{LLij} = \hat{v}_{outL}\left(\frac{sRC_L + sb_{L1}R_{cL}C_L + b_{L1}}{R(1 + sR_{cL}C_L)}\right) \tag{14}$$

Adding equations representing the secondary multimodule group in (12):

$$\sum_{i=1}^{a_{M1}} \sum_{j=1}^{b_{M1}} \hat{\imath}_{LMij} = \hat{v}_{outM} \left(\frac{sRC_M + sb_{M1}R_{cM}C_M + b_{M1}}{R(1 + sR_{cM}C_M)} \right) \tag{15}$$

Equation (15) can be written as:

$$\sum_{i=1}^{a_{M1}} \sum_{j=1}^{b_{M1}} \hat{\imath}_{LMij} = \hat{v}_{outM} \left(\frac{sRC_M + sb_{M1}R_{cM}C_M + b_{M1}}{R(1 + sR_{cM}C_M)} \right) \tag{16}$$

Defining the summation terms of the module's input and output voltage appearing after summing up equations representing the primary multimodule group in (11):

$$\sum_{j=1}^{L} \hat{v}_{cdLj} = \gamma_L \hat{v}_{inL} = \gamma_L \beta_{L2} \hat{v}_{in} \tag{17}$$

where:

- $\gamma_L = 1$, if all the modules in the primary group at the input side are connected in series.
- $\gamma_L = \alpha_{L1}$, if all the modules in the primary group at the input side are connected in parallel or connected in both series and parallel.

$$\sum_{j=1}^{L} \hat{v}_{outLj} = c_L \hat{v}_{outL} = c_L b_{L2} \hat{v}_{out} \tag{18}$$

where:

- $c_L = 1$, if all the modules in the primary group at the output side are connected in series.
- $c_L = a_{L1}$, if all the modules in the primary group at the output side are connected in parallel or connected in both series and parallel.

Defining the summation terms of the module's input and output voltage appearing after summing up equations representing the secondary multimodule group in (11):

$$\sum_{j=1}^{M} \hat{v}_{cdMj} = \gamma_M \hat{v}_{inM} = \gamma_M \beta_{M2} \hat{v}_{in} \tag{19}$$

where:

- $\gamma_M = 1$, if all the modules in the secondary group at the input side are connected in series.
- $\gamma_M = \alpha_{L1}$, if all the modules in the secondary group at the input side are connected in parallel or connected in both series and parallel.

$$\sum_{j=1}^{M} \hat{v}_{outMj} = c_M \hat{v}_{outL} = c_M b_{M2} \hat{v}_{out} \tag{20}$$

where:

- $c_M = 1$, if all the modules in the secondary group at the output side are connected in series.

- $c_M = a_{M1}$, if all the modules in the secondary group at the output side are connected in parallel or connected in both series and parallel.

2.2.1. Control-to-Output Voltage Transfer Function

The relation between the output voltage and the duty cycle is obtained by performing two steps. The first step is by adding the L equations in (11) to obtain the relation between $\hat{v}_{outL}$ and $\hat{d}_{Lj}$, assuming $\hat{v}_{inL} = 0$, and $\hat{d}_{Lk} = 0$, where $k = 1, 2, \ldots, L$ and $k \neq j$, and substituting (2), (4), (14), (17) and (18). However, the second step is by adding the M equations in (11) to obtain the relation between $\hat{v}_{outM}$ and $\hat{d}_{Mj}$, assuming $\hat{v}_{inM} = 0$, and $\hat{d}_{Mk} = 0$, where $k = 1, 2, \ldots, L$ and $k \neq j$, and substituting (7), (9), (16), (19) and (20).

Adding the L equations in (11) would result in:

$$\frac{D_{eff1}}{K_1} \sum_{j=1}^{L} \hat{v}_{cdLj} + \frac{\beta_{L2}}{\beta_{L1}} \frac{V_{in}}{K_1} \left(\sum_{j=1}^{L} \hat{d}_{iLj} + \sum_{j=1}^{L} \hat{d}_{vLj} + \sum_{j=1}^{L} \hat{d}_{Lj} \right) = sL_L \sum_{j=1}^{L} \hat{i}_{LLj} + \sum_{j=1}^{L} \hat{v}_{outLj} \tag{21}$$

Substituting (2), (4) and (14) would result in:

$$\frac{D_{eff1}}{K_1} \sum_{j=1}^{L} \hat{v}_{cdLj} + \frac{\beta_{L2}}{\beta_{L1}} \frac{V_{in}}{K_1} \left(\begin{array}{l} -\frac{\beta_{L1} K_1 R_{dL}}{\beta_{L2} V_{in}} \hat{v}_{outL} \left(\frac{sRC_L + s b_{L1} R_{cL} C_L + b_{L1}}{R(1 + sR_{cL} C_L)} \right) + \\ \sum_{j=1}^{L} \frac{a_{L2} b_{L1} \beta_{L1} R_{dL} D_{eff1}}{a_{L1} b_{L2} \beta_{L2} RV_{in}} \hat{v}_{cdLj} + \hat{d}_{L1} \end{array} \right)$$
$$= sL_L \left(\frac{sRC_L + s b_{L1} R_{cL} C_L + b_{L1}}{R(1 + sR_{cL} C_L)} \right) \hat{v}_{outL} + \sum_{j=1}^{L} \hat{v}_{outLj} \tag{22}$$

Substituting (17) and (18) in (22) results in:

$$\frac{D_{eff1}}{K_1} \gamma_L \hat{v}_{inL} + \frac{\beta_{L2}}{\beta_{L1}} \frac{V_{in}}{K_1} \left(\begin{array}{l} -\frac{\beta_{L1} K_1 R_{dL}}{\beta_{L2} V_{in}} \hat{v}_{outL} \left(\frac{sRC_L + s b_{L1} R_{cL} C_L + b_{L1}}{R(1 + sR_{cL} C_L)} \right) + \\ \frac{a_{L2} b_{L1} \beta_{L1} R_{dL} D_{eff1}}{a_{L1} b_{L2} \beta_{L2} RV_{in}} \gamma_L \hat{v}_{inL} + \hat{d}_{L1} \end{array} \right)$$
$$= sL_L \left(\frac{sRC_L + s b_{L1} R_{cL} C_L + b_{L1}}{R(1 + sR_{cL} C_L)} \right) \hat{v}_{outL} + c_L \hat{v}_{outL} \tag{23}$$

Simplifying (23) results in (24):

$$G_{vdL} = \frac{\hat{v}_{outL}}{\hat{d}_{Lj}}$$
$$= \frac{\frac{\beta_{L2}}{\beta_{L1}} \frac{V_{in}}{K_1} (1 + sR_{cL} C_L)}{s^2 L_L C_L \left(1 + \frac{b_{L1} R_{cL}}{R} \right) + s \left(\frac{b_{L1} L_L}{R} + R_{dL} C_L \left(1 + \frac{b_{L1} R_{cL}}{R} \right) + c_L R_{cL} C_L \right) + \frac{b_{L1} R_{dL}}{R} + c_L} \tag{24}$$

Performing the second step which is adding the M equations in (11) to obtain the relation between $\hat{v}_{outM}$ and $\hat{d}_{Mj}$, assuming $\hat{v}_{inM} = 0$, and $\hat{d}_{Mk} = 0$, where $k = 1, 2, \ldots, M$, and $k \neq j$, and substituting (7), (9), (16), (19) and (20) would result in:

Adding the M equations in (11) would result in:

$$\frac{D_{eff2}}{K_2} \sum_{j=1}^{M} \hat{v}_{cdMj} + \frac{\beta_{M2}}{\beta_{M1}} \frac{V_{in}}{K_2} \left(\sum_{j=1}^{M} \hat{d}_{iMj} + \sum_{j=1}^{M} \hat{d}_{vMj} + \sum_{j=1}^{M} \hat{d}_{Mj} \right) = sL_M \sum_{j=1}^{M} \hat{i}_{LMj} + \sum_{j=1}^{M} \hat{v}_{outMj} \tag{25}$$

Substituting (7), (9) and (16) would result in:

$$
\frac{D_{eff2}}{K_2} + \frac{\boldsymbol{\beta_{M2}} V_{in}}{\boldsymbol{\beta_{M1}} K_2}\left(-\frac{\boldsymbol{\beta_{M1}} K_2 R_{dM}}{\boldsymbol{\beta_{M2}} V_{in}} \hat{v}_{outM}\left(\frac{sRC_M + sb_{M1}R_{cM}C_M + b_{M1}}{R(1+sR_{cM}C_M)}\right) \right.
$$
$$
\left. + \sum_{j=1}^{M} \frac{a_{M2}b_{M1}\boldsymbol{\beta_{M1}} R_{dM} D_{eff2}}{a_{M1}b_{M2}\boldsymbol{\beta_{M2}} RV_{in}} \hat{v}_{cdMj} + \hat{d}_{M1} \right)
$$
$$
= sL_M\left(\frac{sRC_M + sb_{M1}R_{cM}C_M + b_{M1}}{R(1+sR_{cM}C_M)}\right)\hat{v}_{outM} + \sum_{j=1}^{M} \hat{v}_{outMj}
\tag{26}
$$

Substituting (19) and (20) results in:

$$
\frac{D_{eff2}}{K_2}\boldsymbol{\gamma_M}\hat{v}_{inM} + \frac{\boldsymbol{\beta_{M2}} V_{in}}{\boldsymbol{\beta_{M1}} K_2}\left(-\frac{\boldsymbol{\beta_{M1}} K_2 R_{dM}}{\boldsymbol{\beta_{M2}} V_{in}} \hat{v}_{outM}\left(\frac{sRC_M + sb_{M1}R_{cM}C_M + b_{M1}}{R(1+sR_{cM}C_M)}\right) \right.
$$
$$
\left. + \frac{a_{M2}b_{M1}\boldsymbol{\beta_{M1}} R_{dM} D_{eff2}}{a_{M1}b_{M2}\boldsymbol{\beta_{M2}} RV_{in}} \boldsymbol{\gamma_M}\hat{v}_{inM} + \hat{d}_{M1} \right)
$$
$$
= sL_M\left(\frac{sRC_M + sb_{M1}R_{cM}C_M + b_{M1}}{R(1+sR_{cM}C_M)}\right)\hat{v}_{outM} + c_M\hat{v}_{outM}
\tag{27}
$$

Simplifying (27) would result in (28).

$$
G_{vdM} = \frac{\hat{v}_{outM}}{\hat{d}_{Mj}}
$$
$$
= \frac{\dfrac{\boldsymbol{\beta_{M2}} V_{in}}{\boldsymbol{\beta_{M1}} K_2}(1+sR_{cM}C_M)}{s^2 L_M C_M\left(1+\dfrac{b_{M1}R_{cM}}{R}\right) + s\left(\dfrac{b_{M1}L_M}{R} + R_{dM}C_M\left(1+\dfrac{b_{M1}R_{cM}}{R}\right) + c_M R_{cM}C_M\right) + \dfrac{b_{M1}R_{dM}}{R} + c_M}
\tag{28}
$$

By adding G_{vdL} and G_{vdM} the control-to-output voltage transfer function can be found.

2.2.2. Control-to-Filter Inductor Current Transfer Function

The relation between the filter inductor current and the duty cycle is derived by performing two steps, where the first step considers the L modules in (11) while the second step considers the M modules in (11). The first step is by substituting $\hat{v}_{outL}$ in terms of $\hat{i}_{LLj}$ using (14) in (23) and assuming $\hat{v}_{inL} = 0$, and $\hat{d}_{Lk} = 0$, where $k = 1, 2, \ldots, L$ and $k \neq j$. However, the second step is by substituting $\hat{v}_{outM}$ in terms of $\hat{i}_{LMj}$ using (16) in (27) and assuming $\hat{v}_{inM} = 0$, and $\hat{d}_{Mk} = 0$, where $k = 1, 2, \ldots, M$ and $k \neq j$.

Substituting $\hat{v}_{outL}$ in terms of $\hat{i}_{LLj}$ using (14) in (23):

$$
\frac{D_{eff1}}{K_1}\gamma_L\hat{v}_{inL} + \frac{\beta_{L2} V_{in}}{\beta_{L1} K_1}\left(-\frac{\beta_{L1} K_1 R_{dL}}{\beta_{L2} V_{in}}\sum_{i=1}^{a_{L1}}\sum_{j=1}^{b_{L1}}\hat{i}_{LLij} + \frac{a_{L2}b_{L1}\beta_{L1} R_{dL} D_{eff1}}{a_{L1}b_{L2}\beta_{L2} RV_{in}}\gamma_L\hat{v}_{inL} + \hat{d}_{L1} \right)
$$
$$
= sL_L\sum_{i=1}^{a_{L1}}\sum_{j=1}^{b_{L1}}\hat{i}_{LLij} + c_L\left(\frac{R(1+sR_{cL}C_L)}{sRC_L+sb_{L1}R_{cL}C_L+b_{L1}}\right)\sum_{i=1}^{a_{L1}}\sum_{j=1}^{b_{L1}}\hat{i}_{LLij}
\tag{29}
$$

$$
\frac{\beta_{L2} V_{in}}{\beta_{L1} K_1}\hat{d}_{L1} - R_{dL}\sum_{j=1}^{L}\hat{i}_{LLj} = sL_L\sum_{j=1}^{L}\hat{i}_{LLj} + c_L\left(\frac{R(1+sR_{cL}C_L)}{sRC_L+sb_{L1}R_{cL}C_L+b_{L1}}\right)\sum_{j=1}^{L}\hat{i}_{LLj}
\tag{30}
$$

Simplifying (30) would result in (31).

$$G_{idL} = \frac{\sum_{j=1}^{L} \hat{\imath}_{LLj}}{\hat{d}_{Lj}}$$
$$= \frac{\frac{\beta_{L2}}{\beta_{L1}} \frac{V_{in}}{K_1}\left(sRC_L + sb_{L1}R_{cL}C_L + b_{L1}\right)}{R\left(s^2 L_L C_L\left(1 + \frac{b_{L1}R_{cL}}{R}\right) + s\left(\frac{b_{L1}L_L}{R} + R_{dL}C_L\left(1 + \frac{b_{L1}R_{cL}}{R}\right) + c_{L}R_{cL}C_L\right) + \frac{b_{L1}R_{dL}}{R} + c_{L}\right)} \tag{31}$$

Performing the second step which is substituting v_{outM} in terms of i_{LMj} using (16) in (27) and assuming $v_{inM} = 0$, and $d_{Mk} = 0$, where $k = 1, 2, \ldots, M$ and $k \neq j$ would result in:

$$\frac{D_{eff2}}{K_2}\gamma_M \hat{v}_{inM} + \frac{\beta_{M2}V_{in}}{\beta_{M1}K_2}\left(-\frac{\beta_{M1}K_2 R_{dM}}{\beta_{M2}V_{in}}\sum_{i=1}^{a_{M1}}\sum_{j=1}^{b_{M1}}\hat{\imath}_{LMij} + \frac{a_{M2}b_{M1}\beta_{M1}R_{dM}D_{eff2}}{a_{M1}b_{M2}\beta_{M2}RV_{in}}\gamma_M \hat{v}_{inM} + \hat{d}_{M1}\right)$$
$$= sL_M \sum_{i=1}^{a_{M1}}\sum_{j=1}^{b_{M1}}\hat{\imath}_{LMij} + c_M\left(\frac{R(1 + sR_{cM}C_M)}{sRC_M + sb_{M1}R_{cM}C_M + b_{M1}}\right)\sum_{i=1}^{a_{M1}}\sum_{j=1}^{b_{M1}}\hat{\imath}_{LMij} \tag{32}$$

$$\frac{\beta_{M2}V_{in}}{\beta_{M1}K_2}\hat{d}_{M1} - R_{dM}\sum_{j=1}^{M}\hat{\imath}_{LMj} = sL_M\sum_{j=1}^{M}\hat{\imath}_{LMj} + c_M\left(\frac{R(1 + sR_{cM}C_M)}{sRC_M + sb_{M1}R_{cM}C_M + b_{M1}}\right)\sum_{j=1}^{M}\hat{\imath}_{LMj} \tag{33}$$

Simplifying (33) would result in (34). By adding G_{idL} and G_{idM}, the control-to-filter inductor current transfer function can be found.

$$G_{idM} = \frac{\sum_{j=1}^{M}\hat{\imath}_{LMj}}{\hat{d}_{Mj}}$$
$$= \frac{\frac{\beta_{M2}V_{in}}{\beta_{M1}K_2}\left(sRC_M + sb_{M1}R_{cM}C_M + b_{M1}\right)}{R\left(s^2 L_M C_M\left(1 + \frac{b_{M1}R_{cM}}{R}\right) + s\left(\frac{b_{M1}L_M}{R} + R_{dM}C_M\left(1 + \frac{b_{M1}R_{cM}}{R}\right) + c_M R_{cM}C_M\right) + \frac{b_{M1}R_{dM}}{R} + c_M\right)} \tag{34}$$

By adding G_{idL} and G_{idM} the control-to-output filter inductor current transfer function can be found.

2.2.3. Output Impedance

The generalized converter output impedance for the hybrid ISIP-OSOP multimodule DC-DC converter can be found by considering two groups of equations. The primary group is the L number of KCL equations presented in (12). However, the secondary group is the M number of KCL equations presented in (12).

To find the generalized converter output impedance, the KCL equation in (12) can be rewritten as follows:

$$
\begin{cases}
\hat{i}_{LL11}+\hat{i}_{LL21} + \cdots + \hat{i}_{LL_{a_{L1}}1} + \hat{i}_{outL} = g_L\hat{v}_{outL1} + \dfrac{\hat{v}_{outL}}{R} \\[2mm]
\hat{i}_{LL12} + \hat{i}_{LL22} + \cdots + \hat{i}_{LL_{a_{L1}}2} + \hat{i}_{outL} = g_L\hat{v}_{outL2} + \dfrac{\hat{v}_{outL}}{R} \\
\qquad\qquad\qquad\vdots \\[1mm]
\hat{i}_{LL1b_{L1}}+\hat{i}_{LL2b_{L1}} + \cdots + \hat{i}_{LL_{a_{L1}b_{L1}}} + \hat{i}_{outL} = g_L\hat{v}_{outLL} + \dfrac{\hat{v}_{outL}}{R} \\[2mm]
\hat{i}_{LM11} + \hat{i}_{LM21} + \cdots + \hat{i}_{LMa_{M1}1} + \hat{i}_{outM} = g_M\hat{v}_{outM1} + \dfrac{\hat{v}_{outM}}{R} \\[2mm]
\hat{i}_{LM12} + \hat{i}_{LM22} + \cdots + \hat{i}_{LMa_{M1}2} + \hat{i}_{outM} = g_M\hat{v}_{outM2} + \dfrac{\hat{v}_{outM}}{R} \\
\qquad\qquad\qquad\vdots \\[1mm]
\hat{i}_{LM1b_{L1}} + \hat{i}_{LM2b_{L1}} + \cdots + \hat{i}_{LMa_{M1}b_{L1}} + \hat{i}_{outM} = g_M\hat{v}_{outMM} + \dfrac{\hat{v}_{outM}}{R}
\end{cases}
\tag{35}
$$

where; $g_L = \dfrac{sC_L}{sR_{cL}C_L+1}$ and $g_M = \dfrac{sC_M}{sR_{cM}C_M+1}$.

Accordingly, the KCL equation in (14) can be modified as follows:

$$
\sum_{i=1}^{a_{L1}}\sum_{j=1}^{b_{L1}}\hat{i}_{LLij} + \hat{i}_{outL} = \hat{v}_{outL}\left(\frac{sRC_L + sb_{L1}R_{cL}C_L + b_{L1}}{R(1 + sR_{cL}C_L)}\right)
\tag{36}
$$

The relationship between the output voltage and the output current for the L modules is obtained by adding the L equations in (11), assuming $\hat{v}_{inL} = 0$, and $\hat{d}_{Lj} = 0$, $j = 1, 2, \ldots, L$, and substituting (2), (4), (17), (18) and (36).

$$
\begin{aligned}
&\frac{D_{eff1}}{K_1}\gamma_L\hat{v}_{inL} + \frac{\beta_{L2}}{\beta_{L1}}\frac{V_{in}}{K_1}\left(
\begin{array}{l}
-\dfrac{\beta_{L1}}{\beta_{L2}}\dfrac{K_1R_{dL}}{V_{in}}\left(\hat{v}_{outL}\left(\dfrac{sRC_L+sb_{L1}R_{cL}C_L+b_{L1}}{R(1+sR_{cL}C_L)}\right) - \hat{i}_{outL}\right)+ \\[2mm]
\dfrac{a_{L2}b_{L1}\beta_{L1}R_{dL}D_{eff1}}{a_{L1}b_{L2}\beta_{L2}RV_{in}}\gamma_L\hat{v}_{inL} + \hat{d}_{L1}
\end{array}
\right) \\[3mm]
&= sL_L\left(\left(\frac{sRC_L+sb_{L1}R_{cL}C_L+b_{L1}}{R(1+sR_{cL}C_L)}\right)\hat{v}_{outL} - \hat{i}_{outL}\right) + c_L\hat{v}_{outL}
\end{aligned}
\tag{37}
$$

$$
\begin{aligned}
&-R_{dL}\left(\hat{v}_{outL}\left(\frac{sRC_L+sb_{L1}R_{cL}C_L+b_{L1}}{R(1+sR_{cL}C_L)}\right) - \hat{i}_{outL}\right) \\[2mm]
&= sL_L\left(\left(\frac{sRC_L+sb_{L1}R_{cL}C_L+b_{L1}}{R(1+sR_{cL}C_L)}\right)\hat{v}_{outL} - \hat{i}_{outL}\right) + c_L\hat{v}_{outL}
\end{aligned}
\tag{38}
$$

Simplifying (38) would result in (39).

$$
\begin{aligned}
Z_{outL} &= \frac{\hat{v}_{outL}}{\hat{i}_{outL}} \\[2mm]
&= \frac{b_{L1}(R_{dL}+sL_L)(1+sR_{cL}C_L)}{s^2L_LC_L\left(1+\dfrac{R_{cL}}{R}\right)+s\left(\dfrac{L_L}{R}+R_{dL}C_L\left(1+\dfrac{R_{cL}}{R}\right)+c_L R_{cL}C_L\right)+\dfrac{R_{dL}}{R}+c_L}
\end{aligned}
\tag{39}
$$

Similarly, the KCL equation in (16) can be modified as follows:

$$
\sum_{i=1}^{a_{M1}}\sum_{j=1}^{b_{M1}}\hat{i}_{LMij} + \hat{i}_{outM} = \hat{v}_{outM}\left(\frac{sRC_M + sb_{M1}R_{cM}C_M + b_{M1}}{R(1 + sR_{cM}C_M)}\right)
\tag{40}
$$

The relationship between the output voltage and the output current for the M modules is derived by summing the M equations in (11), assuming $\hat{v}_{inM} = 0$, and $\hat{d}_{Mj} = 0$, $j = 1, 2, \ldots, M$, and substituting (7), (9), (19), (20) and (40).

$$\frac{D_{eff2}}{K_2}\boldsymbol{\gamma_M}\hat{v}_{inM} + \frac{\boldsymbol{\beta_{M2}}V_{in}}{\boldsymbol{\beta_{M1}}K_2}$$

$$= \left(\begin{array}{c} -\frac{\boldsymbol{\beta_{M1}}K_2 R_{dM}}{\boldsymbol{\beta_{M2}}V_{in}}\left(\hat{v}_{outM}\left(\frac{sRC_M + s\boldsymbol{b_{M1}}R_{cM}C_M + \boldsymbol{b_{M1}}}{R(1 + sR_{cM}C_M)}\right) - \hat{\imath}_{outM}\right) + \\ \frac{\boldsymbol{a_{M2}b_{M1}\beta_{M1}}R_{dM}D_{eff2}}{\boldsymbol{a_{M1}b_{M2}\beta_{M2}}RV_{in}}\boldsymbol{\gamma_M}\hat{v}_{inM} + \hat{d}_{M1} \end{array} \right) \tag{41}$$

$$sL_M\left(\left(\frac{sRC_M + s\boldsymbol{b_{M1}}R_{cM}C_M + \boldsymbol{b_{M1}}}{R(1 + sR_{cM}C_M)}\right)\hat{v}_{outM} - \hat{\imath}_{outM}\right) + \boldsymbol{c_M}\hat{v}_{outM}$$

$$- R_{dM}\left(\hat{v}_{outM}\frac{sRC_M + s\boldsymbol{b_{M1}}R_{cM}C_M + \boldsymbol{b_{M1}}}{R(1 + sR_{cM}C_M)} - \hat{\imath}_{outM}\right)$$

$$= sL_M\left(\left(\frac{sRC_M + s\boldsymbol{b_{M1}}R_{cM}C_M + \boldsymbol{b_{M1}}}{R(1 + sR_{cM}C_M)}\right)\hat{v}_{outM} - \hat{\imath}_{outM}\right) + \boldsymbol{c_M}\hat{v}_{outM} \tag{42}$$

Simplifying (42) would result in (43):

$$Z_{outM} = \frac{\hat{v}_{outM}}{\hat{\imath}_{outM}}$$

$$= \frac{\boldsymbol{b_{M1}}(R_{dM} + sL_M)(1 + sR_{cM}C_M)}{s^2 L_M C_M\left(1 + \frac{\boldsymbol{b_{M1}}R_{cM}}{R}\right) + s\left(\frac{\boldsymbol{b_{M1}}L_M}{R} + R_{dM}C_M\left(1 + \frac{\boldsymbol{b_{M1}}R_{cM}}{R}\right) + \boldsymbol{c_M}R_{cM}C_M\right) + \frac{\boldsymbol{b_{M1}}R_{dM}}{R} + \boldsymbol{c_M}} \tag{43}$$

By adding Z_{outL} and Z_{outM}, the output impedance transfer function can be found.

2.2.4. Converter Gain

The generalized relationship between the output voltage and the input voltage of the hybrid ISIP-OSOP DC-DC converter can be found by performing two steps. The first step is adding the L number of KVL equations presented in (11) for the primary multimodule DC-DC converters, assuming $\hat{d}_{Lj} = 0$, $j = 1, 2, \ldots, L$, and substituting (2), (4), (14), (17) and (18) in the added equation. However, the second step is adding the M number of KVL equations presented in (11) for the secondary multimodule DC-DC converters, assuming $\hat{d}_{Mj} = 0$, $j = 1, 2, \ldots, M$, and substituting (7), (9), (16), (19) and (20) in the added equation.

Adding the L number of KVL equations in (11), assuming $\hat{d}_{Lj} = 0$, $j = 1, 2, \ldots, L$, and substituting (2), (4) and (14) would result in:

$$\frac{D_{eff1}}{K_1}\sum_{j=1}^{L}\hat{v}_{cdLj} + \frac{\boldsymbol{\beta_{L2}}V_{in}}{\boldsymbol{\beta_{L1}}K_1}\left(\begin{array}{c} -\frac{\boldsymbol{\beta_{L1}}K_1 R_{dL}}{\boldsymbol{\beta_{L2}}V_{in}}\hat{v}_{outL}\left(\frac{sRC_L + s\boldsymbol{b_{L1}}R_{cL}C_L + \boldsymbol{b_{L1}}}{R(1 + sR_{cL}C_L)}\right) + \\ \sum_{j=1}^{L}\frac{\boldsymbol{a_{L2}b_{L1}\beta_{L1}}R_{dL}D_{eff1}}{\boldsymbol{a_{L1}b_{L2}\beta_{L2}}RV_{in}}\hat{v}_{cdLj} \end{array} \right) \tag{44}$$

$$= sL_L\left(\frac{sRC_L + s\boldsymbol{b_{L1}}R_{cL}C_L + \boldsymbol{b_{L1}}}{R(1 + sR_{cL}C_L)}\right)\hat{v}_{outL} + \sum_{j=1}^{L}\hat{v}_{outLj}$$

Substituting (17) and (18) in (44) would give:

$$\frac{D_{eff1}}{K_1}\gamma_L\left(1+\frac{a_{L2}b_{L1}R_{dL}}{a_{L1}b_{L2}R}\right)\hat{v}_{inL}$$
$$= (sL_L + R_{dL})\left(\left(\frac{sRC_L+sb_{L1}R_{cL}C_L+b_{L1}}{R(1+sR_{cL}C_L)}\right)\hat{v}_{outL}\right) + c_L\hat{v}_{outL} \tag{45}$$

Simplifying (45) would result in (46):

$$G_{vgL} = \frac{\hat{v}_{outL}}{\hat{v}_{inL}}$$
$$= \frac{\frac{D_{eff1}}{K_1}\gamma_L\left(1+\frac{a_{L2}b_{L1}R_{dL}}{a_{L1}b_{L2}R}\right)(1+sR_{cL}C_L)}{s^2L_LC_L\left(1+\frac{b_{L1}R_{cL}}{R}\right)+s\left(\frac{b_{L1}L_L}{R}+R_{dL}C_L\left(1+\frac{b_{L1}R_{cL}}{R}\right)+c_LC_L\right)+\frac{b_{L1}R_{dL}}{R}+c_L} \tag{46}$$

Similarly, Adding the M number of KVL equations in (11), assuming $d_{Mj} = 0$, $j = 1, 2, \ldots, M$, and substituting (7), (9) and (16) would result in:

$$\frac{D_{eff2}}{K_2}\sum_{j=1}^{M}\hat{v}_{cdMj}+\frac{\beta_{M2}V_{in}}{\beta_{M1}K_2}\left(\begin{array}{l}-\dfrac{\beta_{M1}K_2R_{dM}}{\beta_{M2}V_{in}}\hat{v}_{outM}\left(\dfrac{sRC_M+sb_{M1}R_{cM}C_M+b_{M1}}{R(1+sR_{cM}C_M)}\right)+\\[2ex]\displaystyle\sum_{j=1}^{M}\dfrac{a_{M2}b_{M1}\beta_{M1}R_{dM}D_{eff2}}{a_{M1}b_{M2}\beta_{M2}RV_{in}}\hat{v}_{cdMj}+\hat{d}_{M1}\end{array}\right)$$
$$= sL_M\left(\frac{sRC_M+sb_{M1}R_{cM}C_M+b_{M1}}{R(1+sR_{cM}C_M)}\right)\hat{v}_{outM}+\sum_{j=1}^{M}\hat{v}_{outMj} \tag{47}$$

Substituting (19) and (20) in (47) would give:

$$\frac{D_{eff2}}{K_2}\gamma_M\left(1+\frac{a_{M2}b_{M1}R_{dM}}{a_{M1}b_{M2}R}\right)\hat{v}_{inM}$$
$$+ c_M\hat{v}_{outM}\left(\begin{array}{l}-\dfrac{\beta_{M1}K_2R_{dM}}{\beta_{M2}V_{in}}\hat{v}_{outM}\left(\dfrac{sRC_M+sb_{M1}R_{cM}C_M+b_{M1}}{R(1+sR_{cM}C_M)}\right)+\\[2ex]\displaystyle\sum_{j=1}^{M}\dfrac{a_{M2}b_{M1}\beta_{M1}R_{dM}D_{eff2}}{a_{M1}b_{M2}\beta_{M2}RV_{in}}\hat{v}_{cdMj}+\hat{d}_{M1}\end{array}\right)$$
$$= sL_M\left(\frac{sRC_M+sb_{M1}R_{cM}C_M+b_{M1}}{R(1+sR_{cM}C_M)}\right)\hat{v}_{outM}+\sum_{j=1}^{M}\hat{v}_{outMj} \tag{48}$$

Simplifying (48) would result in (49):

$$G_{vgM} = \frac{\hat{v}_{outM}}{\hat{v}_{inM}}$$
$$= \frac{\frac{D_{eff2}}{K_2}\gamma_M\left(1+\frac{a_{M2}b_{M1}R_{dM}}{a_{M1}b_{M2}R}\right)(1+sR_{cM}C_M)}{s^2L_MC_M\left(1+\frac{b_{M1}R_{dM}}{R}\right)+s\left(\frac{b_{M1}L_M}{R}+R_{dM}C_M\left(1+\frac{b_{M1}R_{cM}}{R}\right)+c_MC_M\right)+\frac{b_{M1}R_{dM}}{R}+c_M} \tag{49}$$

By adding G_{vgL} and G_{vgM}, the output impedance transfer function can be found.

3. Hybrid Input-Series Output-Parallel (ISOP) Multimodule DC-DC Converter

In this section, the hybrid ISOP multimodule power converter circuit diagram, as well as the hybrid ISOP multimodule converter small-signal analysis, are discussed. The analysis carried out in this section is not limited to unidirectional power flow and can be applied for bidirectional power flow.

The generalized small-signal modeling presented in Section 2 is used to derive the small-signal model for the eight-module hybrid ISOP power converter.

3.1. ISOP Circuit Diagram

The conventional ISOP converter shown in Figure 5a consists of multiple DAB units that are connected in series and in parallel at the input and the output sides, respectively, where all the modules are assumed identical. However, the concept of the hybrid ISOP power converter shown in Figure 5b is dividing conventional ISOP multimodule DC-DC converters into two groups. The primary group consists of identical ISOP DC-DC converters and is responsible for delivering a large portion of the total required power with lower switching frequency. This is shown in Figure 5c. However, the secondary group consists of another identical ISOP multimodule DC-DC converters. It is responsible for delivering the remaining power with higher switching frequency. This is shown in Figure 5d. This would result in two groups of multimodule converters operating at a different switching frequency, and utilizing different leakage inductance, transformers, filter inductors, and capacitors.

In this paper, the EV UFC specifications are assumed to deliver a total power of 200 kW using a battery voltage of 400 V, and assuming a grid voltage of 10 kV. It is assumed that the primary group handles 80% of the total battery charging power, while the secondary group handles 20% of the total battery charging power. Therefore, the primary group is responsible for delivering 160 kW, which $\frac{4}{5}$ of the total required power. The primary group is assumed to operate at switching frequency f_{sL} of 10 kHz. However, the secondary group is responsible for delivering the remaining 40 kW, which is $\frac{1}{5}$ of the total required power. The secondary group is assumed to operate at switching frequency of f_{sM} 100 kHz. Accordingly, the input voltage of the primary group V_{inL} is 8 kV, which is $\frac{4}{5}$ of the total input voltage V_{in}, while the input voltage of the secondary group V_{inM} is 2 kV, which is $\frac{1}{5}$ of the total input voltage V_{in}. Similarly, the output current of the primary group I_{oL} is 400 A, which is $\frac{4}{5}$ of the total output current I_{out}, while the output current of the secondary group I_{oM} is 100 A, which is $\frac{1}{5}$ of the total output current I_{out}. It is essential to mention that the portions $\frac{4}{5}$ and $\frac{1}{5}$ are denoted as K_L and K_M, respectively.

By ensuring equal IVS for the primary group and secondary group, the input voltage per module in the primary group is reduced to $\frac{V_{inL}}{4}$, while the input voltage per module in the secondary group is $\frac{V_{inM}}{4}$. Besides, by ensuring equal OCS for the primary and secondary group, the output current of each module in the primary group is reduced to $\frac{I_{oL}}{4}$, while the output current of each module in the secondary group is reduced to $\frac{I_{oM}}{4}$. In which, V_{inL}, V_{inM}, I_{oL}, and I_{oM} are the input voltages and output currents of the primary group and secondary group, respectively.

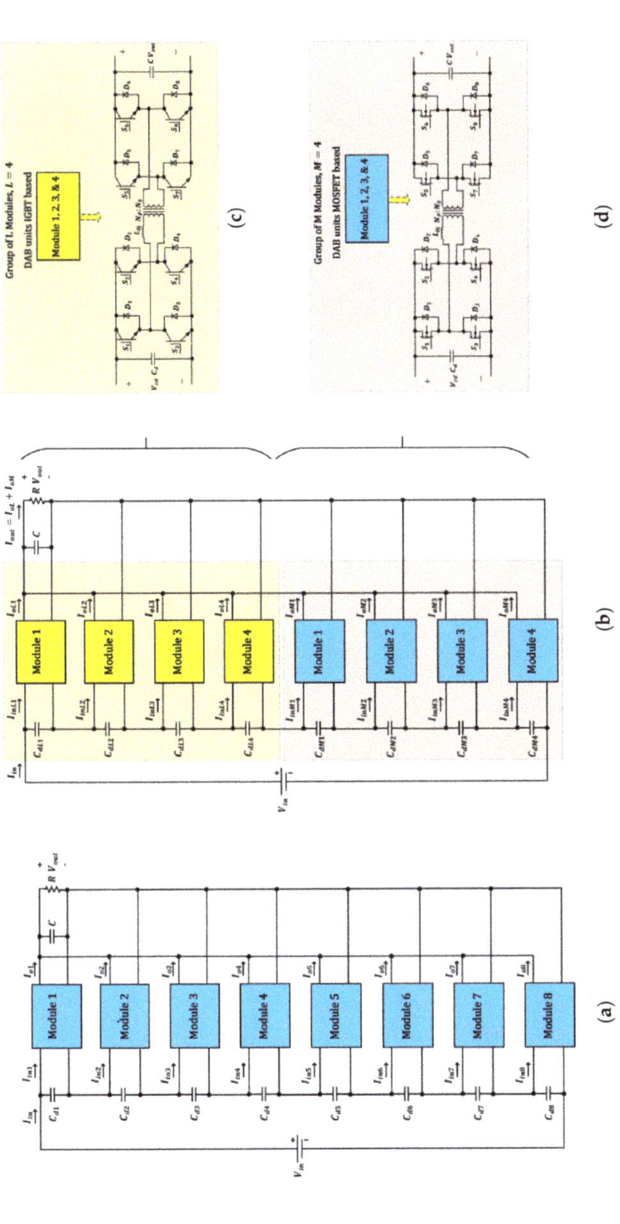

Figure 5. Eight-module ISOP multimodule DC-DC converter circuit diagram; (**a**) Conventional ISOP DC-DC Converter, (**b**) Hybrid ISOP DC-DC Converter, (**c**) DAB Converter based on IGBTs, (**d**) DAB Converter based on MOSFETs.

3.2. Hybrid ISOP Small Signal Analysis

The eight-module hybrid ISOP converter small-signal model shown in Figure 6 is derived using [36]. The generalized model derived in the previous section is used to derive the small-signal functions for the presented converter in Figure 6, as shown in Table 2. The presented transfer functions are used in the control strategy scheme presented in the power balancing section.

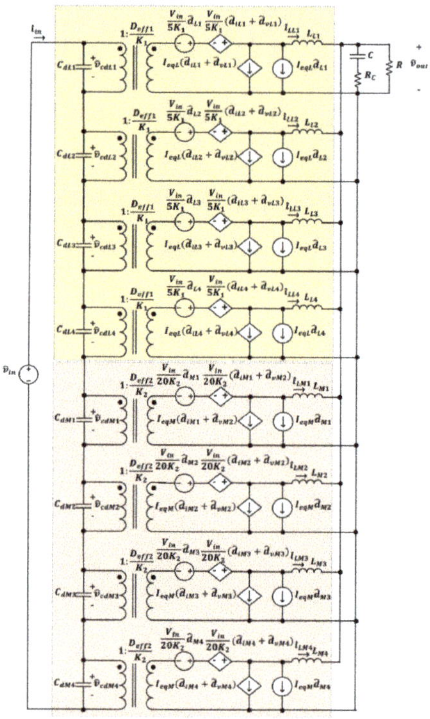

Figure 6. Hybrid ISOP DC-DC converter small-signal model for eight-modules.

Table 2. Generalized model verification with the eight-module hybrid ISOP DC-DC converter.

Transfer Functions for an Eight-Module Hybrid ISOP DC-DC Converter	
G_{vdL}	$\dfrac{\frac{V_{in}}{5K_1}}{s^2 L_L C_L + s(\frac{L_L}{R} + R_{dL} C_L) + \frac{R_{dL}}{R} + 4}$
G_{vdM}	$\dfrac{\frac{V_{in}}{20K_2}}{s^2 L_M C_M + s(\frac{b_{M1} L_M}{R} + R_{dM} C_M) + \frac{R_{dM}}{R} + 4}$
G_{idL}	$\dfrac{\frac{V_{in}}{5K_1}(sRC_L + 1)}{R(s^2 L_L C_L + s(\frac{L_L}{R} + R_{dL} C_L) + \frac{R_{dL}}{R} + 4)}$
G_{idM}	$\dfrac{\frac{V_{in}}{20K_2}(sRC_M + b_{M1})}{R(s^2 L_M C_M + s(\frac{L_M}{R} + R_{dM} C_M) + \frac{R_{dM}}{R} + 4)}$
Z_{outL}	$\dfrac{(R_{dL} + sL_L)}{s^2 L_L C_L + s(\frac{L_L}{R} + R_{dL} C_L) + \frac{R_{dL}}{R} + 4}$
Z_{outM}	$\dfrac{(R_{dM} + sL_M)}{s^2 L_M C_M + s(\frac{L_M}{R} + R_{dM} C_M) + \frac{R_{dM}}{R} + 4}$
G_{vgL}	$\dfrac{\frac{D_{eff1}}{K_1}(1 + \frac{R_{dL}}{R})}{s^2 L_L C_L + s(\frac{L_L}{R} + R_{dL} C_L) + \frac{R_{dL}}{R} + 4}$
G_{vgM}	$\dfrac{\frac{D_{eff2}}{K_2}(1 + \frac{R_{dM}}{20R})}{s^2 L_M C_M + s(\frac{L_M}{R} + R_{dM} C_M) + \frac{R_{dM}}{R} + 4}$

4. Efficiency and Power Density Assessment of the Hybrid Multimodule DC-DC Converter

4.1. Efficiency Assessment

According to the presented ratings of the DAB units, the primary group utilizes power modules that are rated at 40 kW while the secondary group utilizes power modules that are rated at 10 kW. Accordingly, to realize the total desired power, relying only on the secondary group would result in a total number of 20 modules. However, relying on the primary group would result in a total number of five modules. Therefore, it can be said that the primary group results in a lower number of modules but has a limitation in terms of the switching frequency, while the secondary group results in a higher number of modules but with higher switching capability. On the other hand, applying the presented concept would result in a total number of eight modules. Table 3 presents a comparison between the three concepts in terms of the number of employed modules as well as power, voltage, and current ratings.

Table 3. System parameters for conventional and hybrid multimodule DC-DC converters.

Parameters	Multimodule Converter Relying on the Secondary Group	Multimodule Converter Relying on the Primary Group	Hybrid Multimodule Converter	
			Primary Group	Secondary Group
Total rated power	200 kW		160 kW	40 kW
Rated power per module	10 kW	40 kW	40 kW	10 kW
Overall input voltage	10 kV		8 kV	2 kV
Input voltage per module	500 V	2 kV	2 kV	500 V
Total input current		20 A		
Input current per module		20 A		
Overall output voltage		400 V		
Output voltage per module		400 V		
Total output current	500 A		400 A	100 A
Output current per module	25 A	100 A	100 A	25 A
Number of modules	20	5	Total of 8	
			4	4
Switching frequency	100 kHz	10 kHz	100 kHz	10 kHz

The converter efficiency can be presented as in (50):

$$\eta = \frac{P_{in} - P_t}{P_{in}} \tag{50}$$

where, P_t represents the total losses in the employed converter. The semiconductor devices' losses include two types; conduction and switching losses. It is worth mentioning that the following analysis is carried out considering MOSFETs for low power modules and IGBTs for high power modules. The semiconductor conduction losses of the primary and secondary sides can be obtained using the RMS switch currents $I_{S1,RMS}$ and $I_{S2,RMS}$, respectively with the primary and secondary drain-to-source resistances R_{DS1} and R_{DS2} in case of using MOSFETs. The RMS switch currents can be found from the RMS inductor current as follows [42]:

$$I_{S1,RMS} = \frac{I_{L,RMS}}{\sqrt{2}}$$
$$I_{S2,RMS} = n\frac{I_{L,RMS}}{\sqrt{2}} \tag{51}$$

The conduction losses of the primary and secondary sides power switches can be represented as:

$$P_{S1,Cond.} = 4(I_{S1,RMS})^2 R_{DS1}$$
$$P_{S2,Cond.} = 4(I_{S2,RMS})^2 R_{DS2} \tag{52}$$

In case of using IGBTs, the conduction losses can be obtained using the collector-to-emitter voltage $V_{CE(Sat)}$, average IGBT current I_{IGBT} and the duty cycle D. The conduction losses of the primary and secondary IGBTs can be represented as follows:

$$P_{S1,Cond.} = 4V_{CE(Sat)}I_{IGBT1}D$$
$$P_{S2,Cond.} = 4V_{CE(Sat)}I_{IGBT2}D \tag{53}$$

The modulation schemes derived and presented in [29] allow the DAB power converter to operate under ZVS throughout the entire period. Hence the switching losses of the employed power devices can be neglected, assuming that the converter is operating under ZVS [42–45].

The aforementioned power losses are the conduction losses of only one DAB unit. However, the presented converter has a hybrid modular structure, meaning that Equations (52) and (53) are modified according to the configuration of the proposed multimodule converter to include the primary and secondary groups consisting of L and M isolated modules, respectively. Therefore, Equations (52) and (53) are modified to include the conduction losses in the IGBTs and MOSFETs for multiple numbers of DAB units, as shown in (54) and (55). The primary-side conduction losses of the hybrid multimodule converter include the conduction losses in the IGBTs and the conduction losses in the MOSFETs for L and M modules, respectively, and can be represented as follows:

$$P_{S1,Cond.} = 4LV_{CE(Sat)}I_{IGBT1}D + 4M(I_{S1,RMS})^2 R_{DS1} \tag{54}$$

where, $I_{S1,RMS} = \frac{I_{LM,RMS}}{\sqrt{2}}$.

Similarly, the secondary-side conduction losses can be represented as follows:

$$P_{S2,Cond.} = 4LV_{CE(Sat)}I_{IGBT2}D + 4M(I_{S2,RMS})^2 R_{DS2} \tag{55}$$

where, $I_{S2,RMS} = n\frac{I_{LM,RMS}}{\sqrt{2}}$.

To evaluate the losses associated with the hybrid ISOP DC-DC converter and compare it with conventional multimodule DC-DC converter relying on the secondary group and conventional multimodule relying on the primary group, Equations (52)–(55) are used to calculate the conduction losses associated with the semiconductor devices. It is assumed that the turns ratio of the employed transformers is 1 : 1 and that the duty cycle is 0.5 with an RMS inductor current of 25 A for each module, where each converter is rated at 200 kW. The number of L and M modules is specified in Table 3 for the three converters. In the provided assessment, the device parameters of a SiC MOSFET CMF20120D with a drain-to-source resistance of 80 mΩ and the device parameters of an IGBT IKW25N120T2 with a collector-to-emitter voltage of 1.7 V are considered. Considering only the conduction losses of the switching devices in the three multimodule converters, the following can be observed. Conventional multimodule DC-DC converter relying on the secondary group and conventional multimodule DC-DC converter relying on the primary group would results in losses of 5.1 kW (i.e., efficiency of 97.5%) and 680 W (i.e., efficiency of 99.6%), respectively. However, conduction losses in the hybrid multimodule DC-DC converter are found to be 1.6 kW ((i.e., efficiency of 99.2%). Figure 7 presents the efficiency loss curve with respect to power loading for the three converter systems.

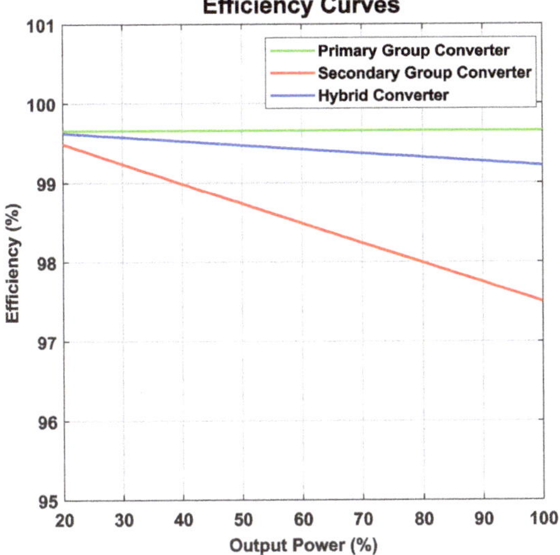

Figure 7. Hybrid Efficiency loss curve with respect to power loading.

4.2. Power Density Assessment

This subsection presents a rough estimation of the power density for the three converters presented in Table 3. The power density can be evaluated in terms of power losses and the volume of the converter components, as defined in (56). The total volume of the converter can be represented by summing up the volume of the utilized power switches, heat sinks, the transformer's winding, and the transformer's core as defined in (57) [46]:

$$\rho = \frac{P_{in} - P_t}{Volume} \tag{56}$$

$$Vol_t = Vol_{sw} + Vol_{HS} + Vol_{Core} + Vol_{Winding} \tag{57}$$

The volume of the overall converters is evaluated, considering the study provided in [47,48]. In which it is assumed that the converter components contribute with the same percentage as presented in [48]. Based on the study provided in [47,48], the volume contribution for the converter components is presented in Figures 8 and 9 considering hard switching and soft switching operation, respectively. It can be observed from Figure 8 that the volume of the heat sinks in the primary group multimodule DC-DC converter is almost the same as the volume of the heat sinks in the secondary group multimodule DC-DC converter. However, the volume of the transformer is higher in the primary group multimodule DC-DC converter due to the lower switching frequency. Accordingly, considering the losses for the three converters presented earlier, the power density of the conventional multimodule DC-DC converter relying on the secondary group is 12.2 kW/L, while the power density of the conventional multimodule DC-DC converter relying on the primary group is 9.9 kW/L. On the other hand, the power density in the hybrid multimodule DC-DC converter is found to be 10.3 kW/L.

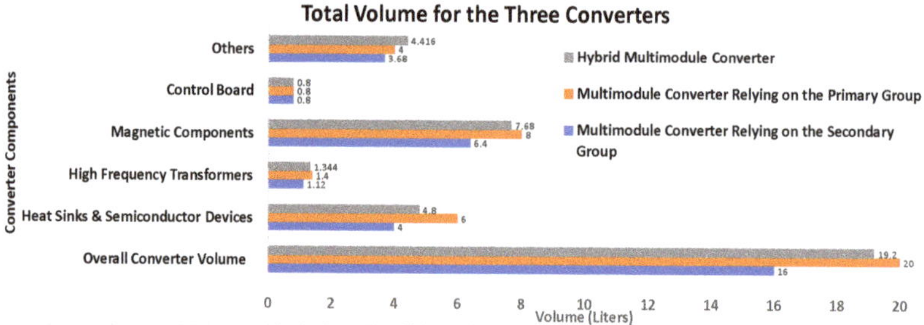

Figure 8. Converter components volume contribution in liters considering hard switching operation.

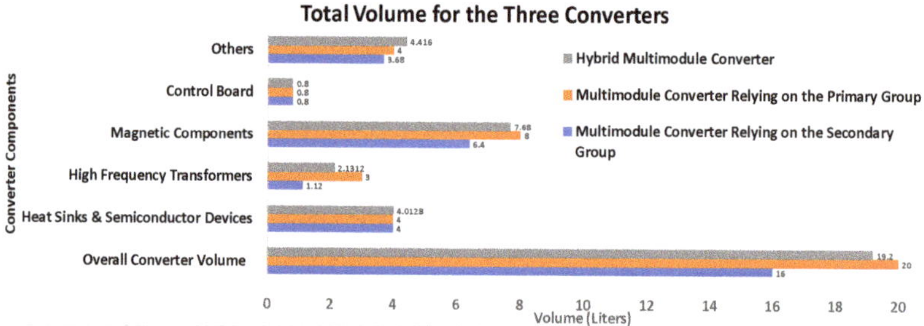

Figure 9. Converter components volume contribution in liters considering soft-switching operation.

5. Power-Sharing in the Eight-Module Hybrid ISOP Fast Charger DC-DC Converter

Since modules in practical applications are not identical, any mismatch in the parameter values can cause unequal power distribution among the modules. Consequently, the voltage of modules is unbalanced, which may cause heavy loading or thermal overstress [39]. Accordingly, a control scheme that ensures uniform power-sharing among the modules is required to achieve reliable operation for the hybrid ISIP-OSOP DC-DC converter.

Since the presented converter is connected in series and parallel at the input and output sides, respectively, a control scheme that ensures IVS and OCS is required. A control strategy for equal power-sharing among the modules is addressed for the eight-module hybrid ISOP DC-DC converter. In other words, a cross-feedback OCS (CFOCS) for ISOP has been presented in [49] to ensure both equal IVS and OCS. The OCS is achieved among the modules and automatically ensuring IVS without the need for an IVS control loop. The presented control strategy in [49] has a fault-tolerant feature even when introducing a mismatch in the module's parameters. In addition, the output voltage regulation for the converter is simplified. The concept of this control strategy is based on applying the feedback control to be the summation of all the other current control loops instead of applying its own current feedback control loop.

In this paper, the CFOCS is applied to the presented hybrid ISOP converter to ensure uniform power-sharing where the system parameters are shown in Table 4. The OCS control, shown in Figure 10, consists of one outer output current loop and eight inner current loops where four are dedicated to the primary multimodule group, and the other four inner loops are dedicated to the secondary multimodule group. The control scheme, shown in Figure 10, for the eight-module hybrid ISOP converter is current-controlled considering a reflex charging technique that is termed as burp charging or negative pulse charging. The control scheme, presented in Figure 10, is tested with reference current relying on the burp charging algorithm to the output current reference signal. The charging cycle starts with a positive sequence from 0.2 s to 0.6 s. After that, a rest period for 0.1 s is applied, then a short discharge sequence for 0.1 s.

Table 4. System parameters used in simulation.

Parameters	Primary Multimodule Group				Secondary Multimodule Group			
	Module 1	Module 2	Module 3	Module 4	Module 1	Module 2	Module 3	Module 4
Overall converter rated power		160 kW		200 kW		40 kW		
Rated power per module		40 kW				10 kW		
Total input voltage		8 kV		10 kV		2 kV		
Input voltage per module		2 kV				500 V		
Overall output voltage				400 V				
Module output voltage				400 V				
DAB units				8				
Turns ratio	1 : 1	1 : 0.95	1 : 0.9	1 : 0.85	4 : 1	4 : 0.95	4 : 0.9	4 : 0.85
Leakage inductance	80 μH	89 μH	99 μH	11 μH	500 nH	554 nH	617 nH	692 nH
Effective duty cycle	0.8	0.84	0.89	0.94	0.8	0.84	0.89	0.94
Input capacitance	50 μF	80 μF	50 μF	80 μF	35 μF	57 μF	35 μF	57 μF
Filter inductance	50 mH	60 mH	60 mH	50 mH	35 mH	42 mH	42 mH	35 mH
Filter capacitance				300 μF				
Resistance				0.8 Ω				
Switching frequency		10 kHz				100 kHz		

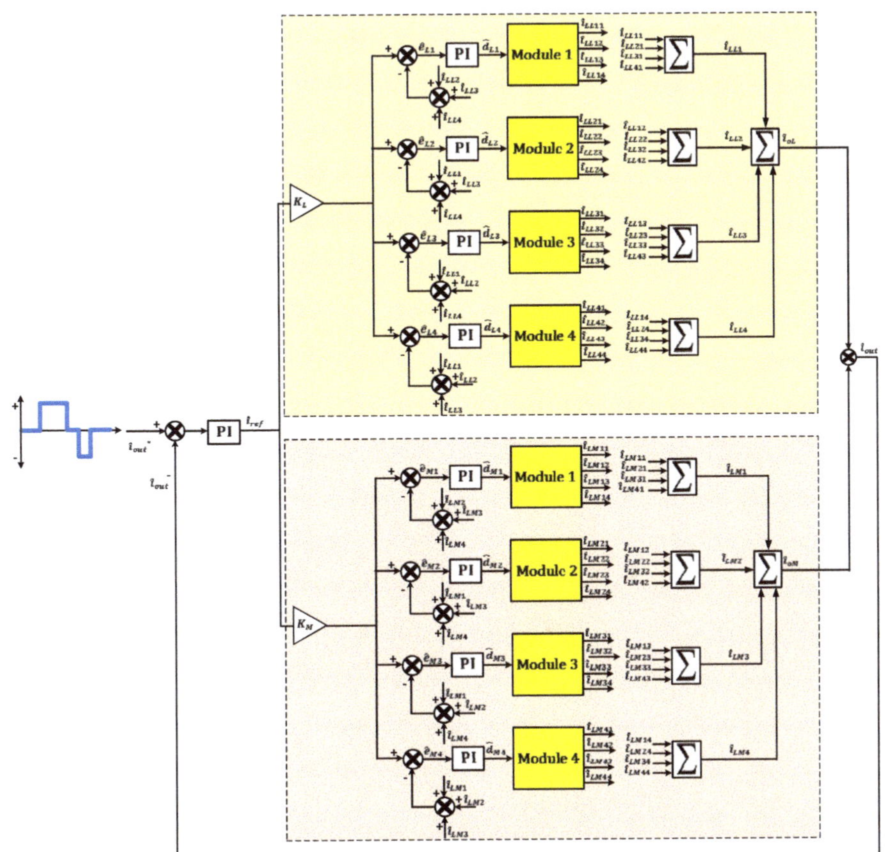

Figure 10. OCS control scheme for the proposed hybrid ISOP DC-DC converter.

6. Discussion

A MatLab/Simulink model is simulated using the small-signal model derived in Section 3 along with the control strategy presented in Figure 10, where the parameter mismatch presented in Table 4 is introduced to the employed modules.

As can be observed from Figure 11, the control strategy for the proposed eight-module hybrid converter can accomplish the requirements. Results, presented in Figure 11, demonstrate that the controller in Figure 10 compensates for the negative influences resulting from the system parameters mismatch. In which the modular input voltages and the modular output currents are equally shared between the four modules. It can be seen from the presented results in Figure 11 that the primary multimodule group is in charge of delivering $\frac{4}{5}$ of the total desired power while the secondary multimodule group is in charge of delivering $\frac{1}{5}$ of the total desired power. Besides, the output current of the proposed power converter tracks the reference current that relies on the burp charging algorithm. Accordingly, it can be concluded that the applied control strategy is reliable and that equal power-sharing is achieved between the employed modules.

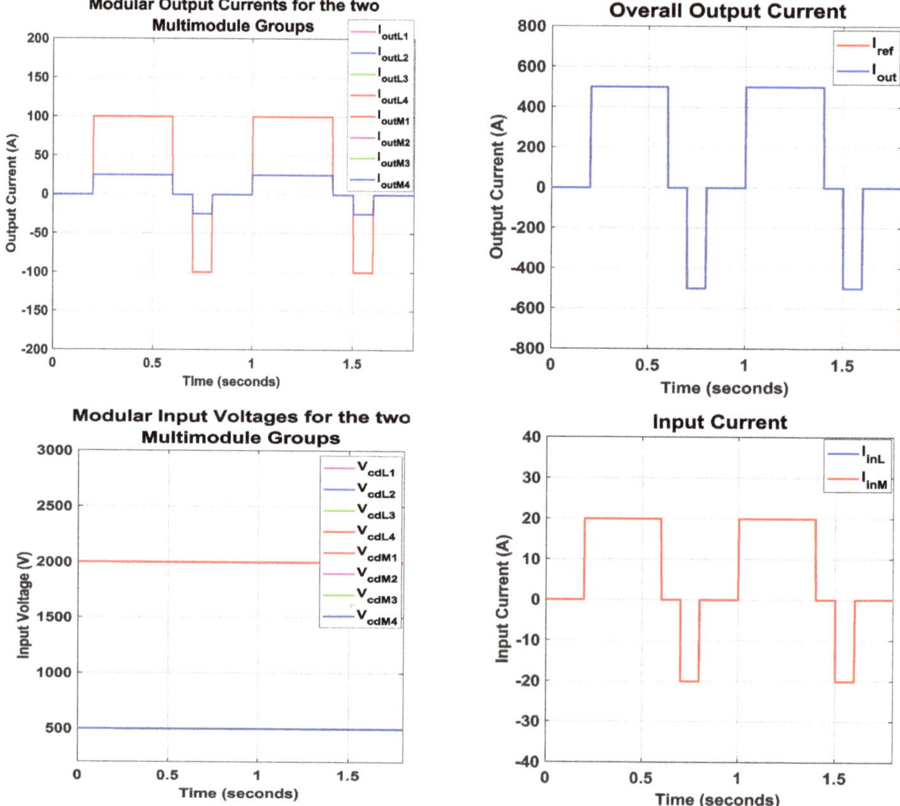

Figure 11. Simulation results for the eight-module hybrid ISOP system.

7. Conclusions

This paper introduces a hybrid multimodule DC-DC converter for EV UFC to achieve both high efficiency and high power density. The hybrid concept is achieved through employing two different groups of multimodule converters. The first is designed to be in charge of a high fraction of the total required power, operating relatively at a low switching frequency. While the second is designed for a small fraction of the total power, operating relatively at a high switching frequency. To support the power converter controller design, a generalized small-signal model for the hybrid multimodule DC-DC converter is studied in detail. This in turn supports the analysis and control design. In addition, the efficiency and power density for the conventional multimodule DC-DC converter based on the primary group, conventional multimodule DC-DC converter based on the secondary group as well as the presented hybrid DC-DC converter are evaluated. In which it has been shown that the presented converter can achieve both high efficiency (99.6%) and high power density (10.3 kW/L), compromising between the two other conventional converters. Since the power switches are the key contributors to the losses and the volume of the overall converter. It is worth mentioning that the assessment provided in this paper takes into account the conduction losses and the volume of the semiconductor switches, assuming that the converters are operating under zero voltage switching (ZVS). Furthermore, cross feedback output current sharing (CFOCS) for the hybrid input-series output-parallel (ISOP) multimodule DC-DC converters to ensure uniform power-sharing among the employed modules and the desired fraction of power handled by each multimodule group is examined. The control scheme for a hybrid eight-module ISOP power converter of 200 kW is investigated. The controller is tested with a reference current that relies on reflex charging scheme. The power loss analysis of the hybrid multimodule converter is provided. Simulation results using the MatLab/Simulink platform are provided to elucidate the presented concept considering parameter mismatches. Simulation results show that the modular input voltage and the modular output current are equally shared among the four modules of each group with the required ratio between the two multimodule groups. Numerical calculation in terms of losses is carried out for the presented converter considering conduction losses of the power switches.

Author Contributions: M.E. and A.M. contributed to the whole research work and analysis tools; M.E. wrote the paper. This work was performed under the supervision with regular and continuous feedback of A.M. All authors have read and agreed to the published version of the manuscript.

Funding: This work was supported by NPRP grant NPRP (10-0130-170286) from the Qatar National Research Fund (a member of Qatar Foundation).

Acknowledgments: This work was supported by NPRP grant NPRP (10-0130-170286) from the Qatar National Research Fund (a member of Qatar Foundation). The statements made herein are solely the responsibility of the authors.

Conflicts of Interest: The authors declare no conflict of interest.

Abbreviations

ICEs	Internal Combustion Engines
EVs	Electric Vehicles
UFC	Ultrafast Charging
DAB	Dual Active Bridge
DHB	Dual Half Bridge
ZVSZ	Zero Voltage Switching
CSC	Zero Current Switching
CFOCS	Cross Feedback Output Current Sharing
ISOP	Input-Series Output-Parallel
ISIP-OSOP	Input-Series Input-Parallel Output-Series Output-Parallel
ICS	Input Current Sharing
IVS	Input Voltage Sharing
OCS	Output Current Sharing
OVS	Output Voltage Sharing

References

1. Chan, C.C. The State of the Art of Electric, Hybrid, and Fuel Cell Vehicles. *Proc. IEEE* **2007**, *95*, 704–718. [CrossRef]
2. Wilson, J.R.; Burgh, G. *Energizing Our Future: Rational Choices for the 21st Century*; John Wiley & Sons Inc.: Hoboken, NJ, USA, 2008.
3. Ehsani, M.; Gao, Y.; Gay, S.E.; Emadi, A. *Modern Electric, Hybrid Electric, and Fuel Cell Vehicles: Fundamentals, Theory, and Design*; CRC Press LLC: Boca Raton, FL, USA, 2005.
4. ElMenshawy, M.; Massoud, A. Modular Isolated DC-DC Converters for Ultra-Fast EV Chargers: A Generalized Modeling and Control Approach. *Energies* **2020**, *13*, 2540. [CrossRef]
5. Wang, X.; He, Z.; Yang, J. Electric vehicle fast-charging station unified modeling and stability analysis in the dq frame. *Energies* **2018**, *11*, 1195. [CrossRef]
6. Yang, X.; Li, J.; Zhang, B.; Jia, Z.; Tian, Y.; Zeng, H.; Lv, Z. Virtual Synchronous Motor Based-Control of a Three-Phase Electric Vehicle Off-Board Charger for Providing Fast-Charging Service. *Appl. Sci.* **2018**, *8*, 856.
7. Lv, Z.; Xia, Y.; Chai, J.; Yu, M.; Wei, W. Distributed Coordination Control Based on State-of-Charge for Bidirectional Power Converters in a Hybrid AC/DC Microgrid. *Energies* **2018**, *11*, 1011. [CrossRef]
8. Zhu, Y.; Wang, T.; Xiong, L.; Zhang, G.; Qian, X. Parallel Control Method Based on the Consensus Algorithm for the Non-Isolated AC/DC Charging Module. *Energies* **2018**, *11*, 2828. [CrossRef]
9. Hõimoja, H.; Vasiladiotis, M.; Grioni, S.; Capezzali, M.; Rufer, A.; Püttgen, H.B. *Toward Ultrafast Charging of Electric Vehicles*; 2012 CIGRE Session: Paris, France, 2012.
10. Tu, H.; Feng, H.; Srdic, S.; Lukic, S. Extreme Fast Charging of Electric Vehicles: A Technology Overview. *IEEE Trans. Transp. Electrif.* **2019**, *5*, 861–878. [CrossRef]
11. Ronanki, D.; Kelkar, A.; Williamson, S.S. Extreme fast charging technology—Prospects to enhance sustainable electric transportation. *Energies* **2019**, *12*, 3721. [CrossRef]
12. Monteiro, V.; Pinto, J.G.; Ferreira, J.C.; Goncalves, H.; Alfonso, J.L. Bidirectional multilevel converter for electric vehicles. In Proceedings of the 2012 Annual Seminar on Automation, Industrial Electronics and Instrumentation, Guimarães, Portugal, 11–13 July 2012; pp. 434–439.
13. Yuan, Z.; Xu, H.; Chao, Y.; Zhang, Z. A novel fast charging system for electrical vehicles based on input-parallel output-parallel and output-series. In Proceedings of the 2017 IEEE Transportation Electrification Conference and Expo, Asia-Pacific (ITEC Asia-Pacific), Harbin, China, 7–10 August 2017; pp. 1–6.
14. Beldjajev, V. Research and Development of the New Topologies for the Isolation Stage of the Power Electronic Transformer. Master's Thesis, Allinn University of Technology, Tallinn, Estonia, 2013.
15. Engel, S.P.; Stieneker, M.; Soltau, N.; Rabiee, S.; Stagge, H.; de Doncker, R.W. Comparison of the Modular Multilevel DC Converter and the Dual-Active Bridge Converter for Power Conversion in HVDC and MVDC Grids. *IEEE Trans. Power Electron.* **2015**, *30*, 124–137. [CrossRef]
16. Sari, H.I. DC/DC Converters for Multi-Terminal HVDC Systems Based on Modular Multilevel Converter. Master's Thesis, Norwegian University of Science and Technology, Trondheim, Norway, August 2016.
17. Yang, H. Modular and Scalable DC-DC Converters for Medium-/High-Power Applications. Master's Thesis, Georgia Institute of Technology, Atlanta, GA, USA, 2017.
18. Fan, H.; Li, H. A high-frequency medium-voltage DC-DC converter for future electric energy delivery and management systems. In Proceedings of the 8th International Conference on Power Electronics—ECCE Asia, Jeju, Korea, 29 May–2 June 2011; pp. 1031–1038.
19. Papadakis, C. Protection of HVDC Grids Using DC Hub. Master Thesis, Delft University of Technology, Delft, The Netherlands, September 2017.
20. Carrizosa, M.J.; Benchaib, A.; Alou, P.; Damm, G. DC transformer for DC/DC connection in HVDC network. In Proceedings of the 2013 15th European Conference on Power Electronics and Applications (EPE), Lille, France, 2–6 September 2013; pp. 1–10.
21. Vasiladiotis, M.; Rufer, A. A Modular Multiport Power Electronic Transformer with Integrated Split Battery Energy Storage for Versatile Ultrafast EV Charging Stations. *IEEE Trans. Ind. Electron.* **2014**, *62*, 3213–3222. [CrossRef]
22. Mortezaei, A.; Abdul-Hak, M.; Simoes, M.G. A Bidirectional NPC-based Level 3 EV Charging System with Added Active Filter Functionality in Smart Grid Applications. In Proceedings of the 2018 IEEE Transportation Electrification Conference and Expo (ITEC), Long Beach, CA, USA, 13–15 June 2018; pp. 201–206.

23. Tian, Q.; Huang, A.Q.; Teng, H.; Lu, J.; Bai, K.H.; Brown, A.; McAmmond, M. A novel energy balanced variable frequency control for input-series-output-parallel modular EV fast charging stations. In Proceedings of the 2016 IEEE Energy Conversion Congress and Exposition (ECCE), Milwaukee, WI, USA, 18–22 September 2016; pp. 1–6.

24. De Doncker RW, A.A.; Divan, D.M.; Kheraluwala, M.H. A three-phase soft-switched high-power-density DC/DC converter for high-power applications. *IEEE Trans. Ind. Appl.* **1991**, *27*, 63–73. [CrossRef]

25. Zahid, Z.U.; Dalala, Z.M.; Chen, R.; Chen, B.; Lai, J.-S. Design of bidirectional dc-dc resonant converter for vehicle-to-grid (v2g) applications. *IEEE Trans. Transp. Electrif.* **2015**, *1*, 232–244. [CrossRef]

26. Cui, T.; Liu, C.; Shan, R.; Wang, Y.; Kong, D.; Guo, J. A Novel Phase-Shift Full-Bridge Converter with Separated Resonant Networks for Electrical Vehicle Fast Chargers. In Proceedings of the 2018 IEEE International Power Electronics and Application Conference and Exposition (PEAC), Shenzhen, China, 4–7 November 2018; pp. 1–6.

27. Srdic, S.; Liang, X.; Zhang, C.; Yu, W.; Lukic, S. A SiC-based high-performance medium-voltage fast charger for plug-in electric vehicles. In Proceedings of the 2016 IEEE Energy Conversion Congress and Exposition (ECCE), Milwaukee, WI, USA, 18–22 September 2016; pp. 1–6.

28. Justino, J.C.G.; Parreiras, T.M.; Filho, B.J.C. Hundreds kW Charging Stations for e-Buses Operating Under Regular Ultra-Fast Charging. *IEEE Trans. Ind. Appl.* **2016**, *52*, 1766–1774.

29. Everts, J. Modeling and Optimization of Bidirectional Dual Active Bridge AC–DC Converter Topologies. Ph.D. Thesis, KU Leuven, Leuven, Belgium, March 2014.

30. Marz, M.; Schletz, A.; Eckardt, B.; Egelkraut, S.; Rauh, H. Power Electronics System Integration for Electric and Hybrid Vehicles. In Proceedings of the 6th International Conference on Integrated Power Electronics Systems (CIPS 2010), Nuremberg, Germany, 16–18 March 2010; pp. 1–10.

31. Emadi, A.; Lee, Y.-J.; Rajashekara, K. Power Electronics and Motor Drives in Electric, Hybrid Electric, and Plug-In Hybrid Electric Vehicles. *IEEE Trans. Ind. Electron.* **2008**, *55*, 2237–2245. [CrossRef]

32. Schoner, H.P.; Hille, P. Automotive power electronics. New challenges for power electronics. In Proceedings of the IEEE 31st Annual Power Electronics Specialists Conference (PESC 2000), Galway, Ireland, 18–23 June 2000; Volume 1, pp. 6–11.

33. Kolar, J.W.; Drofenik, U.; Biela, J.; Heldwein, M.L.; Ertl, H.; Friedli, T.; Round, S.D. PWM Converter Power Density Barriers. In Proceedings of the Power Conversion Conference (PCC 2007), Nagoya, Japan, 2–5 April 2007; pp. 9–29.

34. Kolar, J.W.; Biela, J.; Waer, S.; Friedli, T.; Badstuebner, U. Performance Trends and Limitations of Power Electronic Systems. In Proceedings of the 6th International Conference on Integrated Power Electronics Systems (CIPS 2010), Nuremberg, Germany, 16–18 March 2010; pp. 1–20.

35. Gu, W.-J.; Liu, R. A Study of Volume and Weight vs. Frequency for High-Frequency Transformers. In Proceedings of the IEEE 24th Annual Power Electronics Specialists Conference (PESC 1990), Seattle, WA, USA, 27–30 September 1993; pp. 1123–1129.

36. Vlatkovic, V.; Sabate, J.A.; Ridley, R.B.; Lee, F.C.; Cho, B.H. Small-signal analysis of the phase-shifted PWM converter. *IEEE Trans. Power Electron.* **1992**, *7*, 128–135. [CrossRef]

37. Lian, Y.; Adam, G.; Holliday, D.; Finney, S. Modular input-parallel output-series DC/DC converter control with fault detection and redundancy. *IET Gener. Transm. Distrib.* **2016**, *10*, 1361–1369. [CrossRef]

38. Ruan, X.; Chen, W.; Cheng, L.; Tse, C.K.; Yan, H.; Zhang, T. Control Strategy for Input-Series–Output-Parallel Converters. *IEEE Trans. Ind. Electron.* **2009**, *56*, 1174–1185. [CrossRef]

39. Lian, Y.; Adam, G.P.; Holliday, D.; Finney, S.J. Medium-voltage DC/DC converter for offshore wind collection grid. *IET Renew. Power Gener.* **2016**, *10*, 651–660. [CrossRef]

40. ElMenshawy, M.; Massoud, A. Multimodule ISOP DC-DC Converters for Electric Vehicles Fast Chargers. In Proceedings of the 2019 2nd International Conference on Smart Grid and Renewable Energy (SGRE), Doha, Qatar, 19–21 November 2019; pp. 1–6.

41. ElMenshawy, M.; Massoud, A. Multimodule DC-DC Converters for High-Voltage High-Power Renewable Energy Sources. In Proceedings of the 2019 2nd International Conference on Smart Grid and Renewable Energy (SGRE), Doha, Qatar, 19–21 November 2019; pp. 1–6.

42. Krismer, F.; Kolar, J. Accurate Power Loss Model Derivation of a High-Current Dual Active Bridge Converter for an Automotive Application. *IEEE Trans. Ind. Electron.* **2010**, *57*, 881–891. [CrossRef]

43. Marxgut, C.B. Ultra-Flat Isolated Single-Phase AC-DC Converter Systems. Ph.D. Thesis, Swiss Federal Institute of Technology (ETH Zurich), Power Electronic Systems (PES) Laboratory, Zurich, Switzerland, 2013.
44. Marxgut, C.; Krismer, F.; Bortis, D.; Kolar, J.W. Ultrafast Interleaved Triangular Current Mode (TCM) Single-Phase PFC Rectifier. *IEEE Trans. Power Electron.* **2014**, *29*, 873–882. [CrossRef]
45. Krismer, F.; Kolar, J.W. Closed Form Solution for Minimum Conduction Loss Modulation of DAB Converters. *IEEE Trans. Power Electron.* **2012**, *27*, 174–188. [CrossRef]
46. Barrera-Cardenas, R.; Molinas, M. A Simple Procedure to Evaluate the Efficiency and Power Density of Power Conversion Topologies for Offshore Wind Turbines. *Energy Procedia* **2012**, *24*, 202–211. [CrossRef]
47. Pavlovsky, M.; de Haan, S.W.H.; Ferreira, J.A. Reaching High Power Density in Multikilowatt DC-DC Converters with Galvanic Isolation. *IEEE Trans. Power Electron.* **2009**, *24*, 603–612. [CrossRef]
48. Whitaker, B.; Barkley, A.; Cole, Z.; Passmore, B.; Martin, D.; McNutt, T.R.; Lostetter, A.B.; Lee, J.S.; Shiozaki, K. A High-Density, High-Efficiency, Isolated On-Board Vehicle Battery Charger Utilizing Silicon Carbide Power Devices. *IEEE Trans. Power Electron.* **2014**, *29*, 2606–2617. [CrossRef]
49. Sha, D.; Guo, Z.; Liao, X. Cross-Feedback Output-Current-Sharing Control for Input-Series-Output-Parallel Modular DC-DC Converters. *IEEE Trans. Power Electron.* **2010**, *25*, 2762–2771. [CrossRef]

© 2020 by the authors. Licensee MDPI, Basel, Switzerland. This article is an open access article distributed under the terms and conditions of the Creative Commons Attribution (CC BY) license (http://creativecommons.org/licenses/by/4.0/).

Article

Control of a Fault-Tolerant Photovoltaic Energy Converter in Island Operation

Marino Coppola [1,*], Pierluigi Guerriero [1], Adolfo Dannier [1], Santolo Daliento [1], Davide Lauria [2] and Andrea Del Pizzo [1]

1 Department of Electrical Engineering and Information Technologies, University of Napoli Federico II, Via Claudio 21, 80125 Napoli, Italy; pierluigi.guerriero@unina.it (P.G.); adannier@unina.it (A.D.); daliento@unina.it (S.D.); delpizzo@unina.it (A.D.P.)

2 Department of Industrial Engineering, University of Napoli Federico II, Via Claudio 21, 80125 Napoli, Italy; dlauria@unina.it

* Correspondence: marino.coppola@unina.it; Tel.: +39-081-7683228

Received: 14 May 2020; Accepted: 17 June 2020; Published: 19 June 2020

Abstract: The paper deals with design and control of a fault tolerant and reconfigurable photovoltaic converter integrating a Battery Energy Storage System as a standby backup energy resource. When a failure occurs, an appropriate control method makes the energy conversion system capable of operating in open-delta configuration in parallel with the grid as well as in islanded mode. In case network voltage is lacking due to heavy anomalies or maintenance reasons, the proposed control system is able to quickly disconnect the inverter from the grid while ensuring the energy continuity to the local load and the emergency fixtures by means of the integrated battery packs. In particular, the paper proposes a fast islanding detection method essential for the correct operation of the control system. This specific technique is based on the Hilbert transform of the voltage of the point of common coupling, and it identifies the utility lack in a period of time equal to half a grid cycle in the best case (i.e., 10 ms), thus resulting in good speed performance fully meeting the standard requirements. A thorough numerical investigation is carried out with reference to a representative case study in order to demonstrate the feasibility and the effectiveness of the proposed control strategy.

Keywords: power systems for renewable energy; fault-tolerant photovoltaic inverter; islanding detection; energy storage system

---

1. Introduction

Over the last years, the electric energy generated from renewable energy sources (RES) has grown exponentially. In particular, photovoltaic (PV) energy has now become one of the most relevant parts in energy mix in several geographical areas [1]. In such a scenario, as most of the installed PV generators (PVGs) are grid connected, particular attention must be paid to the system reliability [2] in order to meet the requirements of the electric service in terms of efficiency and power quality. Moreover, the continuous increase in capacity of the PV installed plants makes these resources an important agent in active distribution grids, as they are distributed energy resources (DER) also requiring special control strategies [3]. As a consequence, one of the main purposes of the research is to facilitate the integration of variable renewable sources and distributed generation units within the grid [4].

A particular condition arises when a PV DER feeds a critical local load or emergency fixtures in a grid-tied system; it should be capable of operating in parallel with the grid as well as in island (i.e., regardless of the mains). In fact, the island operating mode becomes unavoidable in case of power outage (e.g., for fault conditions or maintenance purposes) in order to ensure continuous energy supply to the local loads. Obviously, in islanded mode, the inherent intermittent nature of PV generation does not allow fulfillment of the load power demand. For this reason, the integration of a Battery Energy

Storage System (*BESS*) as a standby backup energy resource can be useful to guarantee continuity of the power supply or at least to compensate for the gap between load demand and PV production. Thus, a proper control action must be implemented to ensure the capability of the overall system to work in different operating modes [5]. The islanding detection (ID) and the consequent disconnection from the grid are significant features that the system must implement. In technical literature, different detection techniques are analyzed [6–11]. The ID methods (IDMs) are mainly divided into two categories: local and remote methods. Local IDMs are then classified as passive and active methods. The former methods are based on the monitoring of change or the rate of change in the power system parameters, while the latter rely on the injection of a small perturbation in the output system parameters to identify islanding condition. Comprehensive review and performance evaluations of the several proposed techniques are reported in [10,11]. One of the most relevant figures of merit (FoM) to compare the various IDMs is the time requested to identify the grid trip (i.e., detection time or speed), which should be lower than the standard requirements. In [10,11], active, passive, and modified passive methods based on signal processing are compared in terms of speed. In particular, among the passive methods, the wavelet transform (WT) technique seems to reach the best speed performance but at the cost of increased computational complexity. Recent research works [12–17] propose enhanced IDMs; [12] reports a feedback-based passive islanding detection technique for one-cycle-controlled (OCC) single-phase inverter used in photovoltaic system. This method, as stated by the authors, is not generic and limited to OCC-based inverters while providing a detection time of about 200 ms (i.e., 10 grid cycles). In [13], an IDM based on parallel inductive impedance (PII) switching at a distributed generation (DG) connection point along with monitoring the rate of change of voltage at the DG output is implemented. However, to identify the islanding, this technique needs a two-step procedure that requires at least 300 ms in the worst case, corresponding to a run-on time of 15 grid cycles. In [14], a methodology to detect islanding in a grid-connected photovoltaic system is proposed. A disturbance is injected into the maximum power point tracking (MPPT) algorithm when the absolute deviation of the point of common coupling (PCC) voltage in any phase exceeds a voltage threshold. This determines a shift of the system operating point from its maximum power point (MPP), thus resulting in a relevant output power reduction, and the detection time results within 300 ms (i.e., 15 grid cycles). In [15], the used method for adjustment and evaluation of a voltage relay is based on the combination of the application region (AR) and the power imbalance application region (PIAR) methods, and it leads to a detection time of hundreds of milliseconds (i.e., 100–400 ms, 5–20 grid cycles). In [16], a combination of rate of change of frequency (ROCOF), rate of change of phase angle difference (ROCPAD), rate of change of voltage (ROCOV), and over frequency/under frequency (OF/UF) methods is reported. In such a case, the proposed algorithm represents the merge of different passive ID techniques, thus the detection time is always the minimum among the different used algorithms. As a consequence, it seems to work well (e.g., detection time of few milliseconds) but at the cost of a more complex implementation. A variance in the autocorrelation of the modal current envelope (VAMCE) is used as an islanding detection criterion in [17]. This method employs an autocorrelation function (ACF) of a modal current envelope derived by Hilbert transform, and its detection time is of about two or three grid cycles (i.e., 40–60 ms).

This paper proposes a kind of passive method based on the observation of the envelope of the voltage of the point of common coupling. The envelope of the considered quantity is quickly obtained by means of the Hilbert transform and specifically the proposed algorithm outputs the absolute value of the Hilbert transform that can represent a reliable index of fast change in network behavior. In fact, the grid trip phenomenon causes a sudden variation of the PCC voltage, which results in a spike of the monitored quantity. The latter may no longer fall within a predefined safety range, thus allowing islanding detection according to the requirements of the network code. In principle, the monitoring of the envelope does not need to wait for the PCC voltage crest to verify the boundaries' violation, thus leading to a detection time less than a grid cycle (i.e., 10 ms in the best case), thus showing its benefit in terms of speed with respect to the aforementioned techniques. In addition, the proposed method can be used independently of the specific application.

The paper describes a control strategy that is used to implement grid-connected and intentional islanding operations of a PV inverter when the power circuits are in open-delta configuration as a consequence of a local fault. Under ordinary operating conditions, the PV source delivers the energy to the load and, if the load request is not met, the grid provides the residual part while the battery is in idle mode. In islanded mode, the continuous power supply of the load is ensured due to to the integrated *BESS*, thus overcoming the lack of grid supply and obtaining a flat profile of the inverter output power [18].

The starting architecture for the considered energy conversion system is the conventional double-stage configuration of Figure 1. It represents a centralized solution in which PVG is made up of several strings in parallel in order to achieve the desired rated power.

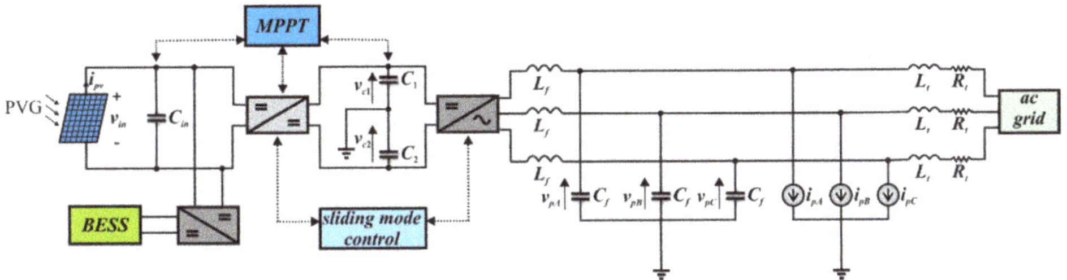

Figure 1. Schematic view of double-stage three-phase photovoltaic (PV) inverter with integrated Battery Energy Storage System (*BESS*).

In a previous paper [19], the authors proposed a solution to enhance fault tolerance of the system and its reliability, introducing a particular control strategy aimed at operating even with only two legs (phases) still being fully functional. In fact, in case of failure of one inverter leg, dedicated switches are capable of short-circuiting the *LC* filter ($L_f - C_f$) of the failing leg, leading to the new open–delta configuration shown in Figure 2, which allows the PV inverter to operate even in the presence of a fault. With reference to the two-leg configuration of the inverter, the effect of *BESS* integration and the effectiveness of the proposed ID method are discussed.

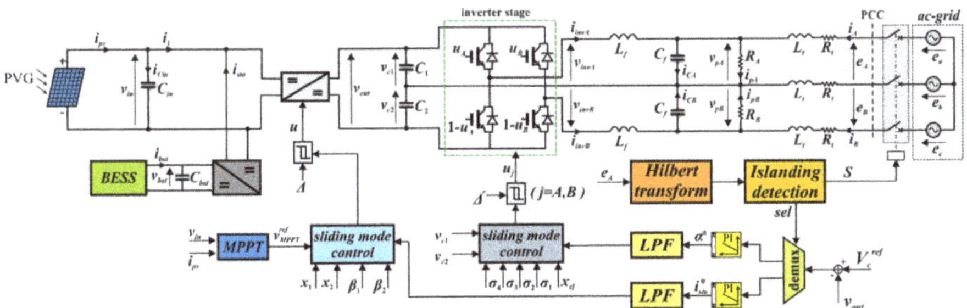

Figure 2. Overall system configuration.

The paper is organized as follows. System modeling is reported in Section 2. Then, Section 3 deals with design and control of the grid-tied PV inverter. In Section 4, the proposed design procedure is validated by carrying out numerical simulations in PLECS environment. Conclusions are summarized in Section 5.

2. System Modeling

The power inverter in the "two legs" configuration shown in Figure 2 is arranged in a double-stage architecture. The first power stage is a step-up DC-DC converter with coupled inductors [20–22] that allows high efficiency and high voltage gain (see Figure 3). It consists of a primary inductor L_1 and a secondary inductor L_2, while resistors R_1 and R_2 account for inductors copper losses.

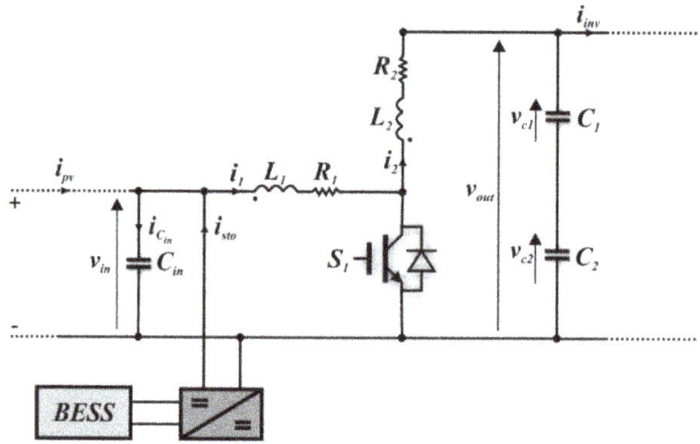

Figure 3. First power stage with coupled-inductors and *BESS*.

The winding ratio of the magnetically coupled inductors is equal to $r_n = N_2/N_1$, where N_1 and N_2 are the turn numbers of the primary and the secondary inductor, respectively. In our analysis, the coupling coefficient k is considered ideal (i.e., $k = 1$), thus the total inductance is $L = (N_1 + N_2)^2 L_0$, with L_0 the inductance of a single winding. The integrated *BESS* is connected to the input DC-link through an auxiliary bidirectional DC-DC converter.

With reference to Figure 3, the mathematical model of the coupled inductors converter can be expressed as:

$$\begin{cases} \dfrac{\mathrm{d}\,i_m}{\mathrm{d}\,t} = \dfrac{v_{in}-R_1\,i_m}{L_1}\,u \;+\; \dfrac{v_{in}-v_{out}}{L_1(1+r_n)}(1-u) \;-\; \dfrac{R_1+R_2}{L_1(1+r_n)^2}\,i_m(1-u) \\[2ex] \dfrac{\mathrm{d}\,v_{in}}{\mathrm{d}\,t} = \dfrac{i_{pv}+i_{sto}}{C_{in}} \;-\; \dfrac{i_m}{C_{in}}u \;-\; \dfrac{i_m}{(1+r_n)C_{in}}\,(1-u) \\[2ex] \dfrac{\mathrm{d}\,v_{out}}{\mathrm{d}\,t} = \dfrac{2\,i_m}{(1+r_n)C}\,(1-u) \;-\; \dfrac{2\,i_{inv}}{C} \end{cases} \tag{1}$$

where $u \in \{0,1\}$. More specifically, $u = 1$ when S_1 is ON; $u = 0$ when S_1 is OFF. Moreover, C is equal to $C_1 = C_2$.

The second power stage is a two-phase DC-AC inverter, where the phase A is supplied by the line-to-line voltage e_A, and the phase B by e_B, while the generated output inverter voltages are v_{invA} and v_{invB}. The two capacitors $C_1 = C_2$ ensure the power decoupling with respect to the first power stage. The inverter supplies the local load modeled by means of two properly sized resistors (i.e., R_A, R_B) through a filter ($L_f - C_f$). The load is clearly unbalanced, but, as demonstrated below, the inverter is always capable of carrying out an effective balancing action. At the network side, the line three-phase transformer is modeled by the equivalent parameters $R_t - L_t$, which also includes a suitable series inductance to decouple the inverter from the grid with the aim of reaching better performance in terms of power quality. Considering the circuit in Figure 2, the following Equations can be written:

$$\begin{cases} \dfrac{\mathrm{d}\,i_j}{\mathrm{d}\,t} = \dfrac{2}{3L_t}\left(e_j - 2\,R_t\,i_j - v_{pj} - R_t\,i_k\right) - \dfrac{1}{3L_t}\left(e_k - 2\,R_t\,i_k - v_{pk} - R_t\,i_j\right) \\[2ex] \dfrac{\mathrm{d}\,i_{invj}}{\mathrm{d}\,t} = \dfrac{1}{L_f}\left(v_{invj} - v_{pj}\right) \\[2ex] \dfrac{\mathrm{d}\,v_{pj}}{\mathrm{d}\,t} = \dfrac{1}{C_f}\left(i_j + i_{invj} - i_{pj}\right) \end{cases} \tag{2}$$

where the subscript indexes $j, k \in \{A, B\}$ with $j \neq k$, while the dynamic behavior of DC-link voltages is given by:

$$\begin{cases} \frac{\mathrm{d}\, v_{c1}}{\mathrm{d}\, t} = -\frac{1}{C_1}\left(u_A\, i_{invA} + u_B\, i_{invB}\right) + \frac{i_m(1-u)}{C_1(1+r_n)} \\ \frac{\mathrm{d}\, v_{c2}}{\mathrm{d}\, t} = \frac{1}{C_2}\left[(1-u_A)\, i_{invA} + (1-u_B)\, i_{invB}\right] + \frac{i_m(1-u)}{C_2(1+r_n)} \end{cases} \tag{3}$$

It should be noted that only the two line-to-line voltages v_{pA} and v_{pB} are controlled in open-delta connection. At the ac network side, a circuit breaker can disconnect the inverter from the grid if this is tripped, thus forcing the operating mode changing from "normal operation" to "islanded mode" in order to meet the requirements of the new circuit configuration.

3. System Control

The controller design plays a very important role in order to guarantee safe and reliable interconnection and interoperability of DER with electric power systems (EPS) as requested by the standard rules [23]. Instead of traditional linear control methods, a sliding mode technique is adopted to control the power stages (i.e., DC-DC converter and inverter) with the aim of obtaining optimal performance in terms of fast dynamic response and robustness against uncertainties and disturbances. The sliding surface of the first power stage is developed to ensure a reduced ripple of the PV voltage, thus limiting fluctuations around the MPP. A suitable control of the inverter is implemented in order to keep balanced the voltages (v_{C1}, v_{C2}) at the inverter DC-link while ensuring higher power quality. In fact, in the two legs configuration (Figure 2), the midpoint at the DC-link capacitors ($C_1 - C_2$) becomes a common-phase of the open-delta system. Consequently, the voltage unbalancing between the two capacitors is a relevant issue to be suitably controlled in order to prevent undesired effects.

The paper does not focus on the control of the power exchange between the *BESS* and the PV sources because of well-established control strategy, and the same applies for the battery management system (BMS). On the contrary, attention is paid to the possibility of enhancing the system reliability due to the *BESS* integration, thus ensuring fulfillment of the load power demand even if the grid is tripped. In fact, the *BESS* does not act during normal operation (i.e., grid connected operation) but only in island mode. In the latter case, the *BESS* can accumulate the excess of power from the PVG (i.e., charging mode) if the power required by the load is less than the generated one.

Otherwise, it can provide the power backup if the load requires more power than generated (i.e., discharging mode). This means that the presence of the storage unit allows a flat profile of the load power regardless of the inherent variability of PV power production. Therefore, *BESS* makes the system able to continuously feed the critical load, thus enhancing the overall reliability.

In normal operation, the control of the DC-AC stage must ensure the energy transfer to the grid with unitary power factor and sinusoidal network currents, also keeping the load and the DC voltages (i.e., v_{C1}, v_{C2}) balanced. Thus, the dynamics of the DC-link voltage v_{out} are relevant to control the displacement angle α (angle between load voltages and network voltages) during normal operation and the storage system during islanding.

As shown in Figure 2, the sliding controllers adapt to circuit configuration due to a de-multiplexer driven by a proper selection signal (*sel*) derived from the islanding detection block, which also activates the circuit breaker by means of the signal S. In other words, the islanding detection block is able to autonomously change the operating mode from normal operation to islanded mode. The two different operating modes are described in the following sub-sections.

3.1. Normal Operation

The normal operation is extensively reported in [19] and here partially recalled for sake of completeness and clarity. The sliding mode control approach features the variable structure nature of DC-DC converters; by means of a proper operation of the switches, the system is forced to reach a suitable selected surface (sliding surface) and to stay on it. As a consequence, the proper choice of the state variables represents a challenge in order to define a state space averaged model of the converter. In particular, in our case, the state variables chosen are the magnetizing current i_m, as defined in [24], and the input voltage v_{in} in order to accomplish the need of MPP tracking by considering the intrinsic variability of PV power generation. The model in Equation (1) can be simplified by neglecting the i_{sto}. The vector $\mathbf{x}$ of the state variables error is:

$$\mathbf{x} = [x_1, x_2]^T = \left[i_m - I_m^{ref}, v_{in} - V_{in}^{ref} \right]^T \tag{4}$$

Thus the following matrix format can be derived:

$$\overset{\bullet}{\mathbf{x}} = \mathbf{A}\,\mathbf{x} + \mathbf{B}\,u + \mathbf{A}\,\mathbf{z} + \mathbf{F} \tag{5}$$

$$\mathbf{A} = \begin{pmatrix} -\dfrac{R_1 + R_2}{L_1(1+r_n)^2} & \dfrac{1}{L_1(1+r_n)} \\[2ex] -\dfrac{1}{(1+r_n)C_{in}} & 0 \end{pmatrix} \tag{6}$$

$$\mathbf{B} = \begin{pmatrix} \dfrac{v_{in}}{L_1} - \dfrac{R_1}{L_1}\,i_m - \dfrac{v_{in}-v_{out}}{L_1(1+r_n)} + \dfrac{R_1+R_2}{L_1(1+r_n)^2}\,i_m + \\[2ex] -\dfrac{i_m}{C_{in}} + \dfrac{i_m}{(1+r_n)C_{in}} \end{pmatrix} \tag{7}$$

$$\mathbf{F} = \begin{pmatrix} -\dfrac{v_{out}}{L_1(1+r_n)} \\[2ex] \dfrac{i_{pv}}{C_{in}} \end{pmatrix} ; \qquad \mathbf{z} = \left[I_m^{ref}, V_{out}^{ref} \right]^T \tag{8}$$

$$I_m^{ref} = \frac{1+r_n}{1-D}\,I_{out}^{ref} = \frac{1+r_n}{1-D}\,\frac{V_{in}^{ref}}{V_{out}^{ref}}\,i_{pv} \tag{9}$$

where the duty cycle reference value is estimated as:

$$D = \frac{1 - \dfrac{v_{in}}{v_{out}}}{1 + r_n \dfrac{v_{in}}{v_{out}}} \tag{10}$$

The chosen sliding surface is a linear combination of the state variables error:

$$S(\mathbf{x}) = \beta_1\,x_1 + \beta_2\,x_2 = \boldsymbol{\beta}^T \mathbf{x} \tag{11}$$

where $\boldsymbol{\beta}^T = [\beta_1, \beta_2]$. The proper choice of the latter coefficients determines the existence conditions of sliding mode [25]:

$$\begin{cases} \overset{\bullet}{S}(\mathbf{x}) < 0 \ \ if \ S(\mathbf{x}) > 0 \\ \overset{\bullet}{S}(\mathbf{x}) > 0 \ \ if \ S(\mathbf{x}) < 0 \end{cases} \tag{12}$$

The respect of the conditions of Equation (12) assures that all the system states near the sliding surface $S(\mathbf{x}) = 0$ are directed towards it for both possible states of the converter switch [26].

With the aim of ensuring that the state of the system remains close to the sliding surface, a suitable operation is necessary for the switch, which links its state with the value of $S(\mathbf{x})$. The latter means that, in a practical case, a discontinuous control law must be defined by using a hysteresis band:

$$u = \begin{cases} 0 \ if \ S(\mathbf{x}) > +\Delta \\ 1 \ if \ S(\mathbf{x}) < -\Delta \end{cases} \tag{13}$$

where 2Δ is the amplitude of the hysteresis band in the sliding surface, being Δ an arbitrary small positive quantity. The reference value (v_{MPPT}^{ref}) of the input voltage is provided by an MPPT algorithm based on a classical perturb and observe (P & O) technique. The voltage reference is sent to the sliding mode controller, thus obtaining an adaptive sliding surface modified at each MPPT step to extract the maximum available power.

The control of the second power stage (i.e., inverter stage) is also based on the sliding approach.

During ordinary operation, the main control goal is to transfer the power generated by PV sources (and not drawn from the load) to the grid by properly adapting the displacement angle α (angle between load voltages and network voltages) to the different operating conditions (i.e., $\alpha > 0$ means that the excess power is transferred to the grid, while $\alpha < 0$ means that the grid provides the difference between load power demand and PV generation). The reference rms value of the line-to-line voltage on the load can be derived by assuming unity power factor at the grid side, thus obtaining:

$$V_p^{ref} = E \frac{\sin \beta}{\sin (\beta - \alpha)} \tag{14}$$

where $\beta = \tan^{-1}(X_t/R_t)$, $X_t = \omega L_t$, ω is the grid angular frequency, and E is the value of both e_A and e_B.

The reference quantities (subscript index 2) for the load line-to-line voltages (i.e., v_{pA}, v_{pB}) and for their derivative (subscript index 1) of the two-phase system are described by the vector:

$$\mathbf{x}_{rj} = \left[x_{rj1}, x_{rj2} \right]^T \tag{15}$$

and the two components of $\mathbf{x}_{rj}$ for $j = A, B$ are:

$$\begin{cases} x_{rA1} = -\sqrt{2}\, \omega\, V_p^{ref} \sin (\omega t + \vartheta + \alpha^*) \\ x_{rA2} = \sqrt{2}\, V_p^{ref} \cos (\omega t + \vartheta + \alpha^*) \\ x_{rB1} = -\sqrt{2}\, \omega\, V_p^{ref} \sin (\omega t + \vartheta + \alpha^* + \pi/3) \\ x_{rB2} = \sqrt{2}\, V_p^{ref} \cos (\omega t + \vartheta + \alpha^* + \pi/3) \end{cases} \tag{16}$$

where ϑ is the phase angle of the grid voltage (at $t = 0$). From Equations (15) and (16), the sinusoidal model of the inverter voltages can be reported in matrix form:

$$\dot{x}_{rj} = \mathbf{A}_r \mathbf{x}_{rj} \qquad j = A, B \tag{17}$$

with:

$$\mathbf{A}_r = \begin{pmatrix} 0 & -\omega^2 \\ 1 & 0 \end{pmatrix} \tag{18}$$

The vector of the state variables is:

$$\mathbf{x}_j = \left[x_{j1}, x_{j2} \right]^T = \left[\dot{v}_{pj}, v_{pj} \right]^T \qquad j = A, B \tag{19}$$

then, the vector $\mathbf{x}_{ej}$ of the state variables error can be derived as follows:

$$\mathbf{x}_{ej} = \mathbf{x}_{rj} - \mathbf{x}_j = \left[x_{rj1} - x_{j1}, x_{rj2} - x_{j2} \right]^T \qquad j = A, B \tag{20}$$

The chosen inverter sliding surface (Equation (20)) is a function of the state variable error $\mathbf{x}_{ej}$, but also of the error of the average and the instantaneous values of the inverter input DC-link voltages in order to meet the requirements for a reliable control action able to avoid DC-link voltages imbalance that could appear for asymmetrical condition during charging transients.

$$S_j\left(\mathbf{x}_{ej}, v_{c1}, v_{c2} \right) = \sigma_1 (x_{rj1} - x_{j1}) + \sigma_2 (x_{rj2} - x_{j2}) + \sigma_3 (\overline{v}_{c1} - \overline{v}_{c2}) + \sigma_4 (v_{c1} - v_{c2}) \qquad j = A, B \tag{21}$$

Moreover, this surface requires a proper hysteresis band, hence the control law becomes:

$$u_j = \begin{cases} 0 & if \quad S_j\left(\mathbf{x}_{ej}, v_{c1}, v_{c2}\right) > +\Delta' \\ 1 & if \quad S_j\left(\mathbf{x}_{ej}, v_{c1}, v_{c2}\right) < -\Delta' \end{cases} \qquad j \in \{A, B\} \tag{22}$$

The choice of the hysteresis band Δ' depends on the maximum switching frequency and on the filter design.

3.2. Islanding Detection

The continuity of power supply to critical local load should be guaranteed also in case of lack of utility grid supply (e.g., fault condition or maintenance purpose). For this purpose, the grid trip must be properly detected to quickly counteract to this event. Several IDMs have been proposed. They can be classified in two main categories: passive and active methods. The former methods are based on the detection of a change or the rate of change in a power system parameter, while the latter are generally based on the introduction of small perturbations at the inverter output, thus generating small changes in a parameter of the power system [6]. As already discussed in the introduction, the proposed method is based on the pure observation of the envelope of the PCC voltage (e.g., e_A in Figure 2) obtained by performing the absolute value of the Hilbert transform, thus it can be classified as a passive IDM. The outcome of the used Hilbert transform algorithm is an analytical complex signal:

$$y(t) = f(t) + j\hat{f}(t), \tag{23}$$

where $f(t)$ is a real function and $\hat{f}(t)$ is its Hilbert transform, which, in the time domain, is a convolution between the Hilbert transformer $1/(\pi\, t)$ and the original signal $f(t)$:

$$\hat{f}(t) = f(t) * \frac{1}{\pi\, t} = \frac{1}{\pi} P \int_{-\infty}^{+\infty} \frac{f(\tau)}{t - \tau} \, d\tau \tag{24}$$

where P represents the Cauchy principal value. By means of some mathematical manipulations, it is easy to verify that a real function and its Hilbert transform are orthogonal. The polar representation of the analytical complex signal is:

$$y(t) = e\,(t)\, e^{j\,\varphi(t)} \tag{25}$$

where $\varphi(t) = \tan^{-1}\left[\hat{f}(t)/f(t)\right]$ is the instantaneous phase, while $e\,(t) = \sqrt{f^2(t) + \hat{f}^2(t)}$ is the instantaneous amplitude or rather the envelope of the original signal. In our case, it is the envelope of the PCC voltage e_A:

$$e\,(t) = \sqrt{e_A^2(t) + \hat{e}_A^2(t)} \tag{26}$$

The grid trip event determines a large spike in the monitored quantity (Equation (26)), which exceeds a suitable permitted range, thus allowing proper and quick detection of the grid fault. In fact, the proposed islanding detection method is carried out by monitoring the envelope of the original sampled voltage e_A (i.e., the line-to-line voltage) at PCC. In particular, a moving window of 20 ms (i.e., one grid cycle) with a time shift of 5 ms is used to evaluate the envelope, but only the center value in the window is considered to detect the possible grid outage with the aim of excluding the edge effects that could produce a false positive.

As a consequence, after a grid trip event, the maximum time requested to detect the islanding is equal to 12.5 ms (i.e., 10 ms due to half of the window length plus half the time shift), while in case the grid trip event results in synchronization with the grid period, the minimum time of 10 ms is required to detect the network failure.

Once this event is detected, the control section rapidly acts in order to disconnect the grid from the PV inverter by means of the proper circuit breaker interposed between the inverter and the utility network (see also Figure 2). Moreover, the selection signal (i.e., *sel* in Figure 2) provided by the islanding detection block is able to commutate the control in order to take into account the new system configuration.

In fact, in islanded mode, the control of the displacement angle α is no longer needed, while the dynamics of DC-link voltage v_{out} are now useful in order to control the *BESS*, which acts to guarantee the energy continuity to the critical local load. As a consequence, an increase of the voltage v_{out} means that the load power request is greater than the PV generated one; the *BESS* should provide the needed amount of power to cover the load demand, while, when the power load request is lower than generated one, the *BESS* can accumulate the excess of power from the PVG.

3.3. Islanded Mode

During islanding operation, the system dynamic behavior consequently changes. In fact, the presence of the *BESS* must be considered in order to properly control the power flow (i.e., $i_{sto} \neq 0$ in Equation (1)). In such a case, by considering the state space averaged model of the step-up converter in Equation (1), the magnetizing current reference can be written as:

$$I_m^{ref} = \frac{1+r_n}{1-D} I_{out}^{ref} = \frac{1+r_n}{1-D} \frac{V_{in}^{ref}}{V_{out}^{ref}} \left(i_{pv} + i_{sto} \right) \tag{27}$$

where the reference storage current is obtained by the PI controller as reported in Figure 2. Then, the same reasoning as in Sub-Section 3.1 can be here repeated in order to obtain a suitable sliding mode control law. Moreover, in the relationship in Equation (2), i_j is zeroed due to activation of the breaker, while the dynamic behavior of DC-link voltages remains the same as described in Equation (3).

The control of the second power stage no longer takes into account the displacement angle α because of grid tripping. Thus, the reference quantities (subscript index 2) for the load line-to-line voltages (i.e., v_{pA}, v_{pB}) and also for their derivative (subscript index 1) of the two-phase system can be described by the vector:

$$\mathbf{x}_{rj} = \left[x_{rj1}, x_{rj2} \right]^T \tag{28}$$

where the two components of $\mathbf{x}_{rj}$ for $j = A, B$ are:

$$\begin{cases} x_{rA1} = - \sqrt{2}\, \omega\, V_p^{ref} \sin\left(\omega\, t + \vartheta\right) \\[6pt] x_{rA2} = \sqrt{2}\, V_p^{ref} \cos\left(\omega\, t + \vartheta\right) \\[6pt] x_{rB1} = - \sqrt{2}\, \omega\, V_p^{ref} \sin\left(\omega\, t + \vartheta + \pi/3\right) \\[6pt] x_{rB2} = \sqrt{2}\, V_p^{ref} \cos\left(\omega\, t + \vartheta + \pi/3\right) \end{cases} \tag{29}$$

Then, the sliding surface can be obtained as in Equation (21) by considering the properly modified circuit variables.

4. Numerical Results

A set of simulations is carried out on the grid-connected PV system (see Figure 2) in PLECS environment. The system under study is implemented by using the circuit parameters listed in Table 1.

Table 1. Used circuit parameters.

Parameters	Values
E (V)	400
C_{in} (mF)	10
L (mH)	20
R_1 (mΩ)	27.14
R_2 (mΩ)	67.86
L_f (mH)	10
C_f (mF)	0.5
R_t (mΩ)	0.267
L_t (mH)	8.46

The PV array is described by a physical model based on the five parameters single diode model, where the unknown parameters are: I_{ph} (photo-generated current), I_0 (reverse saturation current), R_s (parasitic series resistance), R_{sh} (parasitic shunt resistance), and a (diode ideality factor). The aim is the evaluation of these parameters at Standard Test Conditions (STC) and also under varying environmental conditions by using data-sheet information. As a consequence, the parameters of the PV model are estimated, and the used values are reported in Table 2, where $V_t = a\,k\,T_{STC}/q$ is the thermal voltage, T_{STC} is the junction temperature at STC, and n_s is the number of cells.

Table 2. Estimated input parameters of a PV module.

Parameters	Values
I_{ph} (A)	7.362
I_0 (A)	0.351
a	1.2
V_t (V)	0.025
R_s (Ω)	0.204
R_{sh} (Ω)	1168
n_s	60

The considered PVG consists of a PV array with $N_s = 17$ series connected panels (i.e., PV string) and $N_p = 3$ parallel connected strings to reach the desired power level. Figure 4 shows the current–voltage (I–V) and the power–voltage (P–V) characteristics of the considered PV array in Standard Conditions. The corresponding MPP values are $V_{MPP} = 412$ V, $I_{MPP} = 20.4$ A, $P_{MPP} = 8.41$ kW. Moreover, the used control parameters are summarized in Table 3.

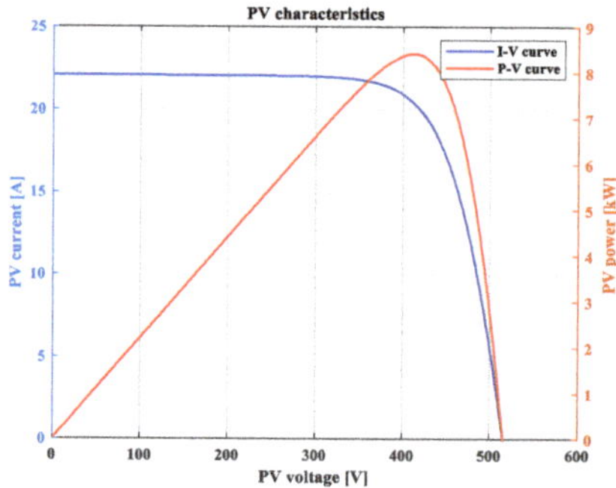

Figure 4. Current–voltage (I–V) and power–voltage (P–V) curves at Standard Test Conditions (STC).

Table 3. Control parameters.

Parameters	Values
T_{MPPT} (ms)	100
ΔV_{MPPT} (V)	1
v_{MPPT}^{ref} (V)	414.2
β_1	0.5
β_2	-1
Δ	1
σ_1	0.001
σ_2	1
σ_3	0.06
σ_4	0.06
Δ'	4

Regarding the sizing of the battery pack, the most relevant parameter to take into account is the battery capacity, which is assumed constant, even in cases of different discharging current rates, in order to simplify the model. The other main parameters are the stored energy and the State of Charge (SOC). The latter is reported in the following Equation:

$$SOC(t) = SOC(t_0) - \frac{1}{Q_0} \int_{t_0}^{t} i_b \, d\tau \tag{30}$$

where Q_0 is the total charge storable in the battery, while i_b is the discharging current.

A simplified circuital model is implemented to describe the behavior of the $BESS$. The equivalent circuit is the series of a voltage source providing the open-circuit voltage V_{oc} and a lumped resistor R_{int}, modeling internal resistance of batteries. The open-circuit voltage is dynamically obtained as function of the SOC by the following non-linear relationship:

$$V_{oc} = E_0 + \frac{R\,T}{F} \, \log\left(\frac{SOC}{1 - SOC}\right) \tag{31}$$

where E_0 is the standard potential of the battery, R is the ideal gas constant, T is the absolute temperature, and F is the Faraday constant. The size of the battery results from the need to support the PV power generation in order to meet the load demand. As a consequence, a storage unit should be able to provide the total rated load power (i.e., 4 kW) for an hour, corresponding to a battery energy of 4 kWh. This choice allows one to overcome a grid trip and also to mitigate the inherent variability of PV production while ensuring a continuous power supply to critical load. Hence, the used $BESS$ presents a capacity of 20 Ah with a rated voltage of about 200 V and total internal resistance of 30 mΩ.

The load is considered pure resistive, and the chosen resistance of each load is equal to the *rms* line-to-line grid voltage divided by the corresponding power $R_A = R_B = E^2/P_{load} = 80 \, \Omega$ (see Figure 2). Moreover, an additional load in parallel with the grid is considered to take into account other loads on the grid network, which may continue to be supplied by the inverter after the grid trip and before the grid disconnection. These additional loads are assumed to be an order of magnitude greater than the local load.

4.1. Normal Operation and Islanding Detection

In this work, we paid particular attention to both islanding detection and $BESS$ control in order to guarantee fault tolerant operation of the inverter. The load demand is globally set at 4 kW (i.e., 2 kW for each load), while the PVG is considered to be operated at STC, thus leading to a PV power production of about 8.4 kW. In such a case, the power difference between PV available power and load demand is transferred to the grid with unity power factor due to the proper control of the displacement angle α, which assumes a positive value. As a consequence, the main issue to be addressed is represented by the islanding detection and the consequent disconnection from the utility network within the time constraint fixed by standard rules [27], which is typically 2 s.

Figure 5 shows the behavior of the PV voltage, which tracks stably the desired MPP in steady-state, thus leading to an MPPT efficiency of 99%. Furthermore, the control strategy allows one to obtain the desired voltage level at the inverter input (see Figure 6a) by means of well-balanced voltages v_{c1}, v_{c2} at the DC-link (see Figure 6b,c).

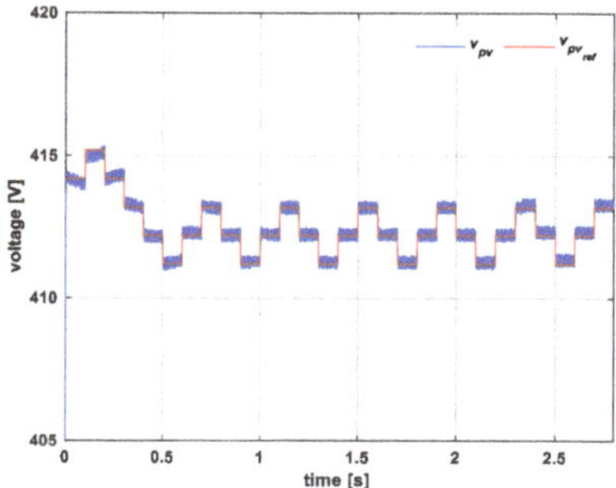

Figure 5. PV voltage (blue line) vs. maximum power point tracking (MPPT) reference voltage (red line).

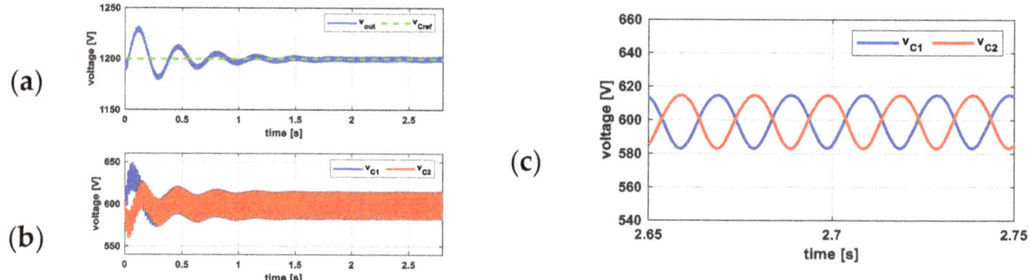

Figure 6. Time behavior of DC-link voltage: (**a**) total DC-link voltage; (**b**) separate DC-link voltage at each capacitor; (**c**) zoom view of separate DC-link voltage at each capacitor.

The time behavior of the displacement angle α is drawn in Figure 7; as expected, it assumes a positive value, which means that the excess power (i.e., the difference between PV power generation and load power request) is transferred to the grid. The voltage and the current behavior of the latter are shown in Figure 8. The grid currents results are sinusoidal and in phase with the grid voltages, thus leading to an almost unity power factor.

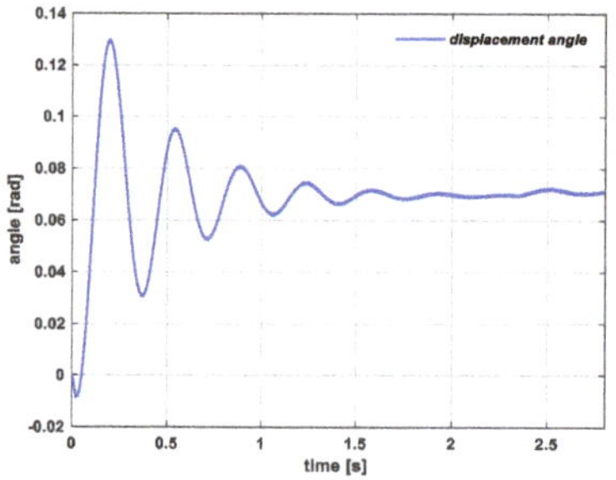

Figure 7. Time behavior of displacement angle α.

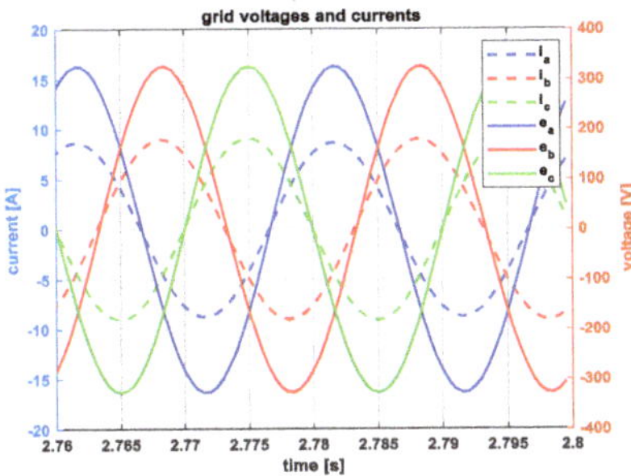

Figure 8. Steady-state behavior of grid currents and voltages.

The steady-state behavior of the load current is in Figure 9a, while Figure 9b shows the power at steady-state. It can be noted that PV excess power with reference to the load power demand is transferred to the grid.

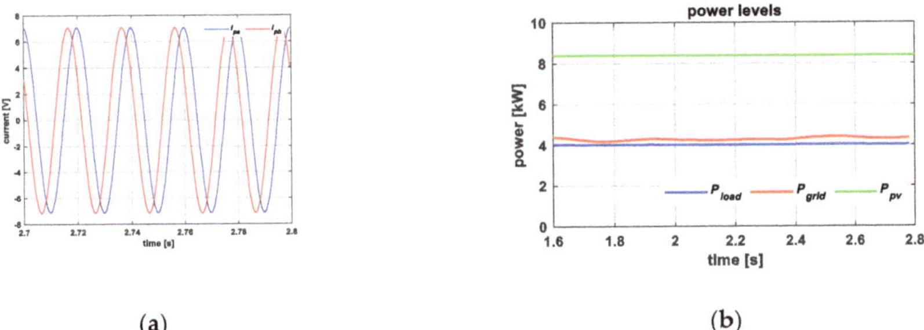

(**a**) (**b**)

Figure 9. Steady-state behavior of load currents (**a**) and of power (**b**): load power (blue line), PV power (green line), and grid power (red line).

Figure 10 shows the results of the proposed ID method. The circles correspond to the envelope value in the middle of the chosen moving window, which has a time duration of 20 ms, as highlighted in the figure and discussed in Sub-Section 3.2. The arrow lines identify the sliding windows, while the colors are the same of the corresponding envelope detected value. The detection happens when the PCC voltage envelope falls outside a suitable safety range, whose typical upper and lower limits are +10%/−15% of the rated value. In our case, in a conservative way, we considered a range of ±10% of the rated value equal to $\sqrt{2}\,E$ (i.e., the voltage interval (509–622 V) defined by the blue and red lines in Figure 10).

The grid fault event occurs at $t = 3$ s (i.e., after 200 ms, it reached steady-state condition). The first value of the envelope that lies outside the safety range is the green circle, which represents the center of the window identified with the green arrow line. Nevertheless, as previously discussed, the grid trip event can be detected only at the end of the window or rather, in such a case, 10 ms after the event itself.

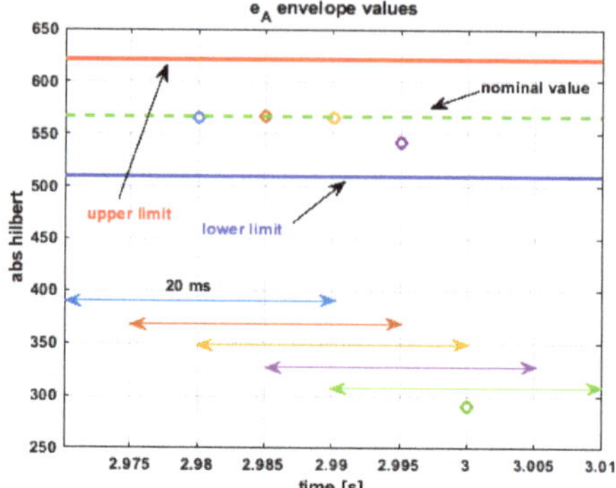

Figure 10. Point of common coupling (PCC) voltage envelope middle values (circles).

4.2. Islanded Mode

Once the grid trip is detected, the control strategy must properly act to prevent supplying the network by disconnecting the inverter, which must continue to feed the critical local load.

The PV and the DC-link voltage behaviors in islanding mode are depicted in Figure 11. The grid trip event occurs at 3 s, while its detection is at 3.01 s, which is the time instant when the inverter is disconnected from the grid and the new control action starts.

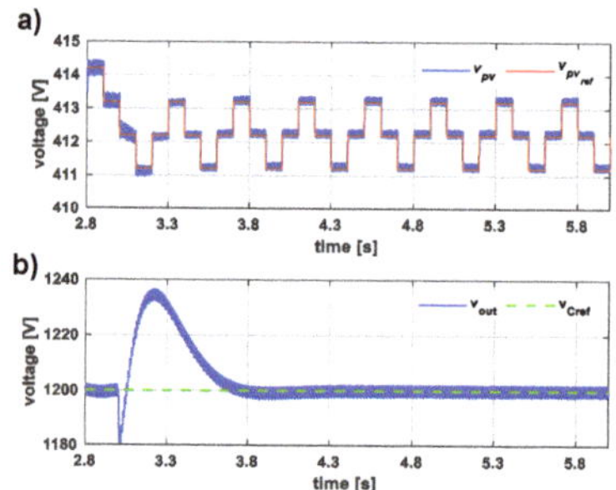

Figure 11. PV and the DC-link voltage behaviors: (**a**) PV voltage (blue line) and maximum power point (MPP) voltage reference (red line); (**b**) DC-link voltage (blue line) and corresponding reference voltage.

The MPP tracking is lost during only one MPPT cycle after the grid fault, while the DC-link voltage diverges from its reference with a maximum overshoot of 0.3% of the reference and recovers the desired behavior in a time interval lower than 1 s.

The time behavior of the *BESS* is drawn in Figure 12; initially (i.e., during normal operation), the battery is in idle mode (battery voltage is equal to the rated value of 200 V, battery current is zero, and the *SOC* is equal to its initial value of 50%). Once the grid trip is detected, the control section activates the battery, which can absorb (i.e., *BESS* charging) the PV power not used by the load; the system operates in island mode without undesired curtailment of PV production. Otherwise, the battery can provide the difference of the PV power with respect to the load demand, ensuring a continuous power supply to the critical load and enhancing system reliability and flexibility.

Finally, in Figure 13, the steady-state load current and the system power behavior are depicted.

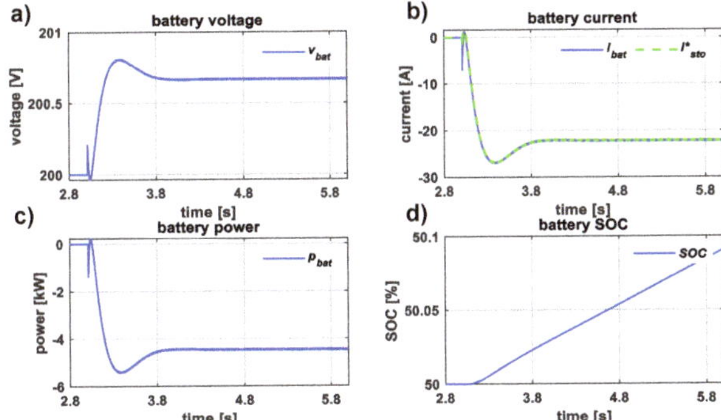

Figure 12. *BESS* time behavior: (**a**) voltage; (**b**) current; (**c**) power; (**d**) State of Charge (*SOC*).

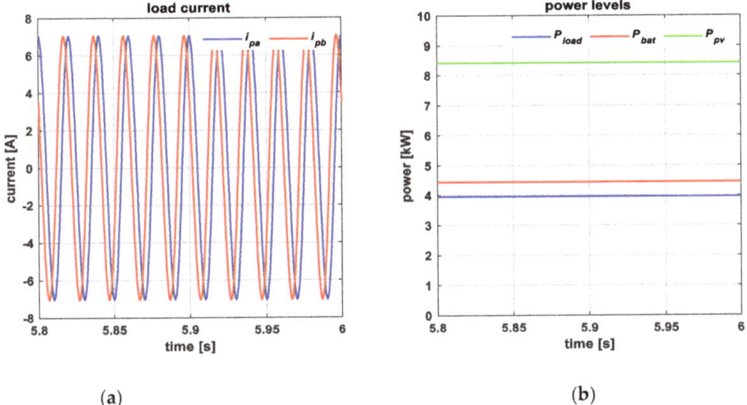

(**a**)   (**b**)

Figure 13. Steady-state behavior of load currents: (**a**) and of power (**b**); load power (blue line), PV power (green line), and battery power (red line).

It can be noted (see Figure 13a) that the load current is the same of Figure 9a, or rather no detrimental effect on the load arises from the grid trip event. In addition, Figure 13b shows how the PV power (i.e., about 8.4 kW) is higher than load demand (i.e., 4 kW). The excess power, which usually would be lost, is provided to the battery, leading to a flat power transfer to the load.

5. Conclusions

This paper is mainly focused on the control of a fault tolerant and reconfigurable grid-connected photovoltaic inverter. The main issue addressed is the possibility to continuously supply critical local load in case of a grid trip event and regardless of the inherent fluctuating nature of the energy generated by PV sources. As a consequence, a suitable islanding detection method is presented and implemented by monitoring the PCC voltage envelope evaluated by means of the Hilbert transform.

A complete set of numerical simulations proved the good performance in normal operation as well as the capability to detect a grid fault event in few milliseconds, thus fully accomplishing the standard requirements. In particular, the minimum time requested to detect the grid outage is equal to 12.5 ms, which can be further reduced to 10 ms in case the grid trip event results in synchronization with the grid period. The proposed IDM appears to be very fast, and it can be compared in terms of speed with respect to different techniques proposed in recent research works, thus showing its benefit in being independent of the specific application.

Moreover, in islanded mode, the system remains fully functional due to a suitable modification of the control strategy, which exploits the integrated *BESS* to continuously provide a flat profile of the load power. Finally, adequate system performance in terms of power quality, power factor, and MPPT efficiency also proves the effectiveness of the proposed control approach.

Author Contributions: Conceptualization, D.L., M.C., and P.G.; methodology, D.L., M.C., and P.G.; software, M.C., P.G., and A.D.; validation, M.C., P.G., and A.D.; formal analysis, D.L. and M.C.; investigation, M.C., P.G., and A.D.; resources, D.L., S.D., and A.D.P.; data curation, M.C.; writing—original draft preparation, M.C., P.G., and A.D.P.; writing—review and editing, A.D.P., D.L., S.D., and A.D.; visualization, M.C., P.G., and A.D.; supervision, A.D.P., D.L., and S.D. All authors have read and agreed to the published version of the manuscript.

Funding: This research received no external funding.

Conflicts of Interest: The authors declare no conflict of interest.

Nomenclature

$C_1; C_2$	DC-link capacitors
C_{bat}	input capacitor of bidirectional DC-DC converter
C_{in}	input capacitor
E	*rms* line-to-line grid voltage
$e_A; e_B$	linetoline voltages
i_{bat}	battery current
i_{pv}	PV current
i_{sto}	storage current
$i_{invA}; i_{invB}$	inverter output currents
$i_{pA}; i_{pB}$	load currents
k	coupling coefficient
L	total inductance of coupled inductors
L_0	inductance of a single winding
$L_1; L_2$	primary and a secondary inductor
$L_f; C_f$	inverter output filter
$N_1; N_2$	turn number of the primary and secondary inductor
$R_1; R_2$	resistors account for inductors copper losses
r_n	winding ratio of the magnetically coupled inductors
$R_A; R_B$	load resistors
$R_t; L_t$	equivalent parameters of line three-phase transformer
S	circuit breaker activation signal
sel	demux selection signal
u	driving signal of coupled inductors DC-DC converter
u_j	driving signal of inverter stage
v_{bat}	battery voltage
v_{C1}, v_{C2}	DClink voltages
v_{in}	PV voltage
$v_{invA}; v_{invB}$	inverter output voltages
$v_{pA}; v_{pB}$	linetoline load voltages
v_{out}	total DClink voltage
v_{MPPT}^{ref}	MPPT voltage reference
α	angle between load voltages and network voltages
$\beta_1; \beta_2$	coefficients of coupled inductors DC-DC converter sliding surface
$x_1; x_2$	statevariables error of coupled inductors DC-DC converter sliding control
$\mathbf{x}_{ej}$	vector of the statevariables error of inverter sliding control
Δ	half the amplitude of the hysteresis band in the sliding surface of coupled inductors DC-DC converter
Δ'	half the amplitude of the hysteresis band in the sliding surface of inverter
$\sigma_1, \sigma_2, \sigma_3, \sigma_4$	coefficients of inverter sliding surface

BESS	battery energy storage system
BMS	battery management system
DER	distributed energy resources
EPS	electric power systems
ID	islanding detection
IDM	islanding detection method
MPP	maximum power point
MPPT	maximum power point tracking
PCC	point of common coupling
P&O	perturb and observe
PV	photovoltaic
PVG	photovoltaic generator
RES	renewable energy sources
SOC	state of charge

References

1. Kouro, S.; Leon, J.I.; Vinnikov, D.; Franquelo, L.G. Grid-Connected Photovoltaic Systems: An Overview of Recent Research and Emerging PV Converter Technology. *IEEE Ind. Electron. Mag.* **2015**, *9*, 47–61. [CrossRef]

2. Romero-Cadaval, E.; Spagnuolo, G.; Franquelo, L.G.; Ramos-Paja, C.A.; Suntio, T.; Xiao, W.M. Grid-Connected Photovoltaic Generation Plants: Components and Operation. *IEEE Ind. Electron. Mag.* **2013**, *7*, 6–20. [CrossRef]

3. Romero-Cadaval, E.; Francois, B.; Malinowski, M.; Zhong, Q. Grid-Connected Photovoltaic Plants: An Alternative Energy Source, Replacing Conventional Sources. *IEEE Ind. Electron. Mag.* **2015**, *9*, 18–32. [CrossRef]

4. European Technology & Innovation Platform. Available online: https://etip-pv.eu/ (accessed on 6 March 2020).

5. Liserre, M.; Sauter, T.; Hung, J.Y. Future Energy Systems: Integrating Renewable Energy Sources into the Smart Power Grid through Industrial Electronics. *IEEE Ind. Electron. Mag.* **2010**, *4*, 18–37. [CrossRef]

6. Petrone, G.; Spagnuolo, G.; Teodorescu, R.; Veerachary, M.; Vitelli, M. Reliability Issues in Photovoltaic Power Processing Systems. *IEEE Trans. Ind. Electron.* **2008**, *55*, 2569–2580. [CrossRef]

7. Zeineldin, H.H.; El-Saadany, E.F.; Salama, M.M.A. Impact of DG interface control on islanding detection and nondetection zones. *IEEE Trans. Power Deliv.* **2006**, *21*, 1515–1523. [CrossRef]

8. Yu, B.; Matsui, M.; Yu, G. A review of current anti-islanding methods for photovoltaic power system. *Sol. Energy* **2010**, *84*, 745–754. [CrossRef]

9. Teodorescu, R.; Liserre, M.; Rodríguez, P. *Grid Converters for Photovoltaic and Wind Power Systems*; John Wiley & Sons Ltd.: Hoboken, NJ. USA, 2011.

10. Kim, M.-S.; Haider, R.; Cho, G.-J.; Kim, C.-H.; Won, C.-Y.; Chai, J.-S. Comprehensive Review of Islanding Detection Methods for Distributed Generation Systems. *Energies* **2019**, *12*, 837. [CrossRef]

11. Ku Ahmad, K.N.E.; Selvaraj, J.; Rahim, N.A. A review of the islanding detection methods in grid-connected PV inverters. *Renew. Sustain. Energy Rev.* **2013**, *21*, 756–766. [CrossRef]

12. Reddy, V.R.; Sreeraj, E.S. A Feedback-Based Passive Islanding Detection Technique for One-Cycle-Controlled Single-Phase Inverter Used in Photovoltaic Systems. *IEEE Trans. Ind. Electron.* **2019**, *67*, 6541–6549. [CrossRef]

13. Rostami, A.; Jalilian, A.; Zabihi, S.; Olamaei, J.; Pouresmaeil, E. Islanding Detection of Distributed Generation Based on Parallel Inductive Impedance Switching. *IEEE Syst. J.* **2020**, *14*, 813–823. [CrossRef]

14. Bakhshi-Jafarabadi, R.; Sadeh, J.; Popov, M. Maximum Power Point Tracking Injection Method for Islanding Detection of Grid-Connected Photovoltaic Systems in Microgrid. *IEEE Trans. Power Deliv.* **2020**. [CrossRef]

15. Babak, S.; Mohamad Esmaeil, H.-G.; Iman, S. Comprehensive investigation of the voltage relay for anti-islanding protection of synchronous distributed generation. *Int. Trans. Elect. Energy Syst.* **2017**, *27*, 1–16.

16. Abyaz, A.; Panahi, H.; Zamani, R.; Haes Alhelou, H.; Siano, P.; Shafie-khah, M.; Parente, M. An Effective Passive Islanding Detection Algorithm for Distributed Generations. *Energies* **2019**, *12*, 3160. [CrossRef]

17. Haider, R.; Kim, C.H.; Ghanbari, T.; Bukhari, S.B.A.; Zaman, M.S.; Baloch, S.; Oh, Y.S. Passive islanding detection scheme based on autocorrelation function of modal current envelope for photovoltaic units. *IET Gener. Transm. Distrib.* **2018**, *12*, 726–736. [CrossRef]

18. Sirico, C.; Teodorescu, R.; Séra, D.; Coppola, M.; Guerriero, P.; Iannuzzi, D.; Dannier, A. PV Module-Level CHB Inverter with Integrated Battery Energy Storage System. *Energies* **2019**, *12*, 4601. [CrossRef]
19. Lauria, D.; Coppola, M. Design and control of an advanced PV inverter. *Solar Energy* **2014**, *110*, 533–542. [CrossRef]
20. Coppola, M.; Lauria, D.; Napoli, E. On the design and the efficiency of coupled step-up dc-dc converters. In Proceedings of the IEEE International Conference on Electrical Systems for Aircraft, Railway and Ship Propulsion (ESARS), Bologna, Italy, 19–21 October 2010.
21. Coppola, M.; Lauria, D.; Napoli, E. Optimal design and control of coupled-inductors step-up dc-dc converter. In Proceedings of the 2011 IEEE International Conference on Clean Electrical Power (ICCEP), Ischia, Italy, 14–16 June 2011; pp. 81–88.
22. Guerriero, P.; Coppola, M.; Cennamo, P.; Daliento, S.; Lauria, D. A single panel PV microinverter based on coupled inductor DC-DC. In Proceedings of the 2017 IEEE International Conference on Environment and Electrical Engineering and 2017 IEEE Industrial and Commercial Power Systems Europe (EEEIC/I & CPS Europe), Milan, Italy, 6–9 June 2017; pp. 1–5.
23. IEEE. *1547–2018-IEEE Standard for Interconnection and Interoperability of Distributed Energy Resources with Associated Electric Power Systems Interfaces*; IEEE: Piscataway, NJ, USA, 2018.
24. Dwari, S.; Jayawant, S.; Beechner, T.; Miller, S.K.; Mathew, A.; Min, C.; Riehl, J.; Sun, J. Dynamics Characterization of Coupled-Inductor Boost DC-DC Converters. In Proceedings of the 2006 IEEE Workshops on Computers in Power Electronics, Troy, NY, USA, 16–19 July 2006.
25. Skvarenina, T.L. *The Power Electronics Handbook*; CRC Press: Boca Raton, FL, USA, 2002; Chapter 8.
26. Mattavelli, P.; Rossetto, L.; Spiazzi, G. Small-signal analysis of DC-DC converters with sliding mode control. *IEEE Trans. Power Electron.* **1997**, *12*, 96–102. [CrossRef]
27. IEEE. *1547.1–2005-IEEE Standard Conformance Test Procedures for Equipment Interconnecting Distributed Resources with Electric Power Systems, IEEE Std.*; IEEE: Piscataway, NJ, USA, 2005; ISBN 0-7381-4736-2 SH95346.

© 2020 by the authors. Licensee MDPI, Basel, Switzerland. This article is an open access article distributed under the terms and conditions of the Creative Commons Attribution (CC BY) license (http://creativecommons.org/licenses/by/4.0/).

Article

Electrochemical Cell Loss Minimization in Modular Multilevel Converters Based on Half-Bridge Modules

Gianluca Brando [1,*], Efstratios Chatzinikolaou [2], Dan Rogers [2] and Ivan Spina [1]

1   Department of Electrical Engineering and Information Technology, University of Naples Federico II, 80125 Naples, Italy; ivan.spina@unina.it
2   Energy and Power Group, Department of Engineering Science, University of Oxford, Oxford OX1 3PA, UK; stratos_hatz@hotmail.com (E.C.); dan.rogers@eng.ox.ac.uk (D.R.)
*   Correspondence: gianluca.brando@unina.it; Tel.: +39-0817683233

Abstract: In the developing context of distributed generation and flexible smart grids, in order to realize electrochemical storage systems, Modular Multilevel Converters (MMCs) represent an interesting alternative to the more traditional Voltage Source Inverters (VSIs). This paper presents a novel analytical investigation of electrochemical cell power losses in MMCs and their dependence on the injected common mode voltage. Steady-state cell losses are calculated under Nearest Level Control (NLC) modulation for MMCs equipped with a large number of half-bridge modules, each directly connected to an elementary electrochemical cell. The total cell losses of both a Single Star MMC (SS-MMC) and a Double Star MMC (DS MMC) are derived and compared to the loss of a VSI working under the same conditions. An optimum common mode voltage injection law is developed, leading to the minimum cell losses possible. In the worst case, it achieves a 17.5% reduction in cell losses compared to conventional injection laws. The analysis is experimentally validated using a laboratory prototype set-up based on a two-arm SS-MMC with 12 modules per arm. The experimental results are within 2.5% of the analytical models for all cases considered.

Citation: Brando, G.; Chatzinikolaou, E.; Rogers, D.; Spina, I. Electrochemical Cell Loss Minimization in Modular Multilevel Converters Based on Half-Bridge Modules. *Energies* **2021**, *14*, 1359. https://doi.org/10.3390/en14051359

Academic Editor: Jinliang Yuan

Received: 21 January 2021
Accepted: 24 February 2021
Published: 2 March 2021

Publisher's Note: MDPI stays neutral with regard to jurisdictional claims in published maps and institutional affiliations.

Copyright: © 2021 by the authors. Licensee MDPI, Basel, Switzerland. This article is an open access article distributed under the terms and conditions of the Creative Commons Attribution (CC BY) license (https://creativecommons.org/licenses/by/4.0/).

Keywords: lithium batteries; los minimization; Modular Multilevel Converters; optimization methods

1. Introduction

The rapid advances in energy storage technologies that have occurred in recent years, together with the urgent issue of environmental pollution, have driven innovative power electronic solutions that, apart from the main function of power conversion, provide additional functionality, such as minimizing system energy losses while assuring the maximum lifetime of the storage devices.

One set of promising converter topologies are based on the series connection of single half-bridge or full-bridge modules [1], each interfaced with a storage device, such as electrochemical cells or super-capacitors. The connection between the power module and the storage device can be either direct or occur via an additional DC/DC converter [2]. The series connection of a fixed number of modules represents an arm [3] of the converter. Based on the number of arms and their relative configuration, different versions of the Modular Multilevel Converter (MMC) can be realized [4,5].

The star connection of n arms can be adopted in order to realize the simplest n phase AC output converter [6], i.e., Single Star MMC (SS-MMC). On the other hand, a Double Star (DS) connection of 2n arms leads to the DS-MMC [7], where each pair of arms of the same phase are interfaced by two buffer inductances, providing both an n phase AC output and a DC interface. Both architectures have been widely proposed in several contexts, including traction applications [8–12] and transformerless grid connected storage systems [13–16].

The inherent modularity of MMCs helps minimize the system maintenance costs and generates fault tolerance features [14,17,18]. In the case of failure, the failed module can be bypassed without interrupting the operation of the system. The improved system reliability allows scaling to high voltages by connecting a very high number of modules in series. Therefore, a direct (transformerless) connection to high-voltage grids is feasible [13] by means of switches characterized by relatively low-voltage ratings, leading to a significant cost reduction and competitive total system efficiency [15]. A large number of modules also results in smaller rates-of-change of voltage (since the output voltage of each module is only a small percentage of the peak output voltage), strongly reducing Electro-Magnetic Interference (EMI)-related issues.

In the context of electrochemical storage devices [19], in order to safeguard the cells and guarantee the highest number of charge/discharge cycles, it is necessary to keep all of the cells balanced in terms of their State of Charge (SoC). While traditional converters are normally equipped with an additional Battery Management System [20], MMC solutions inherently provide this balancing functionality at the module level [12,16,21,22].

As an additional advantage, these modular converter designs lead to a significant reduction of the converter switching losses, especially when a large number of modules (and subsequently, voltage levels) are available [11]. In these cases, high switching frequency Pulse Width Modulation (PWM) schemes can be replaced by Nearest Level Control (NLC) modulation [23]. NLC generates a nearly-sinusoidal AC output by combining the DC voltages of the converter modules while driving the devices with a switching frequency equal to the fundamental, thus producing negligible switching losses. However, since at least one device per module is always in the current path, for large-scale systems, this approach could result in a substantial increase of conduction losses. For this reason, the careful selection of low-voltage Metal Oxide Semiconductor Field Effect Transistors (MOSFETs) optimized for low on-state resistance is necessary.

In most studies to-date where efficiency is considered, MMC solutions are compared with conventional Voltage Source Inverters (VSIs), but only power electronic losses are taken into account, whereas losses occurring in electrochemical cells are ignored [11]. This approach can be misleading; indeed, according to [15], MMC cell losses can form a significant fraction of overall system losses and will disproportionally affect MMC-type designs because the cells are operated under a low-frequency pulsed current (as opposed to a constant DC current in a VSI). On the other hand, this suggests a considerable margin for improvement, since the potential reduction of the MMC cell losses would have a great impact on the total system efficiency.

These considerations underline the need for a more detailed study where the possibility to influence the MMC cell power losses should be investigated. In fact, although many authors recognize the common mode reference [5] as a degree of freedom for the MMCs, no analysis of its effect on the cell power losses has been conducted.

In the context of MMCs comprising a large number of half-bridge modules, each including an elementary electrochemical cell and driven by NLC modulation, this paper presents a novel analytical investigation where the calculation of the cell losses is generalized for any common mode voltage value, so that the dependence of the cell losses from the common mode component is analyzed. Moreover, an optimum common mode voltage injection law is developed, assuring the minimization of cell power losses.

The proposed optimum strategy is compared with traditional approaches, showing a significant decrease in the cell losses: Considering that MMC cell losses can be a considerable fraction of total system losses, their minimization can significantly improve the system efficiency and reduce detrimental cell heating.

The total cell losses of an SS-MMC and DS-MMC are derived in all of the considered cases and compared with the loss associated with an equivalent VSI solution. Finally, all of the mathematical derivations are validated by means of an experimental set-up (see Figure 1) based on a two-arm SS-MMC with 12 modules per arm, each containing an elementary lithium ion cell. The results exhibit good agreement with the analytic model and demonstrate the effectiveness of the proposed strategy.

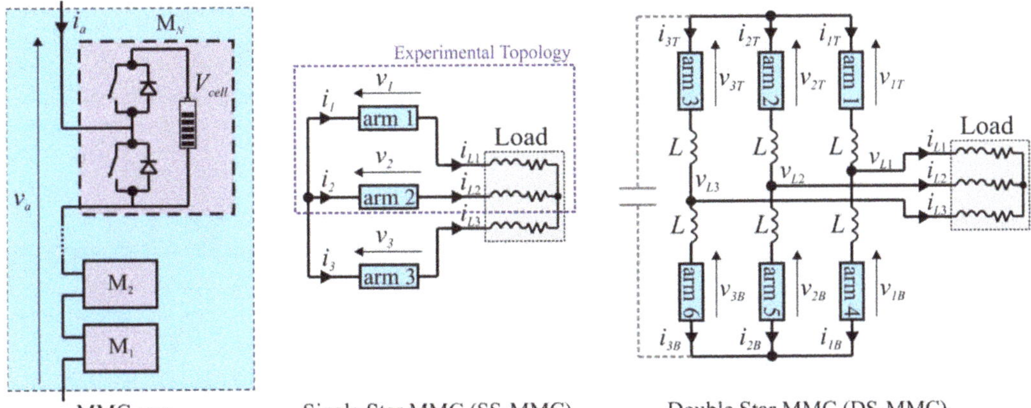

Figure 1. Modular Multilevel Converters (MMCs).

2. Description of the System

Figure 1 shows a single generic converter arm, composed of N half-bridge modules; v_a and i_a represent the total arm voltage and the arm current, respectively. Each module is driven by an electrochemical cell at the voltage V_{cell}, which is the average cell voltage. The SS-MMC and DS-MMC feed symmetrical three-phase loads. Given the possibility of the DS-MMC exchanging power through a DC interface, an optional capacitor (gray dashed line) has also been added. The experimental topology is outlined.

When a module is activated, the corresponding upper switching device is turned ON and its output voltage equals V_{cell}, while the arm current i_a flows in the cell. When a module is de-activated, its output voltage is zero and no current flows in the cell, i.e., the cell is bypassed.

When NLC modulation is implemented, the total number of active modules required to synthesize the instantaneous arm reference voltage v_a^* is given by

$$N_{on} = \text{round}(v_a^* / V_{cell}). \tag{1}$$

In this case, State of Charge (SoC) balancing between the cells can be achieved by choosing which cells contribute to the output arm voltage v_a based on their SoC, i.e., when the system is charging, the N_{on} cells with the lowest SoC will be activated, whereas during discharge, the N_{on} cells with the highest SoC will be placed in the current path.

A second balancing action should be implemented in order to equalize the mean SoCs between the arms through the injection of a proper circulating current value. Since the circulating current is normally a small percentage of the nominal current and is only present during cell balancing [11], its contribution to the losses is not significant and will be neglected in the present discussion.

SS-MMC

With reference to an SS-MMC, denoting the three arm reference voltages as v_1^*, v_2^*, v_3^*, it is possible to define the following common mode component:

$$v_0^* = (v_1^* + v_2^* + v_3^*)/3 \tag{2}$$

Based on Equation (2), the reference voltages can be written as

$$\begin{cases} v_1^* = v_{1,d}^* + v_0^* \\ v_2^* = v_{2,d}^* + v_0^* \\ v_3^* = v_{3,d}^* + v_0^* \end{cases}, \tag{3}$$

where $v^*_{k,d}$ (with $k = 1,2,3$) is, by definition, the differential component of v^*_k. Since the load currents are not affected by v^*_0, it represents an effective degree of freedom in the arm modulation.

As $0 \leq v^*_k \leq NV_{cell}$ ($k = 1,2,3$), the quantity v^*_0 must fulfill the following constrains:

$$\begin{cases} v^*_0 \geq -\min\left(v^*_{1,d}, v^*_{2,d}, v^*_{3,d}\right) = v^*_{0,\min} \\ v^*_0 \leq N\,V_{cell} - \max\left(v^*_{1,d}, v^*_{2,d}, v^*_{3,d}\right) = v^*_{0,\max} \end{cases}. \tag{4}$$

The common mode voltage can be decomposed into its DC and AC components:

$$v^*_0 = v^*_{0,DC} + v^*_{0,AC}.$$

In a normal steady-state condition, the differential component $v^*_{1,d}, v^*_{2,d}, v^*_{3,d}$ will be imposed as a three-phase sinusoidal symmetrical system:

$$\begin{cases} v^*_{1,d} = V_L \sin(\omega t - \varphi) \\ v^*_{2,d} = V_L \sin(\omega t - \varphi - 2\pi/3) \\ v^*_{3,d} = V_L \sin(\omega t - \varphi - 4\pi/3) \end{cases}, \tag{5}$$

where V_L is the load voltage amplitude and ω is the angular frequency.

In the presence of a symmetrical load, given the high number of cells per arm, the three arm currents can be assumed to be sinusoidal and, in particular, since the circulating currents have been neglected, a three-phase sinusoidal symmetrical system can be considered:

$$\begin{cases} i_1 = I_L \sin(\omega t) \\ i_2 = I_L \sin(\omega t - 2\pi/3) \\ i_3 = I_L \sin(\omega t - 4\pi/3) \end{cases}, \tag{6}$$

where I_L is the load current amplitude and φ is the phase shift delay.

By defining $\zeta = 2\,V_L/N\,V_{cell}$ as the arm modulation index, Equation (5) can be rewritten as

$$v^*_{k,d} = N\zeta V_{cell}/2 \sin(\omega t - \alpha_k), \tag{7}$$

where $\alpha_k = 2\pi(k - 1)/3$.

Equivalently, by introducing $\zeta_{DC} = 2v^*_{0,DC}/NV_{cell}$, the DC component of v^*_0 can be written as

$$v^*_{0,DC} = N\zeta_{DC} V_{cell}/2. \tag{8}$$

Based on Equations (7) and (8), Equation (3) can be written as

$$v^*_k = \frac{NV_{cell}\left(\zeta \sin(\omega t - \alpha_k) + \zeta_{DC}\right)}{2} + v^*_{0,AC}. \tag{9}$$

From Equations (1) and (9), the number of modules $N_{on}(t)$ switched ON at instant t is

$$N_{on}(t) = \text{round}\left(\frac{N(\zeta \sin(\omega t - \alpha_k) + \zeta_{DC})}{2} + \frac{v^*_{0,AC}}{V_{cell}}\right). \tag{10}$$

If N is large enough, $N_{on}(t)$ can be reasonably approximated with its continuous equivalent:

$$N_{c,on}(t) = \frac{N(\zeta \sin(\omega t - \alpha_k) + \zeta_{DC})}{2} + \frac{v^*_{0,AC}}{V_{cell}}. \tag{11}$$

DS-MMC

With reference to Figure 1, while, for the SS-MMC, the arm currents are equal to the load currents, in the DS configuration, each k-th phase load current i_{Lk} is given by the difference of a TOP arm current i_{kT} and a BOTTOM arm current i_{kB}. In fact, the DS-

MMC can be seen as being composed of three star-connected TOP arms (with voltages v_{1T}, v_{2T}, v_{3T}) and three star-connected BOTTOM arms (with voltages v_{1B}, v_{2B}, v_{3B}):

$$
\begin{aligned}
\text{TOP}: &\begin{cases} v_{0T}^* = \left(v_{1T}^* + v_{2T}^* + v_{3T}^*\right)/3 = v_{0T,DC}^* + v_{0T,AC}^* \\ v_{kT}^* = v_{kT,d}^* + v_{0T}^* \end{cases} \\
\text{BOT}: &\begin{cases} v_{0B}^* = \left(v_{1B}^* + v_{2B}^* + v_{3B}^*\right)/3 = v_{0B,DC}^* + v_{0B,AC}^* \\ v_{kB}^* = v_{kB,d}^* + v_{0B}^* \end{cases}
\end{aligned}
\tag{12}
$$

with symbols having similar meanings to those in the SS-MMC.

In a normal sinusoidal steady-state condition, and for each k-th phase, the differential component of the BOTTOM arm reference voltage lags the TOP one by an angle of $180°$. Moreover, when the circulating currents are neglected, each BOTTOM arm current is opposed to the TOP one:

$$
\begin{cases} v_{kB,d}^* = -v_{kT,d}^* \\ i_{kT} = -i_{kB} \end{cases}
\tag{13}
$$

Therefore, the DS-MMC is equivalent to two SS-MMCs and all of the previous considerations can be separately applied to both the three TOP arms and the three BOTTOM arms, with the only difference being that, in the presence of a symmetrical load, $i_{kT} = -i_{kB} = i_{Lk}/2$.

3. Conventional Strategies and Cell Loss Calculation

In conventional approaches ([3,4,10,11,24]), the common mode voltage DC component $v_{0,DC}^*$ is equal to $NV_{cell}/2$ (corresponding to $\zeta_{DC} = 1$), so that the arm reference voltages are centered in the range of the possible values $[0, NV_{cell}]$, while the AC component $v_{0,AC}^*$ is set to zero for $\zeta \leq 1$. In order to achieve an overmodulation-free operation in the interval $\zeta \in \left]1, 2/\sqrt{3}\right]$, $v_{0,AC}^*$ can be equal to either v_{SVM} or v_{THI}, with v_{SVM} representing a traditional Space Vector Modulation (SVM) AC common mode component and v_{THI} representing a third harmonic of a proper amplitude:

$$
\begin{aligned}
v_{0,DC}^* &= NV_{cell}/2 \\
v_{0,AC}^* &= \begin{cases} 0 \text{ for } \zeta \leq 1 \\ v_{SVM} \text{or } v_{THI} \text{ for } \zeta > 1 \end{cases}
\end{aligned}
\tag{14}
$$

For the traditional SVM injection,

$$
v_{SVM} = \begin{cases} \dfrac{NV_{cell}\zeta \sin\left(\omega t - \frac{\pi}{3}h\right)}{4} \forall \omega t \in A_{p,h} \\ \dfrac{NV_{cell}\zeta \sin\left(\omega t + \frac{2\pi}{3} - \frac{\pi}{3}h\right)}{4} \forall \omega t \in A_{n,h} \end{cases},
\tag{15}
$$

where $\begin{cases} A_{p,h} \equiv (-\pi/6 + 2\pi/3h, \pi/6 + 2\pi/3h) \\ A_{n,h} \equiv (\pi/6 + 2\pi/3h, \pi/2 + 2\pi/3h) \end{cases}$ and h is an integer number.

For the third harmonic injection,

$$
v_{THI} = \frac{NV_{cell}\zeta \sin 3\omega t}{12}.
\tag{16}
$$

Figure 2 shows the qualitative waveform of $v_{k,d}^* + v_{0,AC}^*$ (normalized with respect to $NV_{cell}/2$) in the presence of the conventional AC common mode injection methods. In particular, starting from the differential component $v_{k,d}^*$ (Figure 2a), the v_{THI} or v_{SVM} common mode AC component (Figure 2b) can be added, producing the waveform of Figure 2c or Figure 2d, respectively.

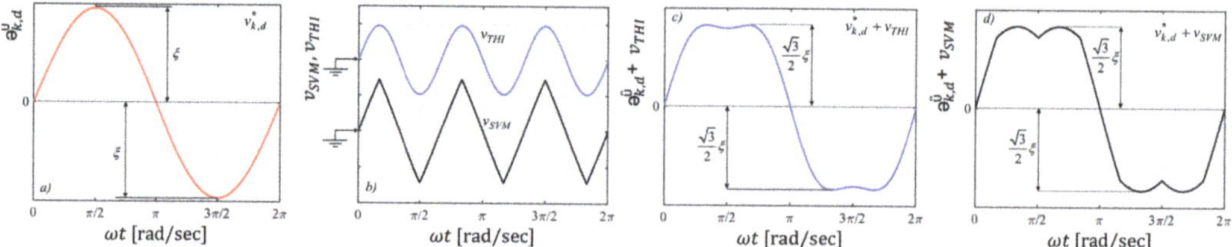

Figure 2. Qualitative waveform of the generic arm reference voltage in the presence of the conventional AC common mode injection: (**a**) Differential component; (**b**) AC conventional common mode components; and (**c,d**) total arm reference voltage AC component.

For the sake of clarity, Figure 2 can refer to either an SS-MMC or the three TOP arms of a DS-MMC, while for the BOTTOM arms, the corresponding waveforms are shifted by 180°.

Thanks to the symmetry of the MMC arms (for both SS and DS configurations), the cell power loss calculation can be performed with reference to a single arm. In particular, if a cell is modeled as a DC voltage source with a series resistance R_{cell}, the instantaneous power dissipated into the arm cells $p_J(t)$ can be expressed as a function of $N_{c,on}$:

$$p_J(t) = N_{c,on}(t)\, R_{cell}\, i_a^2(t), \tag{17}$$

where i_a is the instantaneous arm current. By substituting Equation (11) into Equation (17), the average power losses are given by

$$P_J = \frac{1}{T}\int_0^T N_{c,on} R_{cell}\, i_a^2 dt = R_{cell}\left(\frac{N\zeta_{DC}I_a^2}{4} + \frac{1}{T}\int_0^T \frac{v_{0,AC}^*}{V_{cell}}\, i_a^2 dt\right). \tag{18}$$

By replacing $v_{0,AC}^*$ in Equation (18) with its Fourier-series $\sum_{n=1}^{n=+\infty} V_{ac,h}\sin(n\,\omega t - \varphi_{ac,n})$, P_J is derived:

$$P_J = R_{cell}\left(\frac{N\zeta_{DC}}{4} - \frac{V_{ac,2}}{V_{cell}}\sin(2\varphi - \varphi_{ac,2})\right)I_a^2. \tag{19}$$

According to conventional strategies, considering that $\zeta_{DC} = 1$ and that v_{SVM} and v_{THI} do not contain the second harmonic, P_J becomes

$$P_J = N R_{cell} I_a^2/4, \tag{20}$$

where I_a is the arm current amplitude: $I_a = I_L$ for the SS-MMC and $I_a = I_L/2$ for the DS-MMC.

4. Cell Loss Comparison of MMC and VSI

From Equation (19), it is straightforward to compute the total cell power loss for either a three-phase SS-MMC or DS-MMC characterized by the same number N of cells per arm:

$$\begin{cases} P_{J,SS-MMC} = \frac{3}{4}\, N\, R_{cell,SS}\, I_L^2 \\[6pt] P_{J,DS-MMC} = \frac{3}{8}\, N\, R_{cell,DS}\, I_L^2 \end{cases}. \tag{21}$$

In order to produce the same AC output voltage, the SS-MMC converter requires half as many modules compared to the DS-MMC, where, for each phase, two arms are parallel-connected with respect to the AC side. As a result, for the same total system storage capacity, an SS-MMC cell can be regarded as equivalent to two parallel-connected DS-MMC cells, i.e., $R_{cell,DS} \cong 2R_{cell,SS}$. Therefore, $P_{J,SS-MMC} \cong P_{J,DS-MMC}$.

The mean value of the load active power P_L is naturally independent of the converter configuration:

$$P_L = \frac{3}{2}V_L I_L \cos\varphi = \frac{3}{4}\xi N V_{cell} I_L \cos\varphi. \tag{22}$$

The cell power losses in the MMC (henceforth indicated as $P_{J,MMC}$ for both the SS-MMC and DS-MMC) can be compared to the cell losses $P_{J,VSI}$ occurring in a traditional two-level Voltage Source Inverter (VSI). An equivalent VSI battery pack may be built using three parallel-connected stacks of N cells, as shown in Figure 3. A DC-side LC filter ensures that a smooth DC current i_{dc} is drawn from the pack.

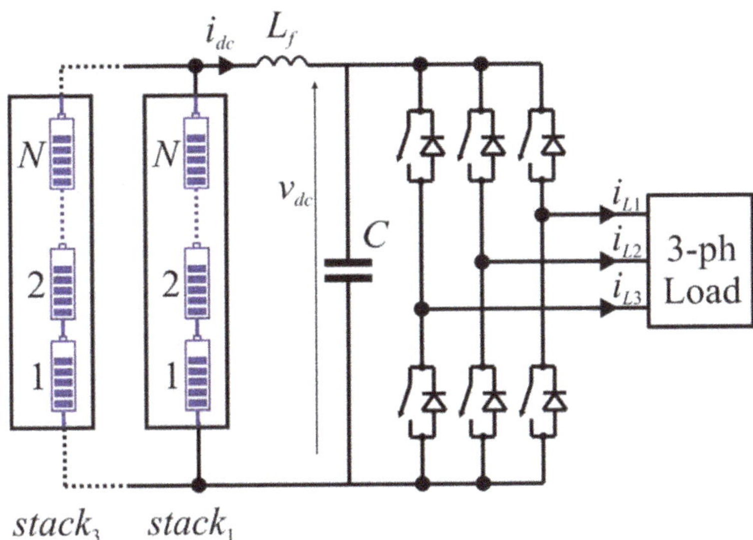

Figure 3. Voltage Source Inverter (VSI).

In this configuration, each stack effectively corresponds to the cells contained in the modules of one SS-MMC arm. If the VSI switching frequency f_s is much greater than the fundamental output frequency f, the current i_{dc} flowing in the battery pack can be assumed to be constant once the DC filter components have been properly chosen. If the converter losses are neglected, i_{dc} only depends on P_L:

$$i_{dc} = \frac{P_L}{N V_{cell}} = \frac{3}{4}\xi I_L \cos\varphi. \tag{23}$$

From Equation (23), the cell losses are

$$P_{J,VSI} = \frac{1}{3}N R_{cell}\, i_{dc}^2 = \frac{3}{16}N R_{cell}(\xi I_L \cos\varphi)^2. \tag{24}$$

This may be directly compared with the first of Equation (21) with $R_{cell,SS} = R_{cell}$ (or with the second of Equation (21) with $R_{cell,DS} = 2R_{cell}$).

As it can be noted, the VSI cell losses depend on both the modulation index and the load power factor (in fact, $P_{J,VSI} \propto P_L^2$). On the contrary, $P_{J,MMC}$ does not depend on ξ or $\cos\varphi$, i.e., at constant I_L, the cell losses do not depend on the output power. Moreover, it is evident that $P_{J,MMC}$ is considerably higher than $P_{J,VSI}$; in the case of the unitary power factor and full modulation index, the cell losses in the MMC are four times those of the VSI (i.e., Equation (21) divided by Equation (24)).

5. Proposed Optimum Injection Strategy

With reference to a three-phase system, $v_{0,AC}^*$ is built on multiples of the third harmonic in order to keep the arm reference voltages symmetrical across the converter phases. In this context, as per Equation (19), P_J does not directly depend on $v_{0,AC}^*$. Nevertheless, a proper $v_{0,AC}^*$ injection law can be formulated in order to decrease the minimum settable value of ξ_{DC}, to which, as stated by Equation (19), the arm cell losses are proportional. ξ_{DC} can be regarded as the shift to be applied to the AC component of the arm reference voltage $(v_{k,d}^* + v_{0,AC}^*)$ in order to guarantee that $v_k^* \in [0, NV_{cell}]$. Therefore, the range of possible values for $\xi_{DC} \in [\xi_{DC,\min}, \xi_{DC,\max}]$ depends on both the modulation index ξ and on $v_{0,AC}^*$. The cell loss minimization problem can thus be reduced to the formulation of an optimal injection law able to minimize $\xi_{DC,\min}$ for each $\xi \in [0, 2/\sqrt{3}]$. The proposed common mode voltage injection law is based on the following AC component:

$$v_{0,AC,OPT} = -\frac{NV_{cell}\xi\left(\sin\left(\omega t - \frac{2\pi}{3}h\right) + \frac{3\sqrt{3}}{2\pi}\right)}{2}\ \forall \omega t \in A_h,\tag{25}$$

where $A_h \equiv (7\pi/6 + 2\pi/3h, 7\pi/6 + 2\pi/3(h+1))$ and h is an integer number.

Figure 4a shows the qualitative waveform of $v_{0,AC,OPT}$, while the total AC arm reference voltage $v_{k,d}^* + v_{0,AC,OPT}$ (normalized with respect to $NV_{cell}/2$) is shown in Figure 4b; as previously mentioned, for the BOTTOM arms of the DS-MMC, a 180° shift must be considered.

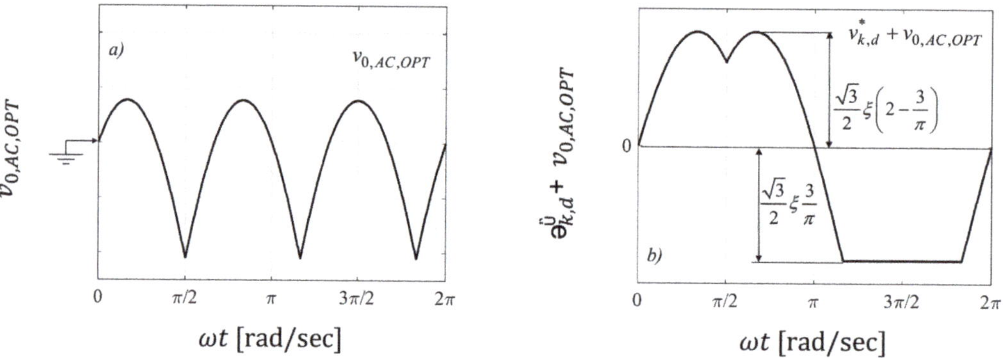

Figure 4. Qualitative waveform of the generic arm reference voltage in the presence of the proposed AC common mode injection: (**a**) AC proposed common mode components, and (**b**) total AC arm reference voltage.

As can be noted from Figure 2a, when $v_{0,AC}^* = 0$, i.e., no injection is used, the minimum ξ_{DC} value, necessary to shift the resultant arm reference voltage such that $v_k^* \geq 0$, is equal to the modulation index ξ (i.e., $\xi_{DC,\min} = \xi$). Instead, in both cases, when $v_{0,AC}^* = v_{SVM}$ or $v_{0,AC}^* = v_{THI}$, it is $\xi_{DC,\min} = \sqrt{3}\xi/2$ (Figure 2c,d). On the other hand, the proposed optimum law gives $\xi_{DC,\min} = 3\sqrt{3}\xi/2\pi$ (Figure 4b), which is the lowest of the three values.

The proposed strategy can be summarized by

$$\begin{cases} v_{0,AC}^* = v_{0,AC,OPT} \\ \xi_{DC} = 3\sqrt{3}\xi/2\pi \end{cases} \Rightarrow P_{J,opt} = \frac{3\sqrt{3}}{8\pi}\xi R_{cell} I_a^2.\tag{26}$$

In order to verify that this is actually the optimum voltage injection law, which guarantees the lowest $\xi_{DC,\min}$, leading to the lowest cell losses of a three-phase MMC, it is convenient to consider the constrains Equation (4). In the possible range $v_{0,\min}^* \leq v_0^* \leq v_{0,\max}^*$, it is evident that the choice $v_0^* = v_{0,\min}^*$ minimizes the quantity ξ_{DC}, thus guaranteeing the lowest cell losses. On the other hand, it is easy to verify that the DC component of $v_{0,min}^*$ can be obtained by Equation (8), with $\xi_{DC}=3\sqrt{3}\xi/2\pi$, while its AC component corresponds to the $v_{0,AC,OPT}$ of Equation (15). It can be pointed out that the lower achievable $\xi_{DC,min}$ is linked to the asymmetrical waveform of $v_{0,AC,OPT}$, which leads to a lower injected absolute minimum value (Figure 4b) with respect to the other methods.

Applying the proposed method, the total cell power losses for an SS-MMC and a DS-MMC characterized by the same number of cells per arm (N) are

$$\begin{cases} P_{J,opt,SS-MMC} = \frac{9\sqrt{3}}{8\pi}\xi N\,R_{cell,SS}\,I_L^2 \\[2mm] P_{J,opt,DS-MMC} = \frac{9\sqrt{3}}{16\pi}\xi\,N\,R_{cell,DS}\,I_L^2 \end{cases}. \tag{27}$$

It should be noted that in the case of a DS-MMC whose DC side is not isolated (see Figure 1), the power loss minimization techniques cannot be employed since ξ_{DC} is constrained to the total DC-link voltage value.

By comparing Equation (27) with Equation (21), the cell energy saving $\left(P_J - P_{J,opt}\right)/P_J$ guaranteed by the proposed method is linear with ξ, and in particular, it is equal to 100% for $\xi = 0$ and around 5% for $\xi = 2/\sqrt{3}$.

As an example, if a grid connected MMC is sized such that SoC = 0 corresponds to the maximum modulation index $\xi = 2/\sqrt{3}$ (worst case for evaluating the benefit of the proposed technique), during a complete charge (or discharge) cycle, and considering the typical lithium cell voltage variation, the modulation index will vary from $2/\sqrt{3}$ to about 0.9, with an average ξ value equal to about 1. The corresponding average energy saving is equal to about 17.5%. Naturally, in an application characterized by a lower value of ξ, the improvement is even more significant. It is important to note, however, that, even with a unitary power factor and full modulation index, an MMC operated with the optimal injection law will still cause approximately 3.3 times the cell loss of an equivalent VSI (i.e., Equation (26) divided by Equation (24)). In practice, the choice between an MMC and VSI design must involve a careful trade-off between the significantly higher cell loss incurred by the MMC and the inherent modularity, BMS flexibility, low EMI, and high-voltage advantages that an MMC has over a VSI.

6. Experimental Validation

To validate the power loss minimization method presented in section V, the experimental setup of Figure 5 was used.

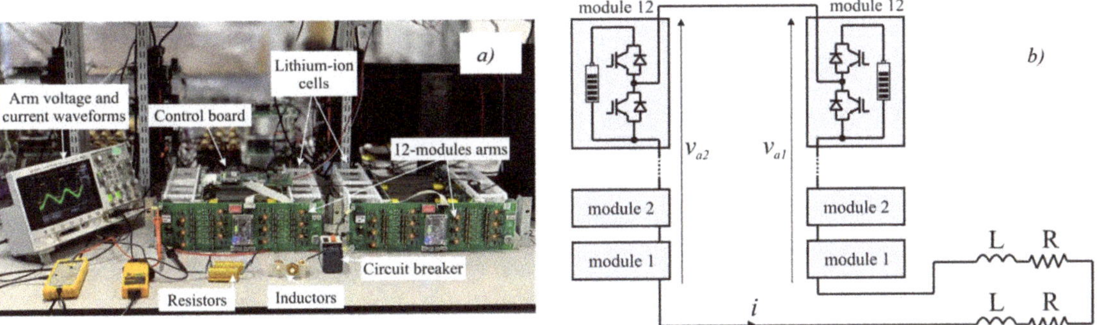

Figure 5. Experimental setup: (**a**) Photograph and (**b**) schematic.

Each arm in Figure 5 comprises twelve modules connected in series. Each module includes a 20 Ah lithium titanate cell that is individually controlled by an H-bridge converter (a detailed description of the control architecture of the experimental hardware is presented in [16]). In these experiments, each H-bridge is operated as a half-bridge, only allowing for a positive or zero voltage output.

Detailed characteristics of the setup and experiments are presented in Table 1. According to the cell loss minimization techniques explained in the previous section and, in particular, by changing the settings of the variables ξ_{DC} and $v^*_{a,0,AC}$, four different tests were performed.

Table 1. Details of the performed tests and results.

N		12		
$V_{cell}(\text{V})$		2.5		
$R_{cell}(\text{m}\Omega)$		5		
ξ		2/3		
$v^*_{a1,d}$		$10\cos(2\pi\,50\,t)$		
$v^*_{a2,d}$		$10\cos(2\pi\,50\,t+2\pi/3)$		
$R(\Omega)$		7.5		
$L(\text{mH})$		3.125		
$P_{arm}(\text{W})$		3.88		
Test n°	**1** **Conventional method**	**2**	**3**	**4** **Proposed method**
ξ_{dc}	1	2/3	$1/\sqrt{3}$	$3\sqrt{3}\xi/2\pi$
$v^*_{a,0,AC}$	0	0	v_{THI}	$v_{0,AC,OPT}$
$P_J(\text{mW})$	14.56	9.75	8.47	7.95
$P'_J(\text{mW})$	**14.67**	9.83	8.43	**8.12**
Error (%) $100\frac{P_J-P'_J}{P'_J}$	−0.76	−0.84	0.48	−2.13

Test 1 refers to the conventional method: $v^*_{0,AC}=0$ and $\xi_{DC}=1$.

In Test 2, no AC voltage injection was used, but ξ_{DC} was set to a lower value ($\xi_{DC}=2/3$ instead of $\xi_{DC}=1$).

In Test 3, a third harmonic AC voltage injection was added and ξ_{DC} was further decreased to $1/\sqrt{3}$.

Test 4 refers to the optimum proposed method with $v^*_{0,AC}=v_{0,AC,OPT}$ and $\xi_{DC}=3\sqrt{3}\xi/2\pi$.

For each test, the current and voltage of one arm were sampled through the oscilloscope at a frequency of 25 MHz (500 k samples per 50 Hz period). Each cell current was calculated as follows:

$$i_{cell,n}(k)=g_n(k)\,i_a(k),\tag{28}$$

where $i_{cell,n}(k)$ is the current of the n-th cell at the sampling time k, $i_a(k)$ is the arm current measured at the sample point k, and $g_n(k)$ is the switching signal:

$$g_n(k)=\begin{cases}0,\ \text{if }v_{a1}(k)<nV_{cell}\ (n-\text{cell bypassed})\\1,\ \text{if }v_{a1}(k)\ge nV_{cell}\ (n-\text{cell online})\end{cases}.\tag{29}$$

For each test, the total cell loss of the considered arm is calculated by

$$P'_J=\frac{R_{cell}}{N_s}\sum_{k=1}^{N_s}\sum_{n=1}^{12}g_n(k)\,i_a^2(k),\tag{30}$$

where N_s is the total number of samples per cycle.

Table 1 reports these results, together with the theoretical calculation of P_J. All the tests were carried out in correspondence with the same modulation index value, such that the arm output power P_{arm} was fixed. The error between the experimental and theoretical results, which is also shown, is less than 2.5% for all cases (within the experimental error, e.g., due to arm current measurements).

As can be noted from the Test 1 results (Figure 6a,e), the arm voltage is centered around 15 V ($NV_{cell}/2$) and its waveform corresponds to a quantized sine wave. The theoretical calculated cell losses (see Table 1) are substantially equal to the experimental ones.

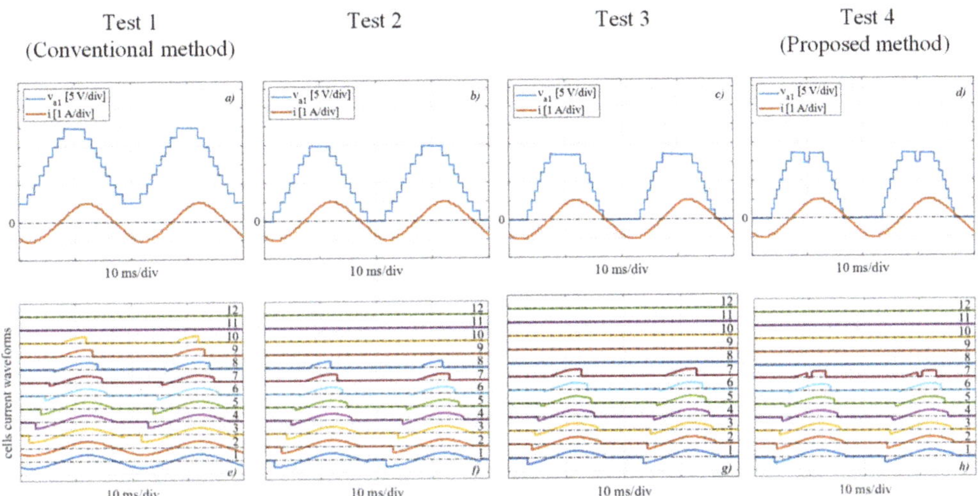

Figure 6. Experimental results: total voltage and current during: (**a**) Test 1, (**b**) Test 2, (**c**) Test 3, (**d**) Test 4; cells currents waveform during: (**e**) Test 1, (**f**) Test 2, (**g**) Test 3, (**h**) Test 4.

As per the theoretical considerations, the Test 2 results testify that the cell losses are dependent on $v_{0,DC}^*$. Indeed, in comparison with Test 1, the losses are lower, proportional to the decrease of ξ_{DC}. From Figure 6b, it can be noted that the arm voltage reaches the zero value. In fact, the ξ_{DC} setting of Test 2 corresponds to the minimum ξ_{DC} value, considering that no AC common mode voltage has been injected.

Test 3 clarifies that even if P_J does not directly depend on $v_{0,AC}^*$, an AC injection can affect the minimum settable value of ξ_{DC}. In this case, indeed, it was possible to set $\xi_{DC} = 1/\sqrt{3}$, without violating $0 \leq v_k^* \leq NV_{cell}$. The third injected harmonic can be recognized from the altered waveform of the arm voltage (Figure 6c), which shows a wider constant value in correspondence with the maximum and minimum point.

Finally, the optimum AC common mode voltage injection of Test 4, which displays visible notches in the arm voltage (Figure 6d), guarantees the minimum settable value of ξ_{DC}, leading to the minimization of cell losses. The decrease of the cell losses is associated with a decrease of the total *rms* current value in the 12 cells, which can be easily noted when compared to Figure 6e,h. In comparison with the conventional method of Test 1, the proposed method assures a cell energy saving of about 45%, which is consistent with the fact that the experimental converter is operating with a mid-range ξ value.

7. Additional Losses Numerical Example

In order to better evaluate the relative impact of the cell loss reduction introduced by the proposed method, other losses occurring in the system should be considered. These are principally switching and conduction losses occurring in the power electronic devices. For an MMC with a large number of modules driven by NLM, the switching losses are negligible, since the power devices switch at the fundamental output frequency. The VSI is instead characterized by both conduction and switching losses. A rapid example can be developed by completing the analysis of [11] (which compares the power electronics losses of a traditional VSI to those of a DS-MMC) with the electrochemical cell losses evaluation of the present work. While the VSI is equipped with FZ300R12KE3G IGBT, the MMC is built upon AUIRFS8409-7P power MOSFET. Both solutions constitute a 80 kW–220 V–250 A converter, fed by a 24 kWh battery system based on 11 Ah Kokam SLPB55205130H cells with an internal resistance of approximately 1.6 mΩ each.

The procedure given in [11] only considers the losses occurring in the power electronic devices; at half full-load, with $\zeta = 0.7$ and $\cos\varphi = 1$, the power electronics losses are as shown in Figure 7a.

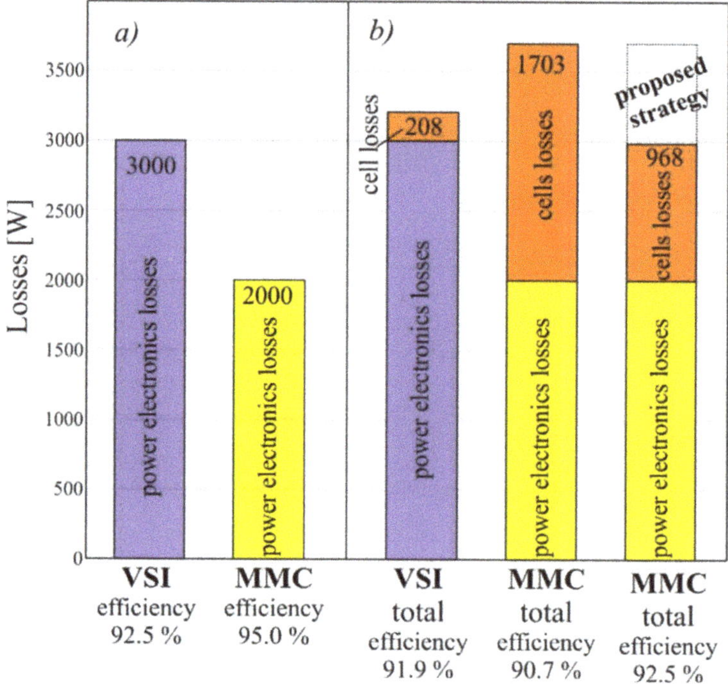

Figure 7. (**a**) Power losses of an 80 kW–24 kWh converter realized by the MMC and VSI solution: (**a**) Power electronics losses and (**b**) all losses.

Cell losses may be added to the analysis of [11], by applying Equations (21), (24) and (27), resulting in the loss breakdown given in Figure 7b. As expected, the MMC cell losses are higher than those of VSI and represent an important fraction of the total losses. In this example, the proposed technique produces a 42% reduction in cell losses, which equates to an overall 19% loss reduction. As a consequence of applying the proposed technique, the total efficiency is increased by 1.8%.

Naturally, the actual increase of the MMC overall efficiency achieved through the proposed technique depends on the particular operating condition. It is important to highlight that loss reduction in electrochemical cells is doubly beneficial as it leads to a direct reduction in detrimental cell heating and increase in the service life.

8. Conclusions

In the present work, the calculation of the cell losses for Modular Multilevel Converters (MMCs) equipped with half-bridge modules, each directly connected to an elementary electrochemical cell and driven by Nearest Level Control (NLC), has been generalized for any common mode voltage value. It was observed that the total cell losses of both a Single Star MMC and a Double Star MMC are significantly higher than those occurring in an equivalent VSI.

The novel analytical investigation allowed the development of an optimal common mode voltage injection law that minimizes the cell power losses under all balanced operating conditions. The proposed method delivers a significant reduction in cell losses, which is dependent on the converter working condition and, in particular, is about 17.5% in the worst case.

All of the analytical derivations have been validated by means of an experimental set-up based on a two-arm SS-MMC with 12 modules per arm, showing an error within 2.5% in all the considered cases and confirming the effectiveness of the proposed technique, with a cell loss reduction of about 45% in the performed tests.

In conclusion, in comparison with conventional VSI solutions, MMC solutions show intrinsic advantages (modularity, BMS functionality, fault tolerance, reduced EMI issue, etc.), but are characterized by higher cell losses, which represent an important fraction of the total losses. The proposed optimum common mode voltage injection law introduces a significant reduction of the converter losses, making MMCs more attractive with respect to conventional power converters.

Author Contributions: Conceptualization, G.B. and I.S.; methodology, D.R. and I.S.; software, E.C.; validation, D.R. and G.B.; formal analysis, D.R.; investigation, E.C. and I.S.; data curation, G.B. and E.C.; writing—original draft preparation, G.B., E.C. and I.S.; writing—review and editing, D.R.; supervision, I.S. All authors have read and agreed to the published version of the manuscript.

Funding: This research received no external funding.

Institutional Review Board Statement: Not applicable.

Informed Consent Statement: Not applicable.

Conflicts of Interest: The authors declare no conflict of interest.

References

1. Jafari, M.; Malekjamshidi, Z.; Li, L.; Zhu, J.G. Performance analysis of full bridge, boost half bridge and half bridge topologies for application in phase shift converters. In Proceedings of the 2013 International Conference on Electrical Machines and Systems (ICEMS), Busan, Korea, 26–29 October 2013; pp. 1589–1595.
2. Coppola, M.; Del Pizzo, A.; Iannuzzi, D. A power traction converter based on Modular Multilevel architecture integrated with energy storage devices. In Proceedings of the 2012 Electrical Systems for Aircraft, Railway and Ship Propulsion, Bologna, Italy, 16–18 October 2012; pp. 1–7.
3. Diab, M.S.; Elserougi, A.A.; Massoud, A.M.; Ahmed, S.; Williams, B.W. A Hybrid Nine-Arm Modular Multilevel Converter for Medium-Voltage Six-Phase Machine Drives. *IEEE Trans. Ind. Electron.* **2019**, *66*, 6681–6691. [CrossRef]
4. Akagi, H. Classification, terminology, and application of the modular multilevel cascade converter (MMC). *IEEE Trans. Ind. Electron.* **2011**, *26*, 3119–3130.
5. Camargo, R.S.; Mayor, D.S.; Miguel, A.M.; Bueno, E.J.; Encarnação, L.F. A Novel Cascaded Multilevel Converter Topology Based on Three-Phase Cells—CHB-SDC. *Energies* **2020**, *13*, 4789. [CrossRef]
6. Lamb, J.; Mirafzal, B.; Blaabjerg, F. PWM Common Mode Reference Generation for Maximizing the Linear Modulation Region of CHB Converters in Islanded Microgrids. *IEEE Trans. Ind. Electron.* **2018**, *65*, 5250–5259. [CrossRef]
7. Cheng, Q.; Wang, C.; Wang, J. Analysis on Displacement Angle of Phase-Shifted Carrier PWM for Modular Multilevel Converter. *Energies* **2020**, *13*, 6743. [CrossRef]
8. Tolbert, L.M.; Peng, F.Z. Multilevel converters for large electric drives. In Proceedings of the Applied Power Electronics Conference and Exposition, Thirteenth Annual, Anaheim, CA, USA, 15–19 February 1998; Volume 2, pp. 530–536.
9. Tolbert, L.M.; Chiasson, J.N.; McKenzie, K.J.; Du, Z. Control of cascaded multilevel converters with unequal voltage sources for HEVs. In Proceedings of the IEEE International Electric Machines and Drives Conference, 2003. IEMDC'03, Madison, WI, USA, 1 June 2003; Volume 2, pp. 663–669.
10. Quraan, M.; Yeo, T.; Tricoli, P. Design and Control of Modular Multilevel Converters for Battery Electric Vehicles. *IEEE Trans. Power Electron.* **2016**, *31*, 507–517. [CrossRef]
11. Quraan, M.; Tricoli, P.; D'Arco, S.; Piegari, L. Efficiency Assessment of Modular Multilevel Converters for Battery Electric Vehicles. *IEEE Trans. Power Electron.* **2017**, *32*, 2041–2051. [CrossRef]

12. Brando, G.; Dannier, A.; Spina, I.; Tricoli, P. Integrated BMS-MMC Balancing Technique Highlighted by a Novel Space-Vector Based Approach for BEVs Application. *Energies* **2017**, *10*, 1628. [CrossRef]
13. Maharjan, L.; Inoue, S.; Akagi, H. A Transformerless Energy Storage System Based on a Cascade Multilevel PWM Converter with Star Configuration. *IEEE Trans. Ind. Appl.* **2008**, *44*, 1621–1630. [CrossRef]
14. Hillers, A.; Biela, J. Fault-tolerant operation of the modular multilevel converter in an energy storage system based on split batteries. In Proceedings of the 2014 16th European Conference on Power Electronics and Applications, Lappeenranta, Finland, 26–28 August 2014; pp. 1–8.
15. Chatzinikolaou, E.; Rogers, D.J. A Comparison of Grid-Connected Battery Energy Storage System Designs. *IEEE Trans. Power Electron.* **2017**, *32*, 6913–6923. [CrossRef]
16. Chatzinikolaou, E.; Rogers, D.J. Cell SoC Balancing Using a Cascaded Full-Bridge Multilevel Converter in Battery Energy Storage Systems. *IEEE Trans. Ind. Electron.* **2016**, *63*, 5394–5402. [CrossRef]
17. Wang, S.; Alsokhiry, F.S.; Adam, G.P. Impact of Submodule Faults on the Performance of Modular Multilevel Converters. *Energies* **2020**, *13*, 4089. [CrossRef]
18. Li, J.; Yin, J. Fault-Tolerant Control Strategies and Capability without Redundant Sub-Modules in Modular Multilevel Converters. *Energies* **2019**, *12*, 1726. [CrossRef]
19. Wang, Z.; Lin, H.; Ma, Y. A Control Strategy of Modular Multilevel Converter with Integrated Battery Energy Storage System Based on Battery Side Capacitor Voltage Control. *Energies* **2019**, *12*, 2151. [CrossRef]
20. Lawder, M.T.; Suthar, B.; Northrop, P.W.; De, S.; Hoff, C.M.; Leitermann, O.; Crow, M.L.; Santhanagopalan, S.; Subramanian, V.R. Battery Energy Storage System (BESS) and Battery Management System (BMS) for Grid-Scale Applications. *Proc. IEEE* **2014**, *102*, 1014–1030. [CrossRef]
21. Chatzinikolaou, E.; Rogers, D.J. Performance Evaluation of Duty Cycle Balancing in Power Electronics Enhanced Battery Packs Compared to Conventional Energy Redistribution Balancing. *IEEE Trans. Power Electron.* **2018**, *33*, 9142–9153. [CrossRef]
22. Uddin, W.; Zeb, K.; Adil Khan, M.; Ishfaq, M.; Khan, I.; Islam, S.u.; Kim, H.-J.; Park, G.S.; Lee, C. Control of Output and Circulating Current of Modular Multilevel Converter Using a Sliding Mode Approach. *Energies* **2019**, *12*, 4084. [CrossRef]
23. Moranchel, M.; Bueno, E.J.; Rodriguez, F.J.; Sanz, I. Implementation of nearest level modulation for Modular Multilevel Converter. In Proceedings of the 2015 IEEE 6th International Symposium on Power Electronics for Distributed Generation Systems (PEDG), Aachen, Germany, 22–25 June 2015; pp. 1–5.
24. Gonçalves, J.; Rogers, D.J.; Liang, J. Submodule Temperature Regulation and Balancing in Modular Multilevel Converters. *IEEE Trans. Ind. Electron.* **2018**, *65*, 7085–7094. [CrossRef]

Article

Performance Analysis of a Full Order Sensorless Control Adaptive Observer for Doubly-Fed Induction Generator in Grid Connected Operation

Gianluca Brando, Adolfo Dannier * and Ivan Spina

Department of Electrical Engineering and Information Technologies, University of Naples Federico II, Via Claudio 21, 80125 Napoli, Italy; gianluca.brando@unina.it (G.B.); ivan.spina@unina.it (I.S.)
* Correspondence: adolfo.dannier@unina.it; Tel.: +39-081-768-3233

Abstract: This paper focuses on the performance analysis of a sensorless control for a Doubly Fed Induction Generator (DFIG) in grid-connected operation for turbine-based wind generation systems. With reference to a conventional stator flux based Field Oriented Control (FOC), a full-order adaptive observer is implemented and a criterion to calculate the observer gain matrix is provided. The observer provides the estimated stator flux and an estimation of the rotor position is also obtained through the measurements of stator and rotor phase currents. Due to parameter inaccuracy, the rotor position estimation is affected by an error. As a novelty of the discussed approach, the rotor position estimation error is considered as an additional machine parameter, and an error tracking procedure is envisioned in order to track the DFIG rotor position with better accuracy. In particular, an adaptive law based on the Lyapunov theory is implemented for the tracking of the rotor position estimation error, and a current injection strategy is developed in order to ensure the necessary tracking sensitivity around zero rotor voltages. The roughly evaluated rotor position can be corrected by means of the tracked rotor position estimation error, so that the corrected rotor position is sent to the FOC for the necessary rotating coordinate transformation. An extensive experimental analysis is carried out on an 11 kW, 4 poles, 400 V/50 Hz induction machine testifying the quality of the sensorless control.

Keywords: doubly-fed induction generator; wind power system; sensorless control; full order observer; field oriented control; grid connected system

Citation: Brando, G.; Dannier, A.; Spina, I. Performance Analysis of a Full Order Sensorless Control Adaptive Observer for Doubly-Fed Induction Generator in Grid Connected Operation. *Energies* **2021**, *14*, 1254. https://doi.org/10.3390/en14051254

Academic Editor: Anibal De Almeida

Received: 1 February 2021
Accepted: 19 February 2021
Published: 25 February 2021

Publisher's Note: MDPI stays neutral with regard to jurisdictional claims in published maps and institutional affiliations.

Copyright: © 2021 by the authors. Licensee MDPI, Basel, Switzerland. This article is an open access article distributed under the terms and conditions of the Creative Commons Attribution (CC BY) license (https://creativecommons.org/licenses/by/4.0/).

1. Introduction

The Doubly Fed Induction Generator (DFIG) is widely employed as a generator, especially in variable speed grid-connected wind energy applications. DFIGs guarantee robust and flexible systems, facilitating electric energy generation in a wide operating range of wind turbines [1–3].

In the grid-connected system, the aim is to maximize the conversion of mechanical input energy from the wind turbine into electric energy, ensuring the minimization of the cost of energy at the same time [4,5]. To meet this goal, the more widespread wind power generation system, for both small and large power, consists of a wind turbine [6,7], usually equipped with pitch control limits, a gearbox, and a DFIG directly connected to the AC grid on the stator side and driven through a power electronic converter on the rotor side. The rating of the power electronic is almost 25% of the rated power, allowing a speed range from nearly 50% to 120% of the rated speed [8].

It is clear that this Wind Energy Conversion System (WECS) is much more efficient than a constant speed squirrel cage induction generator system, but it obviously presents a greater complexity in the control, especially with regard to the knowledge of the rotor position [6].

The traditional control system of the doubly fed induction generator is based on stator flux oriented vector control [9]. Practically, using the rotor position it is possible to decompose the rotor current space vector into two current components in order to decouple control of the flux and electromagnetic torque. A rotor position sensor is required to finalize the transformation of the rotor current space vector. As it is evident, the presence of these sensors on the shaft reduces the robustness of the whole system. In recent years, research has been very active in working to replace the traditional control with a sensorless one, achieving advantages in terms of system robustness, easy installation, and maintenance [10,11]. In the literature, several sensorless methods have been proposed. The different sensorless controls can be collected into the follows categories: open-loop estimation methods, closed-loop estimation methods [12,13], Kalman filter [14], high-frequency signal injections [15,16], Model Reference Adaptive System (MRAS) observers [17,18] with full or reduced order approaches [19–21], and other sensorless methods.

In the open-loop sensorless methods, the rotational speed is obtained via differentiation of the estimated slip angle [22–26]; in the MRAS, the adaptive models are all based on static flux–current relations, and the estimated speed is used as feedback in a vector control system—this approach is very sensitive to machine inductance [27]; in the other sensorless method, it is possible to include all types of sensorless control based on PLLs and similar to MRAS observers. Indeed, in these last cases, the error is driven to zero when the phase shift between the estimated vector and the reference vector is null [28,29]. Ultimately, it is possible to state that in each of these controls, the challenge is to evaluate the position of the rotor by means of an indirect method, preserving at the same time good estimation accuracy against possible parameter deviations.

The performances of the estimation of the rotor position are linked to the implemented control strategy, such as Field Oriented Control (FOC), Direct Torque Control (DTC) [30,31], and Direct Power Control (DPC) [32–34].

This paper deals with an experimental evaluation of the performance of a novel adaptive full-order observer, presented in [35], in order to assess the accuracy of the rotor position estimation by means of an FOC control scheme.

The novelty of this approach is the consideration of the rotor position estimation error as an additional machine parameter. This parameter is accurately tracked by the proposed adaptive law, assuring a good compensation of the projection error from the rotor frame to the stator frame. A prototype system was set up in order to validate the effectiveness of the proposed approach by means of an extensive measurement campaign.

This paper is organized as follows: in Section 2 a description of the system is presented, in Section 3, starting from the mathematical model in a matrix form, the proposed adaptive observer is illustrated, in Section 4 details of the adopted control scheme are highlighted and, finally, in Section 5, an analysis of the experimental results is reported.

2. Description of the System

The present work focuses on the system depicted in Figure 1. In particular, a Doubly Fed Induction Generator (DFIG) was mechanically connected to a wind turbine via a gearbox. The DFIG rotor windings were wound and equipped with slip rings. While the stator windings were directly connected to the three-phase AC grid, the rotor windings were fed through a back-to-back power converter consisting of a common DC link and two three–phase converters: a Voltage Source Inverter (VSI) to the rotor side and a Voltage Source Rectifier (VSR) to the AC grid side. Due to the presence of the bidirectional power converter, the DFIG can operate as a generator or motor in both sub-synchronous and super-synchronous modes. However, in typical applications, the converter can be rated as 25% of the nominal power of the whole system. This limits the sub-synchronous and super-synchronous modes to about 1/3 up and down of synchronous speed.

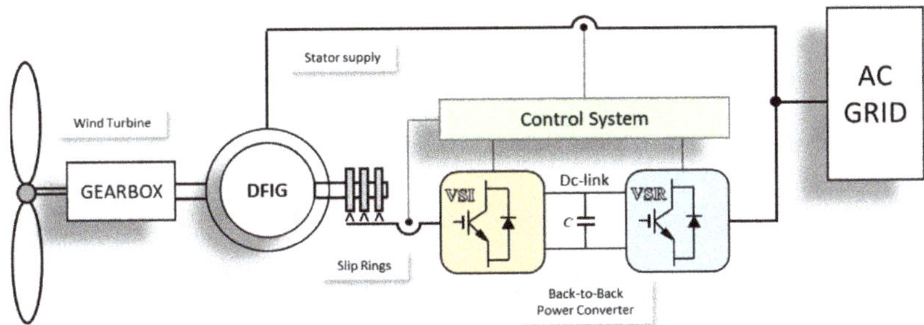

Figure 1. System block scheme.

3. Proposed Adaptive Observer

The mathematical model of the DFIG is well known and can be presented in different forms, depending on the particular choice of the state variables and the reference frame. Adopting the matrix form, the stator current and flux as state variables and the stator reference frame, the DFIG mathematical model is given by (see nomenclature at §.0):

$$\frac{d}{dt}\begin{bmatrix} \mathbf{i}_s \\ \mathbf{\Phi}_s \end{bmatrix} = \begin{bmatrix} A_{11} & A_{12} \\ A_{21} & A_{22} \end{bmatrix}\begin{bmatrix} \mathbf{i}_s \\ \mathbf{\Phi}_s \end{bmatrix} + \begin{bmatrix} B_1 \\ B_2 \end{bmatrix}[\mathbf{v}_s] + \begin{bmatrix} C_1 \\ C_2 \end{bmatrix}[\mathbf{v}_r] \tag{1}$$

It is worth noting that by imposing $\mathbf{v}_r = 0$ in Equation (1), the mathematical model of a traditional induction machine (with short-circuited rotor windings) is obtained.

In Equation (1), the state variables, the inputs, and the outputs are space vectors. By consequence, all the corresponding matrix elements are complex numbers. In particular

$$\begin{cases} A_{11} = -\left(\frac{R_s}{L_{s,eq}} + f_{r,eq}\right) + jp\omega_r \\ A_{12} = \frac{\sigma f_{r,eq}}{L_{s,eq}} - j\frac{p\omega_r}{L_{s,eq}} \\ A_{21} = -R_s \\ A_{22} = 0 \end{cases} \quad \begin{cases} B_1 = \frac{1}{L_{s,eq}} \\ B_2 = 1 \end{cases} \quad \begin{cases} C_1 = -\frac{B_1}{\mu_r} \\ C_2 = 0 \end{cases} \tag{2}$$

where:

$$\mu_r = \frac{L_m}{L_r}; \; \sigma = (1 - \mu_s\mu_r); \; L_{s,eq} = \sigma L_s; \; f_{r,eq} = \frac{R_r}{L_{r,eq}} \tag{3}$$

with $\mu_s = L_m / L_r$ and $L_{r,eq} = \sigma L_r$.

It is possible to derive the space vector of the rotor current from the definition of the stator flux ($\mathbf{\Phi}_s = L_s\mathbf{i}_s + L_m\mathbf{i}_r$):

$$\mathbf{i}_r = \frac{\mathbf{\Phi}_s}{L_m} - \frac{L_s}{L_m}\mathbf{i}_s \tag{4}$$

The following full-order Lueberger observer can be defined based on the DFIG mathematical model (Equation (1)):

$$\frac{d}{dt}\begin{bmatrix} \hat{\mathbf{i}}_s \\ \hat{\mathbf{\Phi}}_s \end{bmatrix} = A\begin{bmatrix} \hat{\mathbf{i}}_s \\ \hat{\mathbf{\Phi}}_s \end{bmatrix} + B[\mathbf{v}_s] + C[\mathbf{v}_r] + \begin{bmatrix} G_1 \\ G_2 \end{bmatrix}[\mathbf{i}_s - \hat{\mathbf{i}}_s] \tag{5}$$

The elements G_1 and G_2 of the observer matrix should be designed with reference to the observer state matrix A_O:

$$A_O = \begin{bmatrix} A_{11} - G_1 & A_{12} \\ A_{21} - G_2 & A_{22} \end{bmatrix} \tag{6}$$

having eigenvalues $p_{O,1}$ and $p_{O,2}$. It should be noted that the above equations relate to a proportional type Lueberger observer. Other more complex architectures (such as proportional–integral and modified integral [36]) may lead to an overall improved noise rejection. However, the proportional type allows for a relatively simple selection of the observer gains. In particular, the observer gains should be selected such as that the error $\mathbf{i}_s - \hat{\mathbf{i}}_s$ quickly converges to zero. In other words, the observer dynamic must be faster than that of the DFIG in all operating conditions and, in particular, for any value of the rotational speed. The faster DFIG dynamic is linked to the DFIG high-frequency pole $p_{D,HF}$ at zero speed. Thus, by introducing an overall observer gain $K_G > 1$, the two observer poles that $p_{O,1}$ and $p_{O,2}$ can be fixed proportional to $p_{D,HF}$ through K_G. Considering that a good approximation by excess of the effective $p_{D,HF}$ value is:

$$p_{D,HF} \cong -\left(\frac{R_s}{L_{s,eq}} + f_{r,eq} \right) \tag{7}$$

the eigenvalues $p_{O,1}$, $p_{O,2}$ can be fixed as:

$$p_{O,1} = p_{O,2} = p_O = -K_G \left(\frac{R_s}{L_{s,eq}} + f_{r,eq} \right) \tag{8}$$

where K_G is the overall observer gain.

In order to preserve system stability, an optimal value for K_G should be eventually fixed by trials on the real system, depending on the measurement equipment and noise; K_G values are typically chosen between 2 and 5. Hence, the elements G_1 and G_2 of the observer matrix are given by:

$$\begin{cases} G_1 = A_{11} - 2p_O \\ G_2 = A_{21} + p_O^2 / A_{12} \end{cases} \tag{9}$$

In the dynamic system (Equation (5)), the space vectors $\mathbf{v}_r, \mathbf{v}_s$ and $\mathbf{i}_s$ represent inputs. In particular, while $\mathbf{v}_s$ and $\mathbf{i}_s$ can be simply measured by voltage and current transducers, knowledge of $\mathbf{v}_r$ would require an additional rotor position sensor. Indeed, $\mathbf{v}_r$ is defined by a rotational transformation of the actual space vector voltage $\mathbf{v}_r^{(r)}$ provided by the rotor power converter through the electrical rotor position ϑ_e:

$$\mathbf{v}_r = \mathbf{v}_r^{(r)} e^{j\vartheta_e} \tag{10}$$

where

$$\vartheta_e = p\vartheta_r \tag{11}$$

with ϑ_r represents the mechanical rotor position and p the number of pole pairs.

On the other hand, in the context of a sensorless control, the observer outputs can be exploited in order to estimate the rotor position and, consequently, the space vector of the rotor voltage in the stator reference frame $\mathbf{v}_r$, without the use of an additional position sensor.

The rotor position can be estimated considering that Relation (10) also applies to the rotor current space vector in the two reference frames:

$$\mathbf{i}_r = \mathbf{i}_r^{(r)} e^{j\vartheta_e} \tag{12}$$

where, naturally, $\mathbf{i}_r^{(r)}$ is the rotor current space vector provided by the power converter and $\mathbf{i}_r$ is the rotor current space vector in the stator frame.

While the quantity $\mathbf{i}_r^{(r)}$ is directly measurable at the converter terminals, an estimated version of $\mathbf{i}_r$ can be calculated as per Equation (4) by substituting the actual flux $\mathbf{\Phi}_s$ with the estimated one $\hat{\mathbf{\Phi}}_s$:

$$\hat{\mathbf{i}}_r = \frac{\hat{\mathbf{\Phi}}_s}{L_m} - \frac{L_s}{L_m} \mathbf{i}_s \tag{13}$$

Once $\mathbf{i}_r^{(r)}$ and $\hat{\mathbf{i}}_r$ are known, the rotor electrical position can be derived as:

$$e^{j\hat{\vartheta}_e} = \frac{\hat{\mathbf{i}}_r}{\mathbf{i}_r^{(r)}} \Rightarrow \hat{\vartheta}_e = \arg(\hat{\boldsymbol{\Phi}}_s - L_s \hat{\mathbf{i}}_s) - \arg\left(\mathbf{i}_r^{(r)}\right) \tag{14}$$

Hence, $\mathbf{v}_r$ can be calculated as per Equation (10).

The knowledge of $\hat{\vartheta}_e$ also allows for the computation of the rotor speed via the numerical time derivative. In order to dampen the oscillations resulting from the time derivative computation, the calculation of the rotor speed should be processed by a properly sized Low Pass Filter (LPF):

$$\hat{\omega}_r = \frac{1}{p} \Im\left(\frac{d\hat{\vartheta}}{dt}\right) \tag{15}$$

where $\Im$ denotes the functional associated to the LPF.

As expected, knowledge of machine parameter values plays an important role in the described estimation procedure. However, the machine parameters cannot be known with absolute precision, and only their estimated values can actually be used in the previous equations. In particular, the symbols A, B, C, $\mathbf{v}_r$ and L_s appearing in Equations (5), (10), and (14) should formally be replaced with $\hat{A}$, $\hat{B}$, $\hat{C}$, $\hat{\mathbf{v}}_r$ and $\hat{L}_s$, this last representing their estimated versions. As a consequence of the possible parameter deviations, the rotor position estimation $\hat{\vartheta}_e$ will be affected by inaccuracy.

Let us define $\Delta\vartheta_e$ as the rotor position estimation error:

$$\Delta\vartheta_e = \vartheta_e - \hat{\vartheta}_e \tag{16}$$

According to Definition (16), the actual rotor position ϑ_e can be expressed as $\hat{\vartheta}_e + \Delta\vartheta_e$. Hence, the actual rotor voltage $\mathbf{v}_r$ can be written as:

$$\mathbf{v}_r = \mathbf{v}_r^{(r)} e^{j(\hat{\vartheta}_e + \Delta\vartheta_e)} = \mathbf{v}_r^{(r)} e^{j\hat{\vartheta}_e} e^{j\Delta\vartheta_e} = \hat{\mathbf{v}}_r e^{j\Delta\vartheta_e} \tag{17}$$

where the quantity $\mathbf{v}_r^{(r)} e^{j\hat{\vartheta}_e}$ corresponds to $\hat{\mathbf{v}}_r$: the estimated version of the rotor voltage.

Equation (17) clarifies that the estimated rotor voltage $\hat{\mathbf{v}}_r$ does not correspond to the actual rotor voltage $\mathbf{v}_r$ due the projection error $e^{j\Delta\vartheta_e}$. At the same time, this suggests that the projection error could be compensated if a $\Delta\vartheta_e$ tacking procedure is conceived.

Let us denote the tracking procedure output with $\Delta\hat{\vartheta}_e$. i.e., $\Delta\hat{\vartheta}_e$ is the estimation of the actual $\Delta\vartheta_e$ value. For small $\Delta\hat{\vartheta}_e$ values, the following approximation can be assumed:

$$e^{j\Delta\hat{\vartheta}_e} = \cos\Delta\hat{\vartheta}_e + j\sin\Delta\hat{\vartheta}_e \cong 1 + j\Delta\hat{\vartheta}_e \tag{18}$$

In light of the previous consideration on the parameter deviations, and considering Equation (17) with approximation (18), the Luemberger observer (Equation (5)) can now be re-written:

$$\frac{d}{dt}\begin{bmatrix} \hat{\mathbf{i}}_s \\ \hat{\boldsymbol{\Phi}}_s \end{bmatrix} = \hat{A}\begin{bmatrix} \hat{\mathbf{i}}_s \\ \hat{\boldsymbol{\Phi}}_s \end{bmatrix} + \hat{B}[\mathbf{v}_s] + \hat{C}(1 + j\Delta\hat{\vartheta}_e)[\hat{\mathbf{v}}_r] + G[\mathbf{e}] \tag{19}$$

where $G = \begin{bmatrix} G_1 & G_2 \end{bmatrix}^T$ and $\mathbf{e} = \mathbf{i}_s - \hat{\mathbf{i}}_s$.

The quantity $\Delta\hat{\vartheta}_e$ in Equation (19) can be regarded as an additional machine parameter, and its value can be estimated by the following adaptive law based on the Lyapunov theory:

$$\Delta\hat{\vartheta}_e = K_{\Delta\vartheta} \int_0^t \left(\hat{v}_{ry} e_x - \hat{v}_{rx} e_y\right) dt + \Delta\hat{\vartheta}_{e0} \tag{20}$$

where $K_{\Delta\vartheta}$ is a not negative number.

The model (Equation (19)) and the adaptive law (Equation (20)), together with the Relations (14) and (15), define the proposed adaptive observer, which provides the estimation of rotor speed $\hat{\omega}_r$ and position $\hat{\vartheta}_e$. Moreover, the proposed technique is robust against possible machine parameter uncertainties and is capable of tracking the rotor position estimation error $\Delta\hat{\vartheta}_e$.

4. Control Scheme

In the control scheme depicted in Figure 2, the proposed sensorless adaptive observer provides the necessary rotational transformation angle and the rotor speed value for a traditional stator flux Field Oriented Control (FOC). In particular x, y represent the stator reference frame axis; α, β refer to the rotor reference frame axis; and d, q represents the rotating reference frame aligned with the stator flux $\mathbf{\Phi}_s$.

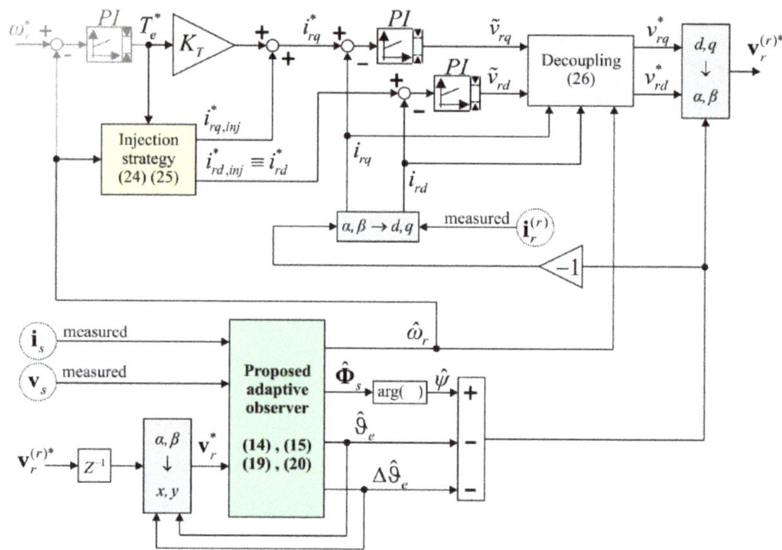

Figure 2. Control diagram.

Since the observer model has been written into the stator reference frame, its inputs ($\mathbf{v}_s$, $\mathbf{i}_s$ and $\mathbf{v}_r$) are referred to as x, y coordinates. In particular, while the first two inputs are measured directly on the stator side connected to the AC grid, the third input needs to be transformed from α, β to x, y coordinates. Indeed, the quantity $\mathbf{v}_r^{*(r)} = v_{r\alpha}^* + j v_{r\beta}^*$, representing the reference voltage for the VSI, is transformed into x, y coordinates by the angle $\hat{\vartheta}_e + \Delta\hat{\vartheta}_e$. In this way, the projection error is compensated by the position estimation error.

The adaptive observer outputs are the estimation of rotor speed $\hat{\omega}_r$, position $\hat{\vartheta}_e$, the rotor position estimation error $\Delta\hat{\vartheta}_e$ and the stator flux space vector $\mathbf{\Phi}_s$. From this last output, it is possible to calculate the angle of the stator flux $\hat{\psi}$:

$$\hat{\psi} = \arg(\mathbf{\Phi}_s) \tag{21}$$

The necessary transformation from α, β to x, y coordinates can thus be operated by the angle $\hat{\vartheta}_e + \Delta\hat{\vartheta}_e - \hat{\psi}$:

$$\mathbf{i}_r^{(\Phi)} = \mathbf{i}_r^{(r)} e^{j(\hat{\vartheta}_e + \Delta\hat{\vartheta}_e)} e^{-j\hat{\psi}} = i_{rd} + j i_{rq} \tag{22}$$

where $\mathbf{i}_r^{(r)}$ is the rotor current space vector, measurable in α, β coordinates.

As per the traditional FOC scheme, the obtained components i_{rd}, i_{rq} are processed by two PI controllers driven by the errors computed with respect to the correspondent reference quantities i_{rd}^*, i_{rq}^*. In particular, while i_{rq}^* is set proportional to the reference torque T_e^* by the torque constant K_T, i_{rd}^* should be set equal to zero. However, it must be considered that the tracking of ϑ_e and $\Delta\vartheta_e$ could prove ineffective during very low torque operating conditions where both current components move toward zero. For this reason, an injection strategy has been implemented:

$$\begin{cases} i_{rd}^* = i_{rd,inj}^* \\ i_{rq}^* = K_T T_e^* + i_{rq,inj}^* \end{cases} \tag{23}$$

where:

$$\begin{cases} i_{rd,inj}^* = \lambda_d A_{inj} \cos\left(2\pi f_{inj} t\right) \\ i_{rq,inj}^* = \lambda_q A_{inj} \cos\left(2\pi f_{inj} t\right) \end{cases} \tag{24}$$

A_{inj} and f_{inj} are the amplitude and the frequency of the injected current components. Activation and deactivation of the injection are operated through λ_d and λ_q, which can assume either 0 or 1 values. The q injection is activated if the required torque is too small. The d injection is activated if the required torque is too small or if the slip angular frequency is too small. Fixing $\Delta T_{e,inj}$ and $\Delta\omega_{inj}$ as threshold values, respectively, for the required torque and the slip angular frequency, λ_d and λ_q can be expressed as (in C language style):

$$\begin{cases} \lambda_d = |p\omega_r - \omega| < \Delta\omega_{inj} \; || \; |T_e^*| < \Delta T_{e,inj} \\ \lambda_q = |T_e^*| < \Delta T_{e,inj} \end{cases} \tag{25}$$

The converter space vector reference voltage components v_{rd}^*, v_{rq}^* are obtained by compensating the current PI regulator outputs $\tilde{v}_{rd}, \tilde{v}_{rq}$ through the decoupling action:

$$\begin{cases} v_{rd}^* = \tilde{v}_{rd} - \hat{\omega}_\sigma L_{s,eq} i_{rq} \\ v_{rq}^* = \tilde{v}_{rq} + \hat{\omega}_\sigma L_{s,eq} i_{rd} + \hat{\omega}_\sigma \Phi_{s,R} \end{cases} \tag{26}$$

with $\hat{\omega}_\sigma = \omega - p\hat{\omega}_r$ being the estimated rotor slip angular frequency, and $\Phi_{s,R}$ being the rated stator flux.

Finally, the obtained space vector $v_{rd}^* + jv_{rq}^* = \mathbf{v}_r^{*(\Phi)}$ is transformed from d, q to α, β coordinates by the angle $\hat{\psi} - \hat{\vartheta}_e - \Delta\hat{\vartheta}_e$:

$$\mathbf{v}_r^{*(r)} = \mathbf{v}_r^{*(\Phi)} e^{-j(\hat{\vartheta}_e + \Delta\hat{\vartheta}_e)} e^{j\hat{\psi}} = v_{r\alpha}^* + jv_{r\beta}^* \tag{27}$$

The α, β components of Equation (27) can be modulated as per a Space Vector Modulation (SVM) in order to drive the VSI at the rotor side. Naturally, as mentioned above, $\mathbf{v}_r^{*(r)}$ is also sent back and, through a unit time delay, constitutes a closed loop for the proposed adaptive law.

5. Experimental Results

The proposed control strategy was validated experimentally by mechanically connecting a three-phase wound rotor, 11 kW, 4 poles, 400 V/50 Hz induction machine to a 27 kW DC machine acting as a prime motor. Both the electric machines were driven by power converters based on Semikron IGBT modules, while the system control was implemented on *dSpace 1103* hardware whose digital outputs were properly routed to the IGBT drivers' inputs. The whole experimental setup is shown in Figure 3. In order to evaluate the sensorless estimation errors, a 4000 pulse/round incremental encoder mechanically connected to the DFIG was used to determine the rotor speed and position.

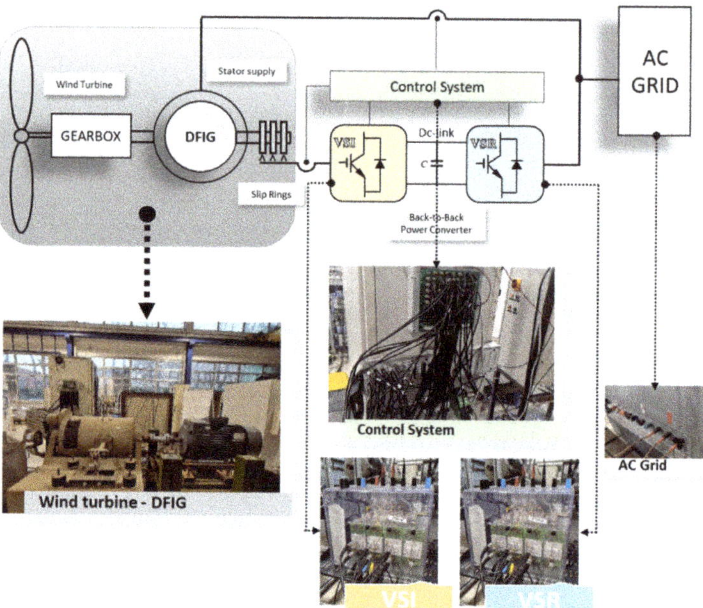

Figure 3. Experimental setup.

The motor parameters used to build the observer matrix were estimated by means of standard blocked rotor/no-load IEEE 112 tests executed on the DFIG. Since no additional parameter tuning was carried out, the experimental results also highlight the robustness of the adaptive observer with respect to substantial parameter deviations in the characterization of the induction machine. Indeed, standard IEEE 112 tests are affected by non-negligible errors, especially when compared to more advanced off-line estimation methods [37]. This approach allows us, therefore, to portray the improved accuracy in rotor position estimation granted by the adaptive observer in comparison to a standard observer.

To validate the effectiveness of the conceived adaptation law both at different speed values and different torque values, the following test was performed:

(a) Initially, a startup procedure (which is not shown) takes the system to the test initial condition, where the rotor speed is set to 70% of the DFIG synchronous speed and the DFIG reference torque is set to 50% of its rated value. Steady state condition is reached at $t = 0$.

(b) After one second of steady state condition, (at $t = 1$) the DFIG reference torque is set to the full rated value and again to half its rated value after one second (at $t = 2$).

(c) At $t = 3$ the reference speed is changed to the DFIG synchronous speed. To achieve a quasi-stationary transition which allows us to check the system response in the whole speed range, the reference speed was processed by a rate limiter filter. Consequently, the speed reached the new reference value through a linear behavior in around 2.2 s.

(d) The two-step torque variation of point (b) was repeated.

(e) The reference speed was changed to 130% of the DFIG synchronous speed.

(f) The two-step torque variation of point (b) was repeated, and the test was concluded.

The experimental results of the whole test are shown in Figure 4 (rotor speed), Figure 5 (rotor axes currents), and Figure 6 (rotor electric angle estimation error). From Figure 4, where both the measured and estimated rotor speed are plotted, it can be deduced that the DFIG observer was able to effectively track the real system speed with a negligible error, which stayed always under 0.5%. Naturally, the observable speed over- and undershoots are linked to the corresponding step variations of the DFIG reference torque. From Figure 5, where both the reference and actual rotor axes currents are plotted, it can be deduced that the control system (driven by the estimated values of the rotor speed and position) was able to effectively drive the rotor currents in the whole speed range. It should be pointed out that the ripples in the rotor axis current are visible only around the DFIG synchronous speed. This is indeed the result of the frequency injection in the rotor currents, which is used to keep the adaptation law sensible to the rotor voltage projection error when the rotor voltages become too small (in this instance, when the DFIG was around the synchronous operation mode). Finally, Figure 6 shows the rotor position error obtained with the proposed adaptation law versus that which would affect the observer when the adaptation law was not engaged. It can be deduced that the performance of the observer is appreciably improved: while the position error was kept between -5 degrees and 8 degrees with the adaptation law, it varied between 10 degrees and 20 degrees when the position error was not compensated. The maximum improvement can be noted at the low speed, where the error is 3 degrees, versus 17–20 degrees at the high speed.

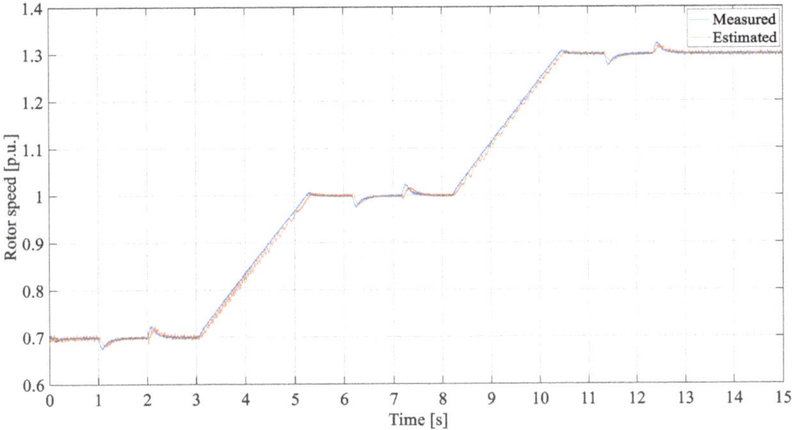

Figure 4. Behavior of the estimated and actual rotor speed in the whole test.

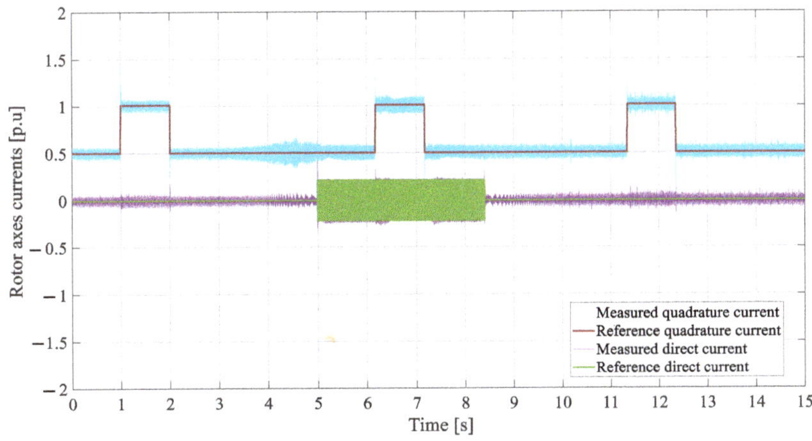

Figure 5. Behavior of the reference and actual rotor axes currents in the whole test.

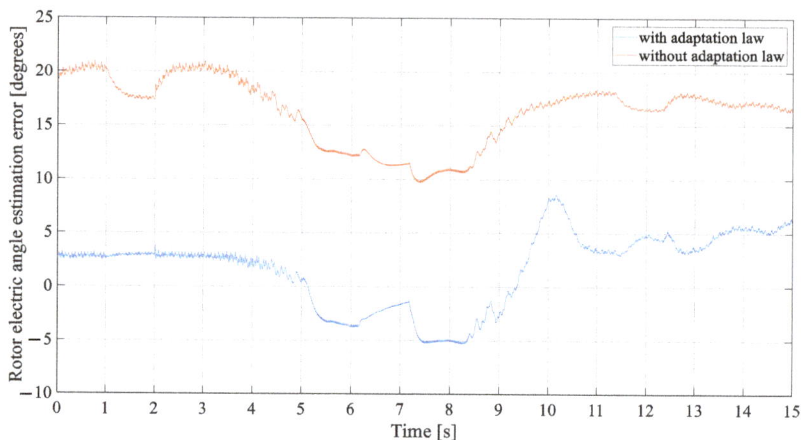

Figure 6. Behavior of the position estimation error with/without adaptation law in the whole test.

Figure 7 shows the behaviors of the first and second phase stator currents (green and violet lines) versus the corresponding rotor currents (blue and red lines) when the rotor speed was equal to 130% of the synchronous speed and the reference torque was set to 100% of the rated value. As expected, while the stator currents oscillated at the grid frequency (50 Hz), the frequency of the rotor currents were linked to the actual rotor speed. Given the value of the rotor speed, the resulting frequency was 15 Hz—this value is coherent with the difference between the synchronous speed and the actual one. It can also be noted that the first phase rotor current lagged after the second phase current—this is also expected since the DFIG was working with a rotor speed higher than the synchronous one.

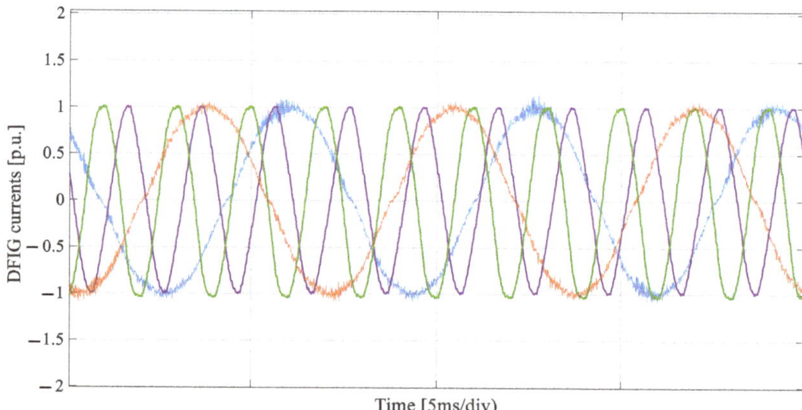

Figure 7. Behavior of the DFIG stator and rotor currents at the rated torque in the high speed region.

6. Conclusions

The aim of this paper is to experimentally evaluate the performance of a novel adaptive full-order observer in order to assess the accuracy of rotor position estimation by means of a Field Oriented Control (FOC) scheme for a Doubly Fed Induction generator (DFIG).

In particular, the work demonstrates that the rotor position evaluated by the observer estimated flux can be affected by significant errors due to parameter inaccuracies. The novelty of the proposed approach is linked to the fact that the rotor position estimation error is considered as an additional machine parameter. Thus, an adaptive law based on the Lyapunov theory was proposed for the tracking of the rotor position estimation error and, additionally, a current injection strategy was developed in order to ensure the necessary tracking sensitivity around zero rotor voltages. The roughly evaluated rotor position was corrected by means of the tracked rotor position estimation error so that the corrected rotor position was sent to the FOC for the necessary rotating coordinate transformation.

The proposed technique was tested experimentally on an 11 kW DFIG prototype moved by a 27 kW DC machine acting as the prime motor. The experimental results testify to the quality of the sensorless control, which is able to effectively track the real system speed with a negligible error ($< 0.5\%$) in the range of 60–130% of the rated speed value. Moreover, the proposed adaptive law clearly improved observer performance by significantly reducing the rotor position estimation error from 17–20 degrees to 3 degrees in the best case.

Future work will focus on the analytical analysis of the stability of the conceived adaptive observer and on its robustness with respect to parameter variations and electric grid and mechanical perturbances, such as negative voltage sequences or rotor eccentricity.

Author Contributions: G.B. formalized the control algorithm and wrote the relevant section, I.S. implemented the control algorithm on the real time controller and wrote the relevant section, A.D. supervised the experimental tests, wrote the other sections and edited the whole manuscript. All authors have read and agreed to the published version of the manuscript.

Funding: This research received no external funding.

Institutional Review Board Statement: Not applicable.

Informed Consent Statement: Not applicable.

Conflicts of Interest: The authors declare no conflict of interest.

Abbreviations

$\mathbf{i}_s / \mathbf{i}_r$	Space vector of the stator/rotor current
$L_{\sigma s} / L_{\sigma r}$	Stator/rotor leakage inductance
L_m	Air-gap linkage inductance
L_s	Stator inductance $L_s = L_{\sigma s} + L_m$
L_r	Rotor inductance $L_r = L_{\sigma r} + L_m$
R_s / R_r	Stator/rotor resistance
$\mathbf{v}_s / \mathbf{v}_r$	Space vector of the stator/rotor voltage
$\vartheta_r / \vartheta_e$	Mechanical/electrical rotor position
$\mathbf{\Phi}_s / \mathbf{\Phi}_r$	Space vector of the stator/rotor flux
ω_r / ω	Rotor angular speed/angular frequency
$\hat{\ }$	Superscript to indicate estimated quantity
$*$	Superscript to indicate reference quantity
(r)	Superscript to indicate rotor reference frame

References

1. Carrasco, J.; Franquelo, L.; Bialasiewicz, J.; Galvan, E.; PortilloGuisado, R.; Prats, M.; Leon, J.; Moreno-Alfonso, N. Power-Electronic Systems for the Grid Integration of Renewable Energy Sources: A Survey. *IEEE Trans. Ind. Electron.* **2006**, *53*, 1002–1016. [CrossRef]
2. Cardenas, R.; Pena, R.; Alepuz, S.; Asher, G. Overview of Control Systems for the Operation of DFIGs in Wind Energy Applications. *IEEE Trans. Ind. Electron.* **2013**, *60*, 2776–2798. [CrossRef]
3. Li, C.; Hang, Z.; Zhang, H.; Guo, Q.; Zhu, Y.; Terzija, V. Evaluation of DFIGs' Primary Frequency Regulation Capability for Power Systems with High Penetration of Wind Power. *Energies* **2020**, *13*, 6178. [CrossRef]
4. Qiao, W.; Zhou, W.; Aller, J.; Harley, R. Wind Speed Estimation Based Sensorless Output Maximization Control for a Wind Turbine Driving a DFIG. *IEEE Trans. Power Electron.* **2008**, *23*, 1156–1169. [CrossRef]
5. Baran, J.; Jąderko, A. An MPPT Control of a PMSG-Based WECS with Disturbance Compensation and Wind Speed Estimation. *Energies* **2020**, *13*, 6344. [CrossRef]
6. Polinder, H.; Ferreira, J.A.; Jensen, B.B.; Abrahamsen, A.B.; Atallah, K.; McMahon, R.A. Trends in Wind Turbine Generator Systems. *IEEE J. Emerg. Sel. Top. Power Electron.* **2013**, *1*, 174–185. [CrossRef]

7. Guo, J.; Lei, L. Flow Characteristics of a Straight-Bladed Vertical Axis Wind Turbine with Inclined Pitch Axes. *Energies* **2020**, *13*, 6281. [CrossRef]
8. Muller, S.; Deicke, M.; De Doncker, R.W. Doubly fed induction generator systems for wind turbines. *IEEE Ind. Appl. Mag.* **2002**, *8*, 26–33. [CrossRef]
9. Tapia, A.; Tapia, G.; Ostolaza, J.X.; Saenz, J.R. Modeling and control of a wind turbine driven doubly fed induction generator. *IEEE Trans. Energy Convers.* **2003**, *18*, 194–204. [CrossRef]
10. Holtz, J. Sensorless control of induction motor drives. *Proc. IEEE* **2002**, *90*, 1359–1394. [CrossRef]
11. Holtz, J. Sensorless Control of Induction Machines—With or Without Signal Injection? *IEEE Trans. Ind. Electron.* **2006**, *53*, 7–30. [CrossRef]
12. Malakar, M.K.; Tripathy, P.; Krishnaswamy, S. A predictor-corrector based rotor slip-position estimation technique for a DFIG. In Proceedings of the 2017 7th International Conference on Power Systems (ICPS), Pune, India, 21–23 December 2017; pp. 424–429.
13. Soares, E.L.; Rocha, F.V.; De Siqueira, L.M.S.; Rocha, N. Sensorless Rotor Position Detection of Doubly-Fed Induction Generators for Wind Energy Applications. In Proceedings of the 2018 13th IEEE International Conference on Industry Applications (INDUSCON), São Paulo, Brazil, 12–14 November 2018; pp. 1045–1050.
14. Abdelrahem, M.; Hackl, C.; Kennel, R. Sensorless control of doubly-fed induction generators in variable-speed wind tur-bine systems. In Proceedings of the 2015 International Conference on Clean Electrical Power (ICCEP), Taormina, Italy, 16–18 June 2015; pp. 406–413.
15. Reigosa, D.D.; Briz, F.; Charro, C.B.; di Gioia, A.; García, P.; Guerrero, J.M. Sensorless Control of Doubly Fed Induc-tion Generators Based on Rotor High-Frequency Signal Injection. In Proceedings of the 2012 IEEE Energy Conversion Congress and Exposition (ECCE), Raleigh, NC, USA, 15–20 September 2012; pp. 2268–2275.
16. Xu, L.; Inoa, E.; Liu, Y.; Guan, B. A New High-Frequency Injection Method for Sensorless Control of Doubly Fed Induc-tion Machines. *IEEE Trans. Ind. Appl.* **2012**, *48*, 1556–1564. [CrossRef]
17. Cárdenas, R.; Peña, R.; Proboste, J.; Asher, G.; Clare, J.; Wheeler, P. MRAS observers for sensorless control of doubly-fed induction generators. In Proceedings of the 4th IET Conference on Power Electronics, Machines and Drives, York, UK, 2–4 April 2008; pp. 568–572.
18. Pattnaik, M.; Kastha, D. Adaptive speed observer for a stand-alone doubly fed induction generator feeding nonlinear and unbalanced loads. In Proceedings of the 2013 IEEE Power & Energy Society General Meeting, Vancouver, BC, Canada, 21–25 July 2013.
19. Bhattarai, R.; Gurung, N.; Thakallapelli, A.; Kamalasadan, S. Reduced-Order State Observer-Based Feedback Control Methodolo-gies for Doubly Fed Induction Machine. *IEEE Trans. Ind. Appl.* **2018**, *54*, 2845–2856. [CrossRef]
20. Forchetti, D.G.; Garcia, G.O.; Valla, M.I. Adaptive Observer for Sensorless Control of Stand-Alone Doubly Fed Induction Generator. *IEEE Trans. Ind. Electron.* **2009**, *56*, 4174–4180. [CrossRef]
21. Yang, S.; Ajjarapu, V. A Speed-Adaptive Reduced-Order Observer for Sensorless Vector Control of Doubly Fed Induc-tion Generator-Based Variable-Speed Wind Turbines. *IEEE Trans. Energy Convers.* **2010**, *25*, 891–900. [CrossRef]
22. Abolhassani, M.; Niazi, P.; Toliyat, H.; Enjeti, P. A sensorless integrated doubly-fed electric alternator/active filter (IDEA) for variable speed wind energy system. In Proceedings of the 38th IAS Annual Meeting on Conference Record of the Industry Applications Conference, Salt Lake City, UT, USA, 12–16 October 2003; pp. 507–514.
23. Datta, R.; Ranganathan, V.T. A simple position-sensorless algorithm for rotor-side field-oriented control of wound-rotor induction machine. *IEEE Trans. Ind. Electron.* **2001**, *48*, 786–793. [CrossRef]
24. Morel, L.; Godfroid, H.; Mirzaian, A.; Kauffmann, J. Double-fed induction machine: Converter optimisation and field oriented control without position sensor. *IEE Proc. Electr. Power Appl.* **1998**, *145*, 360. [CrossRef]
25. Krzeminski, Z. Sensorless control of a double-fed machine for wind power generators. In Proceedings of the Power Conversion Conference, Osaka, Croatia, 2–5 April 2002. [CrossRef]
26. Hopfensperger, B.; Atkinson, D.; Lakin, R. Stator-flux-oriented control of a doubly-fed induction machine: With and without position encoder. *IEE Proc. Electr. Power Appl.* **2000**, *147*, 241–250. [CrossRef]
27. Schauder, C. Adaptive speed identification for vector control of induction motors without rotational transducers. *IEEE Trans. Ind. Appl.* **1992**, *28*, 1054–1061. [CrossRef]
28. Cardenas, R.; Pena, R.; Clare, J.; Asher, G.; Proboste, J. MRAS observers for sensorless control of doubly-fed induction generators. *IEEE Trans. Power Electron.* **2008**, *23*, 1075–1084. [CrossRef]
29. Castelli-Dezza, F.; Foglia, G.; Iacchetti, M.F.; Perini, R. An MRAS observer for sensorless DFIM drives with direct estima-tion of the torque and flux rotor current components. *IEEE Trans. Power Electron.* **2012**, *27*, 2576–2584. [CrossRef]
30. Mondal, S.; Kastha, D. Improved Direct Torque and Reactive Power Control of a Matrix-Converter-Fed Grid-Connected Doubly Fed Induction Generator. *IEEE Trans. Ind. Electron.* **2015**, *62*, 7590–7598. [CrossRef]
31. Arbi, J.; Ghorbal, M.J.; Slama-Belkhodja, I.; Charaabi, L. Direct Virtual Torque Control for Doubly Fed Induction Genera-tor Grid Connection. *IEEE Trans. Ind. Electron.* **2009**, *56*, 4163–4173. [CrossRef]
32. Hu, J.; Zhu, J.; Zhang, Y.; Platt, G.; Ma, Q.; Dorrell, D.G. Predictive Direct Virtual Torque and Power Control of Doubly Fed Induction Generators for Fast and Smooth Grid Synchronization and Flexible Power Regulation. *IEEE Trans. Power Electron.* **2012**, *28*, 3182–3194. [CrossRef]

33. Amiri, N.; Madani, S.M.; Lipo, T.A.; Zarchi, H.A. An Improved Direct Decoupled Power Control of Doubly Fed Induc-tion Machine Without Rotor Position Sensor and With Robustness to Parameter Variation. *IEEE Trans. Energy Convers.* **2012**, *27*, 873–884. [CrossRef]
34. Abad, G.; Rodriguez, M. Ángel; Poza, J. Three-Level NPC Converter-Based Predictive Direct Power Control of the Doubly Fed Induction Machine at Low Constant Switching Frequency. *IEEE Trans. Ind. Electron.* **2008**, *55*, 4417–4429. [CrossRef]
35. Brando, G.; Dannier, A.; Spina, I. A Full Order Sensorless Control Adaptive Observer for Doubly-Fed Induction Generator. In Proceedings of the 2019 International Conference on Clean Electrical Power (ICCEP), Otranto, Italy, 2–4 July 2019; pp. 464–469.
36. Białoń, T.; Pasko, M.; Niestrój, R. Developing Induction Motor State Observers with Increased Robustness. *Energies* **2020**, *13*, 5487. [CrossRef]
37. Wu, R.-C.; Tseng, Y.-W.; Chen, C.-Y. Estimating Parameters of the Induction Machine by the Polynomial Regression. *Appl. Sci.* **2018**, *8*, 1073. [CrossRef]

MDPI
St. Alban-Anlage 66
4052 Basel
Switzerland
Tel. +41 61 683 77 34
Fax +41 61 302 89 18
www.mdpi.com

Energies Editorial Office
E-mail: energies@mdpi.com
www.mdpi.com/journal/energies